PETER E. RIJTEMA has been involved in environmental research since 1972 and is a specialist in the field of the transport, fate and behaviour of agrochemicals in soils. A leading figure in the development of models on both the field and on the regional scale, which describe the leaching of nutrients and pesticides into groundwater and surface water, he also specialises in scenario analysis for the evaluation of the effects of different regulations in agricultural practice in the reduction of environmental pollution.

PIET GROENENDIJK was responsible for the development and application of deterministic simulation models in relation to water management, hydrological systems analysis, ecology, groundwater chemistry, emission of nutrients to groundwater and surface water systems. An expert in the modelling of hydrological, chemical and biological processes for various levels of integration, he has lately been involved in research on agricultural pollution, as well as in the initiation and supervision of a number of applied research projects on the emission of nutrients and pesticides from agricultural soils to groundwater and surface water systems.

JOOP G. KROES was involved in the development and application of water quality simulation models, the pesticide model TRANSOL and the nutrient model ANIMO. He is an expert in the modelling of solute flow for various levels of integration and his publications have dealt largely with aspects of solute modelling, including developments in information technology. He is currently working on various projects on solute transport.

SERIES ON ENVIRONMENTAL SCIENCE AND MANAGEMENT

Series Editor: Professor J.N.B. Bell
  Centre for Environmental Technology, Imperial College

Published

Vol. 1   Environmental Impact of Land Use in Rural Regions
         P.E. Rijtema, P. Groenendijk and J.G. Kroes

Forthcoming

Vol. 2   Numerical Ocean Circulation Modelling
         D.B. Haidvogel and A. Beckmann

SERIES ON ENVIRONMENTAL
SCIENCE AND MANAGEMENT **VOL. 1**

ENVIRONMENTAL IMPACT OF LAND USE IN RURAL REGIONS

The Development, Validation and Application
of Model Tools for Management and Policy Analysis

Peter E. Rijtema
Piet Groenendijk
Joop G. Kroes

*DLO Winand Staring Centre for Integrated Land,
Soil and Water Research, The Netherlands*

Imperial College Press

Published by

Imperial College Press
203 Electrical Engineering Building
Imperial College
London SW7 2BT

Distributed by

World Scientific Publishing Co. Pte. Ltd.
P O Box 128, Farrer Road, Singapore 912805
USA office: Suite 1B, 1060 Main Street, River Edge, NJ 07661
UK office: 57 Shelton Street, Covent Garden, London WC2H 9HE

Library of Congress Cataloging-in-Publication Data
Rijtema, P.E.
　　The environmental impact of land use in rural regions : the development, validation, and application of model tools for management and policy analysis / Peter Emile Rijtema, Pieter Groenendijk, Johannes Gerardus Kroes.
　　　　p.　cm. -- (Series on environmental science and management : vol. 1)
　　Includes bibliographical references and index.
　　ISBN 1-86094-041-2 (alk. paper)
　　1. Soil moisture -- Computer simulation.　2. Soils -- Solute movement -- Computer simulation.　3. Soil percolation -- Computer simulation. 4. Groundwater flow -- Computer simulation.　I. Groenendijk, Pieter. 1959-　.　II. Kroes, Johannes Gerardus, 1952-　.　III. Title. IV. Series.
S594.R548　1999
628.1'684--dc21　　　　　　　　　　　　　　　　　　　　　　　　98-45843
　　　　　　　　　　　　　　　　　　　　　　　　　　　　　　　　　　CIP

British Library Cataloguing-in-Publication Data
A catalogue record for this book is available from the British Library.

Copyright © 1999 by Imperial College Press

All rights reserved. This book, or parts thereof, may not be reproduced in any form or by any means, electronic or mechanical, including photocopying, recording or any information storage and retrieval system now known or to be invented, without written permission from the Publisher.

For photocopying of material in this volume, please pay a copying fee through the Copyright Clearance Center, Inc., 222 Rosewood Drive, Danvers, MA 01923, USA. In this case permission to photocopy is not required from the publisher.

Printed in Singapore by Uto-Print

PREFACE

This book is written for those who are concerned and responsible for the impacts of human activities in the development of the environment. Although the policy objectives for future rural development differ considerably from region to region, there is a remarkable similarity in objectives concerning the environmental impacts of land use on groundwater and surface water resources. Formulation of rational and effective management systems to improve the use of inputs could already provide a major contribution to the prevention of pollution. Technical innovations and adaptations in economic activities can lead to reductions in the environmental impact of rural land use. The environmental effects of different planned policy measures must be evaluated before implementation, using suitable models.

This book gives a description of the scientific backgrounds of a set of water quality models, developed in the DLO Winand Staring Centre for Integrated Land, Soil and Water Research (SC-DLO), in the framework of regional and national environmental studies for policy analysis and water management. The authors were the leading scientists in the development of these models on field scale, as well as on regional scale describing contaminant leaching to groundwater and surface waters. These models are used in scenario studies for the analysis of environmental aspects of national policies, for the determination of environmental effects of water management and in the evaluation of the reduction in pollution by different measures in agricultural practice.

The models SIMGRO, ANIMO, TRANSOL and EPIDIM are available on request. Requests should be addressed to SC-DLO, P.O. Box 125, Wageningen, The Netherlands.

CONTENTS

PREFACE	v
LIST OF FIGURES	xiii
LIST OF TABLES	xxi
1 INTRODUCTION	1
1.1 Problem and purpose	1
1.2 General aspects of modelling	3
1.3 Scope of the book	6
1.4 Outline of the water quality models ANIMO and TRANSOL	8
1.5 Reading guide	9
References	11
2 WATER TRANSPORT IN SOILS	15
2.1 Vertical transport in the unsaturated domain	16
2.1.1 Physical properties of soils	17
2.1.2 Swelling and shrinking soils	19
2.1.3 Schematization of vertical transport	24
2.1.3.1 General remarks	24
2.1.3.2 The module BALANCE	25
2.2 Transport in the saturated domain	32
2.2.1 Perfect drains	33
2.2.1.1 Theory	33
2.2.1.2 Model schematization	35
2.2.2 Line drains in a soil profile with infinite thickness	36
2.2.2.1 Theory	36
2.2.2.2 Model schematization	38
2.2.3 Line drains in a soil profile with finite thickness	40
2.2.3.1 Theory	41
2.2.3.2 Model schematization	44
2.3 Model approach for regional transport	49
References	56
3 TRANSPORT OF SOLUTES	59
3.1 Transport and mass conservation equation	61
3.1.1 Theory	61
3.1.2 Model schematization	63
3.1.2.1 Single component transport	64

3.1.2.2 Multi-component transport	69
3.1.3 Numerical analysis	70
3.1.3.1 Numerical and physical dispersion	70
3.1.3.2 Stability of the mathematical solution	73
3.1.4 Model verification with an analytical solution	74
3.2 Upper soil storage and surface runoff	76
3.3 Emission of pollutants from the soil to surface water	80
3.3.1 Concentration in drainage water released to perfect drains	80
3.3.2 Soil heterogeneity and the concentration in water released to line drains	82
3.3.3 Effect of buffer zones on surface water pollution	85
3.4 Model schematization for regional groundwater pollution	85
References	87
4 PHYSICAL-CHEMICAL PROCESSES	**89**
4.1 Chemical equilibria and ion speciation	89
4.2 Adsorption	93
4.2.1 Cation exchange	94
4.2.2 Sorption isotherms	98
4.2.2.1 Linear adsorption	98
4.2.2.2 Langmuir adsorption equation	99
4.2.2.3 Freundlich adsorption equation	99
4.2.3 Non-equilibrium sorption	100
4.2.3.1 First order rate adsorption	100
4.2.3.2 Intra-aggregate sorption in structured media	102
4.3 Precipitation and dissolution of minerals	104
4.3.1 Time dependent multi-component model	106
4.3.2 Transfer of gases	107
4.3.3 Schematization in the mass conservation and transport model	110
4.3.3.1 Non-equilibrium precipitation and dissolution	110
4.3.3.2 Instantaneous precipitation and dissolution	111
4.4 Redox reactions	111
4.5 Complexation with organic acids	114
4.6 Model schematization of physical and chemical processes	115
4.6.1 Instantaneous precipitation excluded	116
4.6.2 Instantaneous precipitation included	117
4.7 Some special applications	117
4.7.1 Organic micro-pollutants	117
4.7.2 Heavy metals	119
4.7.3 Phosphate sorption and precipitation	123
References	128

5 BIO-CHEMICAL PROCESSES — 131
  5.1 The model TRANSOL — 132
    5.1.1 Kinetic models describing biological decomposition — 133
    5.1.2 Model schematization in TRANSOL — 134
  5.2 Organic matter models — 135
    5.2.1 Multi-component models — 135
    5.2.2 Models with a time dependent decomposition rate factor — 136
  5.3 Biomass production and decomposition — 138
  5.4 Schematization in the nutrient model ANIMO — 140
    5.4.1 Decomposition of fresh organic materials — 142
    5.4.2 Additions of fresh materials — 146
    5.4.3 Additions of root materials — 146
    5.4.4 Production and decomposition of dissolved organic material — 148
    5.4.5 Biomass production and consumption — 148
    5.4.6 Nutrient availability in organic materials — 151
    5.4.7 Nitrogen processes — 153
      5.4.7.1 Net mineralization — 154
      5.4.7.2 Nitrification — 156
      5.4.7.3 Denitrification — 156
      5.4.7.4 Nitrogen fixation — 158
      5.4.7.7 Volatilization — 160
    5.4.8 Phosphorus processes — 160
  5.5 Additions to the soil and ploughing — 162
  References — 163

6 ENVIRONMENTAL INFLUENCES ON PROCESSES — 167
  6.1 Soil moisture — 167
    6.1.1 Schematization in TRANSOL — 168
    6.1.2 Schematization in ANIMO — 168
  6.2 Aeration and bio-chemical processes — 170
    6.2.1 General — 170
      6.2.1.1 Schematization in TRANSOL — 170
      6.2.1.2 Schematization in ANIMO — 171
    6.2.2 Oxygen requirement in soils — 172
    6.2.3 Vertical oxygen transport in soils — 173
    6.2.4 Diffusion in the water phase — 179
      6.2.4.1 Non-aggregated sandy soils — 180
      6.2.4.2 Swelling and shrinking soils — 184
      6.2.3.3 Rainfall and aeration — 187
  6.3 Influence of temperature — 188
    6.3.1 Temperature and biological processes — 188
    6.3.2 Calculation of soil temperature — 190
    6.3.3 Thermal properties of soils — 194

6.4 Influence of clay content	198
6.5 Influence of pH	199
References	202

7 WATER, NUTRIENT UPTAKE AND CROP PRODUCTION — 205

7.1 Transport to roots	205
7.1.1 Nutrient transport to plant roots	206
7.1.2 Nutrient uptake by plant roots	212
7.2 Gross photosynthesis	215
7.2.1 Gross photosynthesis of a standard crop	216
7.2.2 Gross photosynthesis under non-standard conditions	217
7.2.2.1 Maximum assimilation rate	218
7.2.2.2 Leaf area index	218
7.2.2.3 Soil moisture stress	219
7.3 Dry matter production of arable crops	220
7.3.1 Distribution of dry matter	220
7.3.2 Production of roots and exudates	223
7.3.4 Crop residues at harvest	227
7.4 Dry matter production of grassland	228
7.4.1 Grazing and dry matter production	229
7.4.2 Root dry matter production	230
7.5 Nutrient uptake by crops	231
7.5.1 General aspects	231
7.5.2 Nitrogen requirements of arable crops	234
7.5.3 Nitrogen requirement and gross production of grass	235
7.5.4 Phosphorus uptake by crops	237
7.5.5 Gross production and nitrogen requirement of forest plantations	238
References	239

8 MODEL VALIDATION AT FIELD SCALE — 243

8.1 Validation of TRANSOL	243
8.2 Validation of ANIMO	246
8.2.1 Introduction	246
8.2.2 Forage maize and catch crops on a sandy soil	249
8.2.2.1 Method	249
8.2.2.2 Results	251
8.2.3 Non-grazed grassland on a sandy soil	252
8.2.3.1 Method	252
8.2.3.3 Results	253
8.2.4 Winter wheat on silty loam soil	255
8.2.4.1 Method	255
8.2.4.2 Results	255
8.2.5 Grazed grassland on clay soil	258

8.2.5.1 Method	258
8.2.5.2 Results	259
8.2.6 Phosphorus leaching from grassland on a sandy soil	260
8.2.6.1 Method	260
8.2.6.2 Results	261
8.2.7 Flower bulbs on calcareous sandy soils	262
8.2.7.1 Method	263
8.2.7.2 Results	265
References	267
9 REGIONAL MODEL APPLICATIONS	**271**
9.1 Regional application of ANIMO	271
9.1.1 Leaching of nitrogen and phosphorus from rural areas to surface waters in the Netherlands	272
9.1.1.1 Methodology	272
9.1.1.2 Results	274
9.1.2 Simulation of phosphate leaching in catchments with phosphate saturated soils	278
9.1.2.1 Methodology	278
9.1.2.2 Results	278
9.1.3 Beerze, Reusel and Rosep catchments	280
9.1.3.1 Methodology	280
9.1.3.2 Results	284
9.2 Regional application of TRANSOL	288
9.2.1 Introduction	288
9.2.2 Beerze-Reusel-Rosep catchments	290
9.2.2.1 Methodology	290
9.2.2.2 Results	292
References	298
LIST OF SYMBOLS	**301**
INDEX	**313**

LIST OF FIGURES

1.1. Survey of data files, models and their interactions as used in regional studies of water pollution. Data and models surrounded by the dotted lines, dealing with pollution transport in the soil system are discussed in this publication. ... 7

2.1. Relation between $fr_{linex,wp}$ and $fr_{cl} + 2.5fr_{om}$ between saturation and wilting point. The curve is calculated with Eq. (2.7). a: for different classes of $CaCO_3$; b: for different classes of solid volume at saturation; c: comparison of measured and calculated values of $fr_{linex,wp}$. ... 20

2.2. General form of the shrinkage characteristic; void ratio = volume of pores/volume of solids; moisture ratio = volume of water/volume of solids. ... 21

2.3. Relation between the moisture content of aggregates and dry bulk density for Egyptian soils, with a clay fraction fr_{cl} varying between 0.3 and 0.6. ... 23

2.4. Relation between $dz/d\psi$ and ψ for medium coarse sand. 1: $a = 0.192$ ψ from 85 to 16 000 cm; 2: $\varsigma_1 = 0.026000$, $\varsigma_2 = 0.011700$, $\varsigma_3 = 0.003568$ ψ from 85 to 400 cm; 3: $\varsigma_1 = 0.003800$, $\varsigma_2 = 0.003100$, $\varsigma_3 = 0.000556$ ψ from 400 to 1 500 cm; 4: $\varsigma_1 = 0.000590$, $\varsigma_2 = 0.000875$, $\varsigma_3 = 0.000097$ ψ from 1 500 to 5 000 cm; 5: $\varsigma_1 = 0.000105$, $\varsigma_2 = 0.000300$, $\varsigma_3 = 0.000022$ ψ from 5 000 to 1 6000 cm. ... 26

2.5. Soil moisture distribution in the unsaturated subsoil as function of depth below the rootzone for different soils with the upper boundary at wilting point and a steady state capillary rise of 0.01 cm.d^{-1}. ... 27

2.6. Definition sketch of the distribution of soil moisture below the root zone as used for different situations in the module balance. ... 28

2.7. Stream line distribution as function of the point of infiltration and the distance travelled. ... 34

2.8. a: Model profile with vertical flux q_v, leakage flux q_l and drainage flux q_d; b: Vertical flux q_v as function of depth; c: Residence time as function of depth. ... 35

2.9. Streamline pattern in a groundwater body with infinite thickness. ... 37

2.10. Fraction of total drain discharge which will never be found below z/L. ... 38

2.11. The relation between $F(z/L)$ and z/l. ... 39

2.12. The relation between the residence time in groundwater per m drain distance and the relative distance x/L for a line drain in a soil profile with infinite thickness and a perfect drain with $H_d/L = 0.25$. ... 40

2.13. Cross section showing the regions for horizontal flux between $0 \leq 2x \leq L - 2H_d$ and radial flux for the region $L - 2H_d \leq 2x \leq L$. ... 41

2.14. The effect of the slope δ_r of the phreatic water level near the drains as additional cross-section on residence time. ... 45

2.15. The relation between the relative vertical flux distribution (q_z/q_d) in soil columns with finite thickness and the relative depth (z/L). ... 45

2.16. The relation between q_z^*/q_d and H_d/L calculated with a) Eq. (2.89) and b) Eq. (2.90). ... 47

2.17. The relation between residence time in groundwater per m drain distance and the relative distance x/L; - - - horizontal + radial flow; · · · · horizontal + infinite field flow; -· -· - perfect drain. ... 48

2.18. Schematized illustration of the regional drainage concept for a region with 4 different drainage systems. The different drainage systems are field drains, ditches, canals and discharge to the regional flow system. ... 51

2.19. Schematic presentation of coupling columns in a regional transport concept. ... 52

2.20. a) Relation between regional discharge and phreatic water level according to Ernst (1978); b) schematized relation with 4 levels of drainage. ... 53

2.21. Bottom of model discharge layers as a function of transmissivity in a heterogeneous soil profile. ... 55

3.1. Nutrient balance of the topsoil with respect to N and P. ... 60

3.2. Transport of dissolved compounds to and from a soil layer as influenced by water flow in both the unsaturated and saturated zone. ... 61

3.3. Model transport scheme with layer selection procedure. ... 69

3.4. Dimensionless numerical dispersion as a function of the number of pore water refreshments. ... 73

3.5. Solute concentration (mg dm^{-3}) as a function of depth simulated with TRANSOL and an analytical solution for 3 different substances: a) sorption and conversion, b) conversion and no sorption and c) sorption and no conversion. ... 75

3.6. Schematic presentation of the flux distribution through the surface reservoir in combination with surface runoff. ... 78

3.7. Schematic presentation for the combination of horizontal and radial flux into 2 columns; left: tile drains; right: open field drains. ... 83

3.8. Spatial schematization of the top-system (ANIMO, TRANSOL) and the aquifer system (AQUIMIX). ... 86

4.1. The relation between A_γ and temperature (left) and B_γ and temperature (right). ... 91

4.2. Four possibilities of the concentration course with time within a time step Δt establishing an eventual exceedance of the equilibrium concentration c_{eq}. ... 112

List of Figures

4.3. Phosphate precipitation rate as a function of soil moisture concentration, characterized by a number of equilibrium concentration levels c_{eq} and corresponding rate constants k_{pr}
1) $t_0 <$ time $\leq t$: $c < c_{eq}$, $X_p = 0$, $\partial X_p/\partial t = 0$;
2) $t_0 \leq$ time $\leq \tau$: $c < c_{eq}$, $X_p = 0$, $\partial X_p/\partial t = 0$;
   $\tau \leq$ time $\leq t$: $c = c_{eq}$, $X_p \geq 0$, $\partial c/\partial t = 0$;
3) $t_0 \leq$ time $< \tau$: $c = c_{eq}$, $X_p > 0$, $\partial c/\partial t = 0$;
   $\tau \leq$ time $\leq t$: $c \leq c_{eq}$, $X_p \geq 0$, $\partial X_p/\partial t = 0$;
4) $t_0 \leq$ time $< t$: $c = c_{eq}$, $X_p > 0$, $\partial c/\partial t = 0$. ... 128

5.1. Simplified presentation of the carbon cycle in soil. ... 141

5.2. Decomposition of different types of peat calculated with three different models. ... 143

5.3. Decomposition of different organic materials calculated with three different models. ... 144

5.4. The relation between the apparent age of the organic material after Janssen (1986) and the fraction distribution for the various decomposition rates. The points are the data from Table 5.4; the curves are calculated with Eq. (5.14). ... 146

5.5. Schematic presentation of the nitrogen cycle in soils. ... 153

5.6. Fertilization dependent activity coefficient for N-fixation after Janssen (1992). ... 159

5.7. Schematic presentation of the phosphate cycle in soils. ... 161

6.1. The relation between the reduction factor and the soil moisture suction for different processes. ... 169

6.2. Atmospheric oxygen and nitrate oxygen related processes in the ANIMO model. ... 171

6.3. Schematic representation of diffusive oxygen transport in the aeration module. ... 176

6.4. Schematization of an aerobic and an anaerobic zone around an air filled pore. ... 180

6.5. The relation between the relative distance $R.\bar{r}_p^{-1}$ and the relative oxygen production $D_{sw}^{ox} c_{we}^{ox}.(S_{ox} \bar{r}_p^2)^{-1}$ for non-aggregated soils. ... 183

6.6. The relation between the relative distance R/r_{sbl} and the relative oxygen production $D_w^{ox} c_{we}^{ox}/(S_{ox} r_{sbl}^2)$ for aggregated soils for different aggregate shapes. ... 187

6.7. The relation between the relative biological activity and temperature after data presented by Kolenbrander (pers. comm.) The curves are calculated using Eq. (6.61) and Eq. (6.62). ... 189

6.8. The effect of temperature on different biological processes presented by different authors as collected by van Huet (1983). ... 190

6.9. Comparison between measured temperature and temperature simulated with Eq. (6.66). ... 192

6.10. The relation between damping depth and the sum of 2.5 times organic matter fraction plus clay fraction of the soil; the curve has been calculated ignoring the data at wilting point. 196

6.11. Relation between the function ($\lambda_T.D_m^{-1}$) and the sum of 2.5 times organic matter content and the clay content; the curve is calculated for saturated soils. 197

6.12. Relation between the reduction function and relative water saturation for unsaturated soils. 198

6.13. Relation between the function $\lambda_T.D_m^{-1}$, calculated from the original soil data given by Rijtema et al. (1997), and the value calculated using Eq. (6.84). 198

6.14. The relation between the reduction factor for decomposition of organic material and the clay content of the soil after Hansen et al. (1990). 199

6.15. The effect of pH on the reduction of the rate coefficients for different processes: Ammonification: $\xi_1 = 1.24$; $\xi_2 = 3.7$; Nitrification: $\xi_1 = 2.00$; $\xi_2 = 5.5$; Denitrification: $\xi_1 = 1.50$; $\xi_2 = 5.7$. 200

6.16. The fraction of mineralization of organic material by denitrification under anaerobic conditions in relation to pH. 201

7.1. The relation between relative uptake rate $U_{cr}.\bar{c}^{-1}$ and the concentration ratio $c_r.\bar{c}^{-1}$ for different conditions of root density and transpiration. a_1, b_1 and c_1: wet soils; a_2, b_2 and c_2 dry soils. 210

7.2. The relation between the soil-plant coefficient and the concentration ratio $c_r.\bar{c}^{-1}$ with transpiration rates of 2.5 and 5.0 mm.d^{-1} and different soil moisture conditions; the root densities are given for π_r equalling 1 000, 8 000 and 200 000 m^{-2}. 211

7.3. The relation between Υ_{cpl} and the ratio $c_r^{min}.\bar{c}^{-1}$ for different conditions of root density, transpiration rate and soil moisture, and a rooting depth of 0.5 m. 213

7.4. The relation between the transpiration stream concentration uptake rate σ_{cpl}^{max} and the plant diffusion resistance Υ_{cpl} with different soil moisture, root density and transpiration conditions. 214

7.5. Photosynthesis of a standard crop for both perfectly clear and completely overcast skies as a function of latitude and day number of the year. The markers are the data given by De Wit (1965). 217

7.6. Relation between soil moisture suction and dry matter content of grass shoots. 220

7.7. The relation between the distribution of dry matter production and crop stress. 222

7.8. The relation between depth of rooting and the number of days after planting for different crops. 224

7.9. The relation between the dry matter present in roots and depth of rooting. 225

7.10.	Relation between dry matter present in grass roots and day number of the year for different treatments (Schuurman, 1973).	230
7.11.	The relation between total nitrogen uptake by grass and the accumulation of NO_3-N in the shoots.	236
7.12.	The relation between N- fraction and P-fraction in dry matter of some crops derived from data given by Bosch and de Jonge (1989), Westerdijk et al. (1991) and Asijee (1993).	237
8.1.	Aeric mass of Ethoprophos in the soil layer 0-30 cm, from 1 December 1989 until 30 November 1990.	244
8.2.	Concentration profiles of Ethoprophos, 103 days (a), 214 days (b) and 278 days (c) after application.	244
8.3.	Maximum concentration in the soil layer of 1.0 to 2.0 m below surface (a) and leaching percentage passing through the depth of 1 m below surface (b), as simulated by TRANSOL and PESTLA for 168 different combinations of degradation and adsorption.	245
8.4.	Location of field experiments used to validate the ANIMO model.	247
8.5.	Calibration Heino: a) measured and simulated mineral N in 0-60 cm-surface; b) concentration of nitrate at 1 m-surface; c) uptake of N by forage maize without catch crop and with a fertilizer treatment of 200 kg ha^{-1} N.	250
8.6.	a) Measured and simulated mineral N in 0-60 cm-surface; b) concentration of nitrate at 1m-surface; c) uptake of N by crop; forage maize with different catch crops and fertilizer treatments.	251
8.7.	Grassland with sub-surface injection of cattle slurry 80 ton/ha and artificial fertilizer level of 400 kg/ha (total fertilizer level of 800 kg ha^{-1} N). a) Measured and simulated mineral N in the layer 0-50 cm -surface; b) concentration of nitrate at 1 m-surface and c) uptake of N by crop during the period 1980-1984.	253
8.8.	Validation Ruurlo: a) measured and simulated mineral N in 0-50 cm-surface; b) concentration of nitrate at 1 m-surface and c) uptake of N by crop; average values for period 1980-1984; grassland with different fertilizer management.	254
8.9.	Measured and simulated N-uptake of winter wheat at Nagele; a) fertilizer levels of 110 kg ha^{-1} N; b) fertilizer level of 150 kg ha^{-1} N.	256
8.10.	Validation with winter wheat at Nagele: Soil mineral-N; fertilizer level 110 kg ha^{-1} N; soil layers: a) 0-40 cm, b) 0-100 cm. Nitrate (c) and soil mineral ammonium (d), soil layer 0-100 cm. Soil mineral-N, fertilizer level 150 kg ha^{-1} N; soil layers: e) 0-40 cm, f) 0-100 cm.	257
8.11.	Validation Lelystad: Measured and simulated nitrate drain discharge from a field covered by a mixture of grass and clover; daily (a) and cumulative (b); period January 1994-April 1994.	259

8.12. Calibration Putten: measured and simulated fractions (-) of phosphate-occupation (a) and ortho-phosphate concentrations (mg l^{-1} P) in the liquid phase (b) in the upper 80 cm of the soil. ... 261

8.13. Validation Putten: simulated (a) and measured loads (b) of ortho-phosphate (kg ha^{-1} P) discharged by a field ditch. ... 262

8.14. Breakthrough curves of phosphate leaching experiments conducted with two soil samples originating from 0-25 cm depth (bottom) and 52-72 cm depth (top); + = measurement, full drawn line simulation. ... 264

8.15. Measured and simulated ortho-phosphate and dissolved organic phosphate concentrations in soil water at 40 cm, 60 cm, 100 cm and 140 cm depth. ... 266

8.16. Measured and simulated total-P concentrations in drainwater at a St.Maartensbrug flower bulb field plot. ... 266

9.1. Four regional applications of the model ANIMO in the Netherlands: nutrient leaching in the regions Bergambacht, Schuitenbeek and Beerze-Reusel-Rosep (right); phosphorus leaching from sandy soils (left). ... 271

9.2. Model instrument (between the dotted lines) and main data flow. ... 273

9.3. N concentrations (g.m^{-3} N) in water leaching during the winter period from the soil to surface waters (simulated) and concentrations in surface waters (measured) in the region 'Schuitenbeek'. ... 274

9.4. Nitrogen (kg N.ha^{-1}.a^{-1}) and phosphorus (kg P$_2$O$_5$.ha^{-1}.a^{-1}) leaching from soils of rural areas to surface water systems in the Netherlands as calculated for the scenarios: *present*, *policy* and *zero*. ... 275

9.5. Nitrate concentrations in groundwater in 1985 as a result of fertilization according to the scenaior *present*. ... 276

9.6. Nitrate concentrations in groundwater in 2015 as a result of fertilization according to the scenario *policy*. ... 277

9.7. Phosphate leaching (kg.ha^{-1}.a^{-1}) from sandy soils of rural areas to surface water systems as resulting from a scenario with 50 years of phosphate surplus of 10 kg.ha^{-1}.a^{-1} P$_2$O$_5$. ... 279

9.8. Survey of data files, models and their interactions as used in the regional study of water pollution in the Beerze, Reusel and Rosep catchments. ... 281

9.9. Linking of the ANIMO model to the groundwater quality model AQUIMIX to account for regional interactions between infiltration areas and exfiltration areas. ... 283

9.10. Mean groundwater level during spring (left) raise of spring level resulted from the raise of third order drainage level and the weir management during summer (middle) and the additional raise of the spring level caused by the restoration of the historical flow regime of streams. ... 285

9.11. Nitrate concentration in the reference situation predicted for areas covered by grassland (left) decrease of nitrate concentration as resulted from the raise of third order drainage level and the weir management

	during summer (middle) and the additional decrease of the nitrate concentration caused by the restoration of the historical flow regime of streams (right).	286
9.12.	P-leaching to surface water in the reference situation (left), P-discharge as resulted from the raise of third order drainage level and the weir management during summer (middle) and the raise of the spring level caused by the restoration of the historical flow regime of streams (right).	287
9.13.	Survey of data files, models and their interactions as used in the regional study of water pollution by pesticides in the Beerze, Reusel and Rosep catchments.	291
9.14.	The distribution of the load of bentazone on surface water after 20 years of application (left) and the distribution of the bentazone load to groundwater after 20 years of application (right) in the catchments of the rivers Beerze, Reusel and Rosep. The loads are expressed in $kg.ha.a^{-1}$ of the area where the pesticide was used.	297

LIST OF TABLES

2.1. Values of k_0, α, ψ_b and a for the relations between capillary conductivity and suction.   18

2.2. Soil moisture content as volume percentage in relation to suction for standard soils.   19

2.3. Critical values of soil moisture and depth, as used in the model BALANCE.   32

4.1. Optimized ion-dependent parameters d and C for Eq. (4.6) after Ritsema (1993).   92

4.2. Description of the stoichiometry of a system containing four cations exchanging according to the hetero-valent exchange equations.   96

4.3. Description of the stoichiometry of a system containing three cations exchanging according to either the Gaines–Thomas equation or the Vanselow equation.   97

4.4. Description of the stoichiometry of a precipitation and dissolution reaction.   107

4.5. The stoichiometry of a system, describing the transport of CO_2 to the water phase.   108

4.6. Definition of the chemical system of aqueous species considered in calcareous equilibria in aquifers and alkalinization in (semi-)arid regions.   109

4.7. Definition of the chemical system of aqueous species considered in the research on the acidification potential of acid sulphate soils (Bronswijk and Groenenberg, 1994).   113

4.8. Adsorption characteristics for two herbicides in a loamy sand soil.   118

4.9. Logarithmic values of formation constants of complexes at 25 °C and ionic strength = 0, after de Vries and Bakker (1996).   120

4.10. Values for the Freundlich exponent (N) and the coefficients ς_n in the transfer function with the Freundlich adsorption coefficient K_F (after de Vries and Bakker 1996).   120

4.11. Values of pK_a (- log K_a) and log K_c^M describing the dissociation and complexation of organic acids with heavy metals by a mono-protic acid after de Vries and Bakker (1996).   121

4.12. Values of the coefficients ς_n used in Eq. (4.89) for the relationship between the partition coefficient K_d^{tot} and soil properties after de Vries and Bakker (1996).   123

4.13. Parameters describing the rate dependent phosphate sorption for a wide range of Dutch sandy soils (after Schoumans, 1995).   125

5.1.	Start decomposition rates k_i and apparent age of different organic materials after Hendriks (1991).	138
5.2.	Biomass carbon turnover rates as collected from literature by Dendooven (1990).	139
5.3.	Decay rates for soil biomass as given in literature (see Dendooven,1990).	140
5.4.	Constants ς_1 and ς_2 of Eq. (5.7), the apparent age of organic materials (Hendriks, 1991) and fractions of different organic materials as defined by their decomposition rates, at an average soil temperatrure of 10 °C.	143
5.5.	Fraction distribution of the hypothetical organic material, with an apparent age 0, as the basis for the fraction distribution of organic materials as a function of the apparent age.	145
5.6.	The apparent age of organic manures (Janssen, 1995) and fractions of different organic materials as defined by their decomposition rates.	145
5.7.	Fraction of living biomass under steady state conditions for different fresh materials with an assumed assimilation factor of 0.3 and different biomass death rates.	150
5.8.	Decomposition rates, nutrient fractions and organic material fractions of different materials used in ANIMO.	152
5.9.	Emission fractions for ammonium volatilization of different land spreading techniques for animal slurry, after Roelsma (1997).	160
6.1.	Relation between the reduction factor for stress conditions and pF.	170
6.2.	Relative diffusion coefficient D_s/D_a for different soils after Bakker et al. (1987).	174
6.3.	Relation between temperature and the oxygen diffusion coefficient D_w^{ox} in water.	181
6.4.	Thermal properties of soil materials after de Vries (1963).	195
6.5.	Values of the constants ξ_1 and ξ_2 of Eq. (6.86) for different processes.	200
6.6.	Transfer functions for assessment of pH-values from soil chemical properties.	201
7.1.	Values of ψ_l^c in bars for a number of field crops derived from the analysis by Rijtema and El Guindi (1986).	222
7.2.	Relative production respiration and relative root production of some crops.	223
7.3.	Crop residues in kg dry matter.ha^{-1} (de Jonge, 1981).	227
7.4.	Default values for dry matter present in grass swards for different types of grassland utilization.	228
7.5.	N-content of crop residues at harvest in %.	235
7.6.	Dry matter production and nitrogen uptake of stemwood, foliage and fine roots by deciduous trees and spruce trees in the Netherlands after Rijtema and De Vries (1994).	238

List of Tables

8.1.	Validated variables of ANIMO model at different locations in The Netherlands.	247
8.2.	Main characteristics of field experiments used to validate the ANIMO model.	248
8.3.	Parametrization of the Langmuir equation describing the equilibrium sorption of phosphate in calcareous sandy soils.	263
9.1.	Groundwater regime classes, as defined by the mean highest groundwater table, the mean lowest groundwater table and the groundwater fluctuation in regions with pleistocene sands, marine clay soils and dune sands.	279
9.2.	Average nitrate concentration of groundwater in areas covered by grassland, classified by the average annual sprinkling requirement.	285
9.3.	The distribution of land use and pesticide use in the catchments of the Beerze, Reusel and Rosep. Total area of the three catchments is 44 000 ha. The use of pesticides is given in kg active compound per ha (Aarnink *et al.* 1996).	293
9.4.	The physical-chemical properties of the selected representative pesticides.	294
9.5.	The concentration of representative pesticides in shallow groundwater after 20 years of application in relation to the cumulative fraction of the area where these pesticides have been applied in the catchments of the rivers Beerze, Reusel and Rosep, derived from Aarnink *et al.* (1996).	295
9.6.	The load of representative pesticides to surface water after 20 years of application in relation to the cumulative fraction of the area where these pesticides have been applied in the catchments of the rivers Beerze, Reusel and Rosep, derived from Aarnink *et al.* (1996).	296

ENVIRONMENTAL IMPACT OF LAND USE IN RURAL REGIONS

CHAPTER 1

INTRODUCTION

1.1 Problem and purpose

Groundwater is born as rainwater or as water infiltrating from rivers and lakes with a water level above the piezometric head in the aquifer below. This origin, however, has only limited influence on the groundwater composition. During the fall through the air, rainwater will pick up atmospheric pollution. Due to the use of fossil fuels and the increase in CO_2 content of the atmospheric air, the pH of rainwater shows a drop to a value of 5.6 at the moment, while the discharge of sulphur and nitrogen oxides by industry and traffic has caused the pH locally to fall to values as low as 4.0.

Groundwater quality changes by natural causes, as groundwater recharge and chemical weathering of the soil substrate, but not as quickly as the quality of surface waters due to large differences in residence times. Our understanding of nature is seriously hampered by a lack of information on reactions typically encountered in natural waters. The distribution of chemical species in waters and sediments is strongly influenced by an interaction of mixing water from different sources and biological cycles. Often mass balance models may be used to describe the dynamics of some processes.

Aquatic chemistry is of practical importance because water is a necessary resource for humans. Generally, we are not concerned with the quantity of water, as water as a substance is abundantly available, but with the quality of the water and the distribution of fresh water. Water represents one of the most essential constituents for the human environment; its protection is imperative for access to safe drinking water and sustainable water sources for other purposes.

In recent years, an increasing threat of groundwater pollution due to human activities has become of great importance for many countries. Considering the effects on man and on the environment, the key role of groundwater as the main source for drinking water supply has to be taken into account. Conservation of aquatic resources can only be achieved when human interference with the aquatic systems is taken into account. Water pollution control cannot solely consist of waste water treatment.

When dealing with groundwater pollution, it should be kept in mind that there is an important difference between groundwater and surface water pollution and treatment, respectively. While the decision to restore surface water is always made realizing that the quality soon will be restored after the sources of pollution are removed, this does not pertain to groundwater which will remain polluted for decades or even centuries, due to its specific feature of being a slow motion, slow response resource. Restoration of groundwater quality is generally a very costly operation.

For the major part, the adverse effects on groundwater quality are the results of man's activity at the soil surface. In addition to natural causes, water quality changes reflect the evolution of agricultural practices, atmospheric deposition and urban and industrial waste inputs as well as the success of waste water treatment programmes.

By evaluating the strength of significant emission sources and by comparing natural and pollutant fluxes, it is necessary to assess more quantitatively the influences of human activities on the pollution of fresh waters. It is necessary to develop the ability to modify and manipulate the aquatic environment in order to maintain its quality as a life preservation system and a reservoir for genetic diversity.

Growing world population demands an increasing agricultural production. Though the volume of agricultural output has increased considerably over the past few decades, there is an imbalance between population growth and increase in agricultural production per inhabitant in different regions (Rijtema 1993, 1994). Both good agricultural practice and regulations by authorities must lead to optimization of agricultural production with minimization of environmental impacts. Management supporting models are useful tools in predicting long term effects of emission reducing measures. Models incorporating production relations can also be used to support risk analysis of expanding agricultural activities.

Large quantities of treated or untreated waste water can be used in agricultural irrigation, landscape irrigation and groundwater recharge. Agricultural irrigation and landscape irrigation are the largest current and projected uses of treated waste water.

Although human activities have a large impact on the environment, it should be realized that technology as well as energy and resource utilization in themselves are bad. However, for the maintenance of our civilization and culture and for the enhancement of the quality of life, and especially for food production for the increasing world population, we will have to continue to depend on technology and energy utilization. Social criteria and growing pressures for social equality must co-determine scientific and technological development.

In a broad sense, pollution has been characterized as an alteration of our surroundings in such a way that they become unfavourable to us and our life. This characterization implies that pollution is not synonymous with the addition of contaminants or pollutants to the environment but can also result from other direct or indirect consequences of human action, such as the changes in water management on sulphate-acid soils and the effects on processes as a result of changes in the hydrological field due to abstraction of groundwater for municipal water supply.

Point sources of pollution are differentiated primarily from diffuse sources by the precision with which the source of contamination can be identified. Although point sources of pollution may, undoubtedly, threaten the quality of groundwater resources, it might be expected that the impact of leachate migrating from dilute and dispersive sites may be reduced to acceptable levels by proper site selection, execution of preventive protection measures and waste management. In contrast, pollution originating from farming practice, the application of agrochemicals, application of animal slurries and the use of sewage sludge as fertilizer supplements represent diffuse pollution sources. Pollution due to diffuse sources is probably the most difficult pollution to model because the loads are usually non-homogeneous and governed by spatially inhomogeneous and dynamic processes of the chemical and biochemical phenomena, which are often not well known or represented. Although there exist numerous models of point pollution transfer in porous media, only few models have been developed specifically for the prediction of groundwater pollution by diffuse sources.

Concern on groundwater contamination has focused attention on the processes that affect chemical fate in soil-water systems. Independent of the mode of introduction of the chemicals, a major concern with respect to groundwater contamination is the passage of these chemicals through the relatively thin layers of soil that cover many terrestrial surfaces. This water unsaturated zone of the soil profile, extending from the soil-atmosphere interface to the groundwater table, including the root zone of most plants, is the chemically and biologically most active region of the biosphere. The residence time of chemicals within this region, and the processes operative within it, ultimately determine the degree of groundwater contamination. It is now recognized that a better understanding of the basic physical, chemical and biological processes of the unsaturated zone is necessary to induce a better management of human activities, resulting in minimized groundwater as well as surface water contamination and associated risks.

There is a need for decision support systems that can be used for the evaluation of measures for water quality management of groundwater and surface waters, as well as for regional physical planning of land utilization for urban use, agriculture and nature. The increasing complexity of decisions, in terms of the number of alternatives and impacts that must be considered, have increased the demand for such decision supporting systems. Because choice is difficult, part of this demand is negative and expressed as a desire for a technology which yields certainty: a scientific way of taking decisions, so removing the burden from politicians.

More positively, the complexity of decisions requires aids to decision making which reduce the complexity to manageable levels, and clarify the issues involved in the choice. The formulation and implementation of a rational and effective policy to reduce the pollutant load to the environment requires a thorough understanding of the behaviour of pollutants in the soil. This understanding can be increased by using simulation models, which incorporate crop growth, nutrient availability and leaching of pollutants to groundwater and to surface waters.

1.2 General aspects of modelling

Mathematical modelling is used to describe the mechanical, physical-chemical and biochemical phenomena characterizing pollution transport in the soil and the subsoil in order to predict their evaluation under various assumptions.

Structural models use information on the structure of the phenomena, which is as complete as possible, and usually consist of transport and physical-chemical and/or biochemical models. These complete models account for all elementary processes (e.g. mechanical, physical-chemical, biochemical and biological processes), usually by means of partial differential equations. They are generally very complex and their solution most often will be numerical. Structural models consist of physical- and biochemical models to estimate the quantity of the compound available for leaching and they are linked with transport models providing boundary conditions in the form of input water fluxes or of input of concentration values or source terms. They may be difficult to use because they require many parameters which may be hard to measure.

The most detailed models that incorporate the latest technologies and concepts are generally the most technically defensible and have, in theory, the widest range of applicability. Practically, however, these may not be the most useful models for application in farm-management or regional land-use decisions, given the required knowledge of the system and the data requirements.

Structural models can be subdivided into mechanistic and functional models (Bogárdi et al. 1990, Rijtema and van der Bolt 1996):

- Mechanistic implies that the models take into account the most fundamental mechanisms of the processes, as presently known and understood, as for instance Darcy's law for water flow, combination of mass-flow and diffusion-dispersion mechanisms for solute transport, fundamental descriptions of biomass development and biochemical processes, etc. Mechanistic models are primarily research tools.
- The term functional models refers to models that incorporate simplified formulations of processes and treatments of solute and water and make no claim of fundamentality. The functional models act generally as useful guides to support management decisions of soil and water resources.

A structural model, therefore, must be capable of giving a valid representation of, first, the flow system, and second, the physical and geochemical processes acting on the contaminant during its migration along the flow paths.

Various approaches to non-point source contamination in saturated groundwater vary from very simple lumped models to complex models describing flow, transport and chemical reaction in heterogeneous groundwater systems are reviewed by Duffy et al. (1990) and by van Genuchten and Šimůnek (1996). It is usual to distinguish between transport in the unsaturated zone, mainly vertical, and transport in the saturated zone, mainly horizontal.

Many models of varying degree of complexity and dimensionality have been developed during the past decades to quantify the basic physical and chemical processes affecting pollutant transport in the unsaturated zone. Models for variably unsaturated-saturated flow, solute transport, aqueous chemistry and cation exchange were initially developed independently of each other, and only recently there has been significant effort to couple the different models. For example, the processes of adsorption-desorption and cation exchange were often accounted for by using relatively simple linear or non-linear Freundlich isotherms such that all reactions between the solid and liquid phases were forced to be lumped into a single distribution coefficient and possibly a non-linear exponent. Other processes such as precipitation-dissolution and biodegradation were generally simulated by means of simple first and/or zero-order rate processes. These simplifying approaches were needed so as to keep the mathematics relatively simple.

The importance of the unsaturated zone as an inextricable part of the hydrologic cycle has long been recognized. Theoretical and experimental studies on both water flow and solute transport in this zone have been further motivated by attempts to manage the root zone of agricultural soils optimally as well as concerns about soil and groundwater pollution. Reviews of the variety of leaching models of the unsaturated zone and their potential use for management and planning have been given by Vachaud *et al.* (1990) and de Willigen *et al.* (1990). These studies have greatly increased the conceptual understanding of the many complex and interactive physical, chemical and microbiological processes operating in the unsaturated zone. They have also led to a large number of models which vary widely in their conceptual approach and degree of sophistication.

The purpose of physical-chemical and biochemical models is to provide a quantitative estimate of the substances produced from the reactions governing the transformation of different compounds in the soil and subsoil in order to compute the amounts of materials which could possibly be leached and transported in the soil and subsoil. Nitrogen and phosphates can be present in different forms in the soil and the various forms can be transformed through the processes in the nitrogen and phosphate cycle. To quantify these processes in the nitrogen and phosphate cycle it is also necessary to quantify the processes in the carbon cycle because of the many interdependencies between organic matter and nutrients. The problem of coupling models for water flow and solute transport with multi-component chemical equilibrium and non-equilibrium models, combined with more complex biodegradation models, is now increasingly being addressed, facilitated in part by the introduction of more powerful computers and the development of more advanced numerical techniques (Yeh and Tripathi 1989).

The interrelationships between soil, subsoil and surface waters make it unrealistic to treat the saturated and unsaturated zones and the discharge to surface waters separately. If the appropriate model simulations indicate no movement of residues to the groundwater table, then the assessment is complete. However, if the model predicts residues will enter groundwater, additional information is needed about the movement and magnitude of these residues in groundwater. Such information can be obtained by linking the unsaturated zone model to a saturated zone model, considering the outflow at the lower boundary of the unsaturated zone model as input or source term for the saturated model and vice versa.

The use of quasi-three dimensional finite element models, including the unsaturated zone, for the calculation of the water flow is therefore a good choice, using the output of such a model as input in a water quality model for the study of transport of non-point source contaminants.

The unsaturated-saturated flow model SIMGRO (Querner 1988, Querner and van Bakel 1989), using an implicit calculation scheme, is a good example of such a water transport model. The output of this model can be used as input in physical-chemical and biological models for the simulation of water quality in regional studies.

1.3 Scope of the book

Figure 1.1 gives an example of a schematic presentation of the connection between data files, data streams and simulation models that can be used for the evaluation of regional environmental impacts of diffuse sources on groundwater and surface water.

- The model SIMGRO (Querner 1988, Querner and van Bakel 1989) has been developed to simulate flow in the unsaturated and saturated zones. The saturated zone module consists of a quasi-three dimensional finite element model using an implicit calculation scheme. The unsaturated zone is related to land use on a sub-regional level. Subregions are chosen to have relative uniform soil properties and hydrologic conditions. The soil moisture distribution in the unsaturated domain can be calculated with the model CAPSEV.
- The model TREND (van Jaarsveld 1995) determines the long-term atmospheric behaviour of pollutants originating from emissions to the atmosphere by industries, refuse destructors, traffic and agriculture. The model TREND gives the input data for regional spreading of atmospheric deposition both in space and time.
- The input data for the distribution of animal slurries and inorganic fertilizers are obtained from calculations with the model SLAPP, optimizing the nutrient distribution for different forms of land use on basis of input obtained from a GIS-system yielding data on land use and intensity of animal husbandry for districts. The model SLAPP (SLurry APPlication) (van Walsum 1988) translates the animal slurry production per district into actual fertilization data per type of land use per subregion.
- The pesticide information system ISBEST (Merkelbach and Lentjes 1993, Lentjes and Denneboom 1996) calculates per district and per crop species the quantity of each pesticide used. ISBEST is a data base management system in which geographically independent data are stored, such as physical and chemical information of pesticides, application concentration and quantities as well as the number of applications for different crop species under different meteorological conditions.
- The model ANIMO (Rijtema et al. 1997) takes into account the dynamic exchange of nutrients with soil organic matter under influence of biological processes.
- Soil organic matter is considered as an externally determined environmental factor in the model TRANSOL (Kroes and Rijtema 1996). This makes TRANSOL very suitable for the evaluation of the transport of pesticides and other synthetic organic

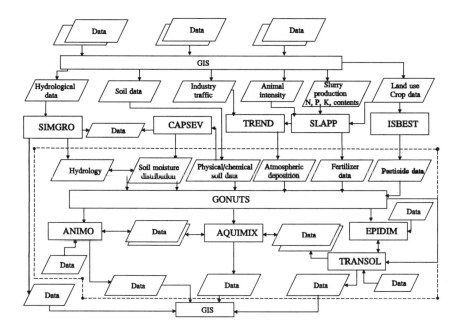

Fig. 1.1: Survey of data files, models and their interactions as used in regional studies of water pollution. Data and models surrounded by the dotted lines, dealing with pollution transport in the soil system are discussed in this publication.

pollutants. Combination of TRANSOL with the chemical equilibrium model EPIDIM (Groenendijk 1997a) is used for the calculation of multi-component transport of inorganic pollutants, as for instance related to acid rain and sulphate-acid soil problems.

- The water quality aspects related to the regional groundwater flow between sub-regions are calculated in the groundwater quality model AQUIMIX (van der Bolt et al. 1996). The relation between ANIMO and TRANSOL and the regional model AQUIMIX is determined by leakage and seepage fluxes between the sub-regional top system and the regional aquifer system. Interactions between sub-regions via the groundwater system are accounted for by linking both models.
- This procedure was formalized for nutrient studies in the programme GONUTS (van der Bolt et al. 1996). The programme GONUTS (Geographical Oriented NUTrient Studies) organizes the data streams between the different models, executes pre-calculations and makes input data files for ANIMO.
- This programme has been extended for the execution of pre-calculations and the creation of input data files for TRANSOL. The programme GONUTS calls the models ANIMO, TRANSOL and AQUIMIX as sub-routines.

The present book does not cover all aspects of regional modelling. The data and models surrounded by the dotted lines in Fig 1.1, dealing with pollution transport in the soil system, are discussed in this publication. In the first chapters of the book a description is given of the several conceptual approaches for modelling solute transport in variably unsaturated-saturated media. The book gives the scientific background of the water quality models that have been developed for the determination of the emission of nutrients, pesticides and other polluting compounds from diffuse sources to groundwater and surface waters. This also includes the description of some pedotransfer functions for the incorporation of soil type dependent physical parameters, affecting the processes. Transport of pollutants, via the soil system, to surface water are considered both as concentration as well as load, but processes in surface water are beyond the scope of the book.

1.4 Outline of the water quality models ANIMO and TRANSOL

The first version of ANIMO was developed by Berghuijs-van Dijk *et al.* (1985). Various versions of the model ANIMO were tested, using measured data from long-term fertilization experiments (Bogárdi *et al.* 1990, Rijtema and Kroes 1991a, 1991b, Jansen 1991, Kroes *et al.* 1996). This resulted in an improved version of the model (Rijtema *et al.* 1997). ANIMO 3.5 makes use of a general module for solute transport, which includes non-equilibrium sorption and precipitation as developed by Groenendijk (1997b).

The first version of TRANSOL was developed by Kroes and Rijtema (1989) and subsequently the model was applied in various projects resulting in an improved version of the model (Kroes and Rijtema 1996). TRANSOL uses the same transport module as is present in ANIMO. Both models must be considered as functional structural models incorporating simplified formulations of processes and treatments of solute and water and it makes no claim of fundamentality. The present publication gives a description of the processes that should be considered in nutrient leaching models.

Both models are operational on field scale as well as on regional scale. The models have been developed as stand alone water quality models, that can be used in combination with different regional models for the quantification of the water fluxes in the hydrological field and the water management in the unsaturated zone. A special hydrological module has been inserted in ANIMO and TRANSOL to adapt the unsaturated soil moisture data, obtained from the hydrological model, to the layers distinguished in the quality model, if necessary. Both models have a few special features which make the models especially attractive for application in regional studies:

- fast semi-analytical solutions of the general transport equations for solute flow;
- solute transport in both the unsaturated and the saturated zone;
- hydrology is not included, which allows linkage to different hydrological models;

- ANIMO calculates transport of NO_3, NH_4, organic and mineral phosphates and dissolved organic matter by surface runoff, drainage to surface waters and leaching to aquifer groundwater systems;
- TRANSOL calculates transport of inorganic compounds as heavy metals, pesticides and other synthetic organic pollutants by surface runoff, drainage to surface waters and leaching to aquifer groundwater systems.

In the water quality model ANIMO, attention is focused on the following processes:

- mineralization of nutrients (N + P) in relation to formation and decomposition of different types of organic matter as organic fertilizers, root material, root exudates, yield losses and native soil organic matter;
- denitrification in relation to partial anaerobiosis and the presence of organic material;
- nitrogen fixation through soil bacteria during growth of leguminosae;
- instantaneous and time dependent adsorption and precipitation of phosphates.

Inputs of both organic and inorganic nitrogen and phosphates in ANIMO can originate from fertilization and soluble N-forms in precipitation. In the soil-water-plant system the different forms can be transformed one to another, and some can be transported to deeper layers. These processes are influenced by environmental factors such as temperature, moisture, aeration and pH. The processes and their influencing factors are quantitatively described in the model. The compounds can leave the topsoil due to harvesting, leaching to deeper layers and volatilization.

In the water quality model TRANSOL, attention is focused on the following processes:

- degradation of organic pollutants, combined with the formation of metabolytes;
- non-equilibrium precipitation and non-equilibrium multi-site adsorption;
- multi-component transport in combination with EPIDIM.

1.5 Reading guide

- Chapter 2 deals with the hydrological schematization in ANIMO and TRANSOL for application of the models on both field scale and regional scale. One-dimensional leaching models generally represent a vertical soil column. Within the unsaturated zone, chemical substances are transported by vertical flows, whereas in the saturated zone the drainage discharge leaves the vertical column sideways In the models, the distribution of lateral drainage fluxes with depth have been used to simulate the response of the load on the surface water system to the inputs in the groundwater system. In the saturated zone of the model approach a number of horizontal layers have been distinguished, so a pseudo two-dimensional transport is considered. The models are stand alone water quality models. They can be used in combination with different hydrological models. The water quality models determine the solute transport following the mixing cell principles. The models can be used to analyse problems on either field, farm or regional scale. They also deal with the pollutant emission to drainage systems and fresh water aquifers.

- Chapter 3 describes the transport and conservation equation, which is the central part of the models ANIMO and TRANSOL. This equation has to be used for all dissolved compounds. With the aid of analytical solutions of this equation the new concentrations of the dissolved compounds in all layers are calculated taking into account simultaneous transport and transformation processes. The soil is divided into a number of horizontal layers, which can differ in thickness. The models work as a pseudo two-dimensional system inside each area.

- In Chapter 4 the physical-chemical processes in the model formulation are treated. Surface waters and soils are open and dynamic systems with variable inputs and outputs of mass and energy for which the state of equilibrium is a construct. Steady-state models reflecting the time-invariant condition of a reaction system may frequently serve as an idealized counterpart of an open natural water system. While the emphasis is on chemical equilibria, kinetic information can often be interwoven with equilibrium information in order to gain understanding of the major factors that influence the chemistry of the aquatic environment.

- In Chapter 5 different approaches to transformations in the carbon cycle are dealt with, resulting in the schematization used in ANIMO. Nutrients can be present in different forms in the soil system and these forms can be transformed into each other, through the processes in the nutrient cycles. To understand some of these processes it is necessary to consider also processes in the carbon cycle, because of the many interdependencies between organic materials and nitrogen and phosphates. The different processes that control the nitrogen mineralization and immobilization in relation to the processes in the organic material are discussed. The main processes taken into consideration are mineralization, immobilization, denitrification and nitrogen fixation.

- In Chapter 6 environmental factors affecting bio-chemical and physical processes are discussed. The main environmental influences on transformation processes considered are soil moisture, oxygen, temperature, pH and clay content.

- Chapter 7 deals with crop production in relation to water and nutrient availability. A description of crop growth in the model is essential, since crops are important sinks in the soil nutrient balance and on the other hand important sources in the organic matter balance of the soil system. High photosynthesis rates can only be obtained when the process in the leaves is not adversely affected by shortage of water or minerals. Photosynthesis is reduced in the model simulation either by moisture stress conditions, resulting in stomatal closure, or by a deficit in nutrient uptake. Reduction in growth due to water stress automatically reduces the nutrient demand by the crop, since it is assumed that the nutrient requirement of the crop is related to the actual growth rate. This means that under growth limiting conditions the effects of water stress on dry matter production are first taken into account and after that the availability of nutrients is considered.

• Chapter 8 gives a number of results obtained during the calibration and validation of the models ANIMO and TRANSOL, using as comparison data obtained from field experiments with different crops and soil types.

• Finally, Chapter 9, deals with some examples of regional and national studies, executed for the evaluation of different developments in land use and agricultural intensity for the support of policy analysis and water quality management in the Netherlands.

References

Berghuijs-van Dijk, J.T., Rijtema, P.E. and Roest, C.W.J. 1985. *ANIMO: agricultural nitrogen model*. Nota **1671**. Institute for Land and Water Management Research: Wageningen, The Netherlands.
Bogárdi, I., Fried J.J., Frind E., Kelly, W.E. and Rijtema, P.E. 1990. Groundwater quality modelling for agricultural non-point sources. In *Proceedings of the International Symposium on Water Quality Modelling of Agricultural Non-point Sources* (ed. Don.G. DeCoursey), ARS-81 Part **1**, 227-252. USA-ARS
Bolt, F.J.E. van der, Groenendijk, P. and Oosterom, H.P. 1996. *Nutriënten belasting van grond- en oppervlaktewater in de stroomgebieden van de Beerze, de Reusel en de Rosep. Simulatie van de nutriëntenhuishouding.* Rapport **306.2**. DLO-Winand Staring Centre: Wageningen, Netherlands.
Duffy, C.J., Kincaid C.T., and Huyakorn, P.S. 1990. A review of groundwater models for assessment and prediction of non-point-source pollution. In *Proceedings of the International Symposium on Water Quality Modelling of Agricultural Non-point Sources*. (ed. Don.G. DeCoursey), ARS-81 Part **1**, 253-275 USA-ARS.
Genuchten, M. Th van and Šimůnek, J. 1996 Evaluation of pollutant transport in the unsaturated zone, In Regional approaches to water pollution in the environment, (ed. P.E. Rijtema and V. Eliáš), NATO ASI SERIES 2 Environment vol. **20**, 139-172. Kluwer: Dordrecht, The Netherlands.
Groenendijk, P. 1997a. *The calculation of complexation, adsorption, precipitation and dissolution in a soil water system with the geochemical model EPIDIM.* Report **70**. DLO Winand Staring Centre: Wageningen, The Netherlands.
Groenendijk, P. 1997b. *Modelling the influence of sorption and precipitation processes on the availability and leaching of chemical substances in soil.* Report **76**. DLO Winand Staring Centre: Wageningen, The Netherlands. (in press)

Jaarsveld, J.A. van, 1995. *Modelling the long-term atmospheric behaviour of pollutants on various spatial scales*. Ph.D. Thesis, University of Utrecht: Utrecht, The Netherlands.

Jansen, E.J. 1991. Results of simulations with ANIMO for several field situations. In *Soil and Groundwater Research Report II: Nitrate in soils*, pp. 269-280 .Comm. Eur. Communities: Brussels, Belgium.

Kroes, J.G. and Rijtema, P. E. 1989. *TRANSOL (TRANsport of a SOLute): User's guide*. Internal Report 5. DLO Winand Staring Centre, Wageningen: The Netherlands.

Kroes, J.G. and Rijtema, P.E. 1996. *TRANSOL, a dynamic simulation model for transport and transformation of solutes in soils*. Report 103. DLO Winand Staring Centre: Wageningen, The Netherlands.

Kroes, J.G., Groot, W.J.M. de, Pankow, J. and Toorn, A.van den. 1996. *Resultaten van onderzoek naar de kwantificering van de nitraatuitspoeling bij landbouwgronden*. Rapport 440. DLO Winand Staring Centre: Wageningen, The Netherlands.

Lentjes P.G. and Denneboom, J. 1996. *Data- en programmabeschrijving ISBEST versie 2.0*. Technical Document 31 DLO Winand Staring Centre: Wageningen, The Netherlands.

Merkelbach, R.C.M. and Lentjes, P.G. 1993. *ISBEST: een informatiesysteem dat het bestrijdings- middelengebruik in Nederland beschrijft*. Internal report. DLO Winand Staring Centre: Wageningen, The Netherlands.

Querner, E.P., 1988. Description of a regional groundwater flow model SIMGRO and some applications. *Agricultural Water Management.* **14**, 209-218.

Querner, E.P. and Bakel, P.J.T. van, 1989. *Description of the regional groundwater flow model SIMGRO*. Report 7. DLO Winand Staring Centre: Wageningen, The Netherlands.

Rijtema, P.E., 1993. Management of nutrient circulation in an expanding world agriculture. In: *On a holistic approach to water quality management; finding life-styles and measures for minimizing harmful fluxes from land to water*. Proceedings Stockholm Water symposium 10-14 august 1992 Publication **2**, 221-233. Stockholm Vatten AB: Stockholm.

Rijtema, P.E., 1994. Landbouw en eerste levensbehoeften. In: *Stofstromen in het landelijk gebied*. (ed. J. Drent) Rapport **365**: 71-95 DLO Winand Staring Centre: Wageningen, The Netherlands.

Rijtema, P.E. and Kroes, J.G. 1991a. Nitrogen modelling on a regional scale. In *Nitrate contamination, Exposure, Consequence and Control* (ed. I. Bogárdi and R.D. Kuzelka), NATO ASI Series, Vol. G **30**, 81-95. Springer-Verlag: New York

Rijtema, P.E. and Kroes, J.G. 1991b. Some results of nitrogen simulations with the model ANIMO. *Fertilizer Research* **27**, 189-198.

Rijtema, P.E. and Bolt, F.J.E. van der, 1996. Regional approaches to water pollution. In *Regional approaches to water pollution in the environment* (ed. P.E. Rijtema and V. Eliáš), NATO ASI SERIES 2 Environment vol. **20**, 1-20. Kluwer: Dordrecht, The Netherlands.

Rijtema, P.E., Groenendijk, P, Kroes, J.G. and Roest, C.W.J. 1997. *Modelling the nitrogen and phosphorus leaching to groundwater and surface water; Theoretical backgrounds and future developments of the ANIMO model*. Report 30. DLO Winand Staring Centre: Wageningen, The Netherlands.

Vachaud, G., Vauclin, M., and Addiscott, T.M. (1990) Solute transport in the vadose zone: a review of models. In *Proceedings of the International Symposium on Water Quality Modelling of Agricultural Non-point Sources* (ed. Don.G. DeCoursey), ARS-81 Part **1**, pp. 81-104 USA-ARS.

Walsum, P.E.V. van, 1988. *SLAPP: een rekenprogramma voor het genereren van bemestingsscenario's (betreffende dierlijke mest and stikstofkunstmest) ten behoeve van milieu-effectonderzoek, versie 1.0*. Nota **1920**. Institute Land and Water Management Research: Wageningen, The Netherlands.

Willigen, P. de, Bergström, L. and Gerritse, R.G. 1990. Leaching models of the unsaturated zone: Their potential use for management and planning. In *Proceedings of the International Symposium on Water Quality Modelling of Agricultural Non-point Sources*. (ed. Don G. DeCoursey), ARS-81 Part **1**, 105-128 USA-ARS.

Yeh, G.T. and Tripathi, V.S. 1989. A critical evaluation of recent developments in hydrogeochemical transport models of reactive multi-chemical components. Water Resources Research **25** (1), 93-108.

CHAPTER 2

WATER TRANSPORT IN SOILS

The solute transport model TRANSOL and the nutrient leaching model ANIMO are stand alone water quality models. They can be used in combination with different hydrological models. The models can be used to analyze problems on either field, farm or regional scale. They also deal with the pollutant load to drainage systems and aquifers. Knowledge of the transport processes in the soil profile is essential for describing the behaviour of solutes in the soil profile. In particular, the transformation and transport processes are strongly influenced by the soil moisture distribution. Five different situations can be considered for the transport of compounds in the soil system. They are:
- recharge of the soil moisture deficit and the distribution of dissolved compounds in the unsaturated zone following precipitation or irrigation;
- redistribution of dissolved compounds in the unsaturated zone as affected by evapotranspiration and capillary rise;
- leaching of dissolved compounds into the saturated zone as influenced by excess precipitation or irrigation water;
- lateral discharge of dissolved compounds to drainage systems in the saturated zone;
- horizontal discharge of dissolved compounds in regional groundwater flow systems.

A good water management model is a prerequisite for correct simulation of the water quality distribution. At field scale much attention will be given to the vertical transport in the unsaturated soil, while in regional studies also the horizontal transport of water and solutes between calculation units plays an important role. The partial differential equation for non-steady state two-dimensional flow of groundwater is written as:

$$\frac{\partial}{\partial x}(k_x H_d \frac{\partial h}{\partial x}) + \frac{\partial}{\partial y}(k_y H_d \frac{\partial h}{\partial y}) = \mu \frac{\partial h}{\partial t} + S \qquad (2.1)$$

where:
$k_x H_d$ = aquifer transmissivity in the x direction in $m^2.d^{-1}$
$k_y H_d$ = aquifer transmissivity in the y direction in $m^2.d^{-1}$
k_x = aquifer hydraulic conductivity in the x direction in $m.d^{-1}$
k_y = aquifer hydraulic conductivity in the y direction in $m.d^{-1}$
h = pressure head in m
μ = storage coefficient in $m^3.m^{-3}$
H_d = head above base of aquifer in m
t = time in days
S = source or sink functions in the saturated zone in $m.d^{-1}$

A numerical solution of the equation can be obtained through a finite difference approach or a finite element approach. This last approach first involves replacing of the continuous aquifer system parameters by an equivalent set of discrete elements. Secondly, the equations governing the flow of groundwater in the discretized model are written in a finite difference form and the resulting set of finite linear equations is solved numerically. An example of such a model is SIMGRO (Querner 1988, Querner and van Bakel 1989). The pseudo-dynamic model DEMGEN (Abrahamse et al. 1982), as used in different policy analyses of the water management in the Netherlands, operates on the basis of series steady state solutions of Eq. (2.1) assuming $\partial h/\partial t = 0$ during the time step under consideration. In these groundwater transport studies simplified functions for the change in water storage in the unsaturated zone are introduced to quantify the storage coefficient µ.

2.1 Vertical transport in the unsaturated domain

The change of water storage in the unsaturated zone with time $\mu \partial h/\partial t$ is expressed as:

$$C_w(h)\frac{\partial h}{\partial t} = \frac{\partial}{\partial z}(k(h)(\frac{\partial h}{\partial z} + 1)) - S_w(h) \qquad (2.2)$$

where:
C_w = differential soil water capacity dθ/dh in m^{-1};
z = vertical distance in m
k_h = hydraulic conductivity in $m.d^{-1}$
S_w = water volume extracted per unit volume of soil per unit of time in $m^3.m^{-3}.d^{-1}$

To solve the implicit second order partial differential equation the flow region is subdivided into a number of finite segments, bounded and represented by a series of nodal points at which a solution is obtained. The solution depends on the solutions of the surrounding segments and on a set of boundary conditions. The top boundary conditions are precipitation, interception, soil evaporation, irrigation and surface runoff. Lateral boundary conditions are root extraction and drainage. In the model SWATRE root extraction is calculated empirically (Feddes et al. 1978, Belmans et al. 1983) as:

$$S_\theta^r(h_\theta) = rf_\theta(h_\theta) S_\theta^{max} \qquad (2.3)$$

where:
$S_\theta^r(h_\theta)$ = water extraction by roots in $m.d^{-1}$
$rf_\theta(h_\theta)$ = prescribed function of pressure head (-)
S_θ^{max} = maximum possible water extraction in $m.d^{-1}$

In the improved version of SWATRE, as introduced in SWAP (van den Broek et al. 1993, van Dam et al. 1997), modifications for variable phreatic water levels, drainage and perched water tables have been introduced. Models like SWATRE and SWAP produce water balances for a freely chosen number of layers. Consequences of irregular water distribution and hysteresis effects are not considered. When the water quality models are used in combination with this type of model, the same layers are used.

2.1.1 Physical properties of soils

In many hydrological investigations a large number of estimates have to be made on the effect of climatological conditions and soil properties in the unsaturated zone on losses of groundwater and on the influence of these factors on groundwater flow. The accuracy by which soil moisture extraction can be forecasted depends on the accuracy by which the soil properties are known. The extraction pattern of soil moisture is not only directly related to the soil physical properties, but is also dependent on the climatological conditions, on the depth of the phreatic water level and on the plant root distribution in the unsaturated zone. Both the capillary properties and the soil moisture characteristics depend on the granular composition, the density and the pore size distribution of the soil. It is apparent that each soil will have its own properties. It is very often impossible, however, to use in regional studies for each soil its specific properties in forecasting procedures, as the physical properties of the soil have to be determined in each specific case. For this reason data of capillary conductivity and soil moisture characteristics available from literature (Richards and Wilson 1936, Wilson and Richards 1938, Christensen 1944, Richards and Moore 1952, Wind 1955, Gardner and Fireman 1958, Nielsen et al. 1960, Butijn 1961, Wind and Hidding 1961, Talsma 1963, Rubin et al. 1964, Rijtema 1965) have been collected. Most measurements cover a limited range from saturation to a suction of 100 to 200 mbar. It appeared from these data that the capillary conductivity in this suction range can be expressed as:

$$k = k_0 \exp[-\alpha \psi] \qquad (2.4)$$

where:
k_0 = capillary conductivity at saturation in cm.d^{-1}
α = soil type dependent constant in mbar^{-1}
ψ = soil moisture suction ($\psi = -h$) in mbar

Only a restricted number of data is available from literature in the suction range between field capacity and wilting point. Rijtema (1965) showed that the relation between capillary conductivity and suction in this range can be given by the expression:

$$k = a\psi^{-n} \qquad (2.5)$$

The value of the exponent *n* was nearly constant for very different soils, covering the range from medium coarse sand to river basin clay. The smallest value of n was 1.35 for basin clay, whereas the highest value was found for a fine sandy loam with $n = 1.46$. For practical application in regional hydrological studies capillary conductivity in the high suction range can be calculated using $\psi^{-1.4}$ as variable. Rijtema (1969) presented on basis of the available data from literature a first series of physical parameters for standard soils. Wösten *et al.* (1994) presented a series of measured data of soil moisture characteristics for Dutch soils and also calculated k-h-θ relations, using the method given by van Genuchten (1980). All these data were averaged for a number of soil groups, resulting in a series of standard soils. When during hydrological investigations in a groundwater basin, a soil survey has been made, the physical properties of these standard soils can be used for forecasting the soil moisture distribution in the unsaturated zone. The mean relation between capillary conductivity and suction of each soil is given in Table 2.1. The constant ψ_b is the boundary value for the use of Eq. (2.4) or Eq. (2.5), respectively. Data of representative soil moisture characteristics are given in Table 2.2.

Table 2.1: Values of k_0, α, ψ_b, and a for the relations between capillary conductivity and suction.

Soil type	k_0 (cm.d^{-1})	α (mbar^{-1})	ψ_b (mbar)	a (cm.mbar$^{-1.4}$.d^{-1})
Coarse sand	38.0	0.1708	71	0.08
Medium coarse sand	10.6	0.0650	85	0.19
Medium fine sand	6.7	0.0617	155	0.54
Fine sand	6.6	0.0482	202	0.66
Loamy medium coarse sand	1.6	0.0578	140	0.49
Loamy medium fine sand	2.3	0.0562	149	0.56
Loamy fine sand	6.4	0.0398	259	0.51
Sandy loam	10.1	0.0667	143	0.76
Loess loam	14.5	0.0490	200	1.34
Fine sandy loam	11.4	0.0478	213	0.78
Silt loam	2.5	0.0365	245	0.71
Loam	1.0	0.0360	220	0.67
Sandy clay loam	8.5	0.0353	285	1.00
Silty clay loam	1.4	0.0285	285	1.15
Clay loam	1.5	0.0268	293	1.69
Light clay	3.5	0.0294	295	1.56
Silty clay	1.3	0.0480	135	1.92
Basin clay	0.5	0.0380	140	2.43
Peat	2.0	0.0567	112	2.75

Table 2.2: Soil moisture content as volume percentage in relation to suction for standard soils.

Soil type	Soil moisture suction in mbar								
	0	10	31	100	250	500	2 500	16 10^3	10^6
Coarse sand	32.1	24.7	13.3	3.2	2.4	1.8	1.5	1.0	0.3
Medium coarse sand	33.9	28.3	22.5	6.1	3.5	2.5	1.6	1.0	0.4
Medium fine sand	35.8	35.1	29.9	13.5	5.6	3.5	2.0	1.5	0.7
Fine sand	37.7	37.0	33.0	17.0	8.2	5.0	2.6	2.2	1.2
Loamy medium coarse sand	33.1	29.9	25.3	13.0	8.3	5.9	2.6	1.2	0.5
Loamy medium fine sand	34.3	32.1	27.9	17.2	10.5	7.2	3.5	1.7	0.7
Loamy fine sand	36.0	34.9	30.8	21.6	12.3	8.8	4.5	2.3	1.3
Sandy loam	37.6	35.5	31.9	24.5	17.6	14.2	9.2	6.1	1.5
Loess loam	41.5	40.6	38.5	34.0	26.6	23.2	15.0	10.0	3.5
Fine sandy loam	40.4	39.2	37.2	32.3	25.3	22.4	13.2	8.7	1.7
Silt loam	41.0	40.7	39.8	34.6	30.2	27.3	17.9	10.8	2.0
Loam	41.0	39.8	37.6	33.7	29.9	27.1	21.3	16.1	2.5
Sandy clay loam	46.6	45.7	43.4	36.9	29.3	23.7	13.9	7.3	4.0
Silty clay loam	46.3	45.8	44.1	37.2	30.5	24.1	14.0	8.5	6.0
Clay loam	49.0	48.4	46.7	41.7	35.3	30.4	19.4	11.6	6.2
Light clay	45.3	44.4	42.5	38.4	32.3	29.4	23.2	17.5	7.5
Silty clay	56.1	55.8	54.4	51.2	46.8	42.9	33.8	25.3	8.5
Basin clay	56.8	56.1	54.5	51.3	47.5	44.5	37.5	30.7	9.9
Peat	87.3	86.1	82.6	72.3	60.0	48.9	29.6	19.1	7.8

2.1.2 Swelling and shrinking soils

Infiltration characteristics of clay soils depend to a large extent on swelling and shrinking behaviour of these soils. Due to swelling and shrinking of clay soils a moisture deficit dependent crack volume must be introduced as a kind of bypass with horizontal infiltration in each layer during recharge of the soil water by precipitation and irrigation. For heavy clay soils these cracks most probably are the major vertical transport pathways for irrigation water and excess precipitation. On a macroscopic level, volume changes of soil aggregates of drying soils result in the occurrence of shrinkage cracks and surface subsidence (Bronswijk 1991). Because of the influence of clay content, clay mineralogy, cation exchange capacity and organic matter on the shrinkage process, there are also many differences in the behaviour of the various soils.

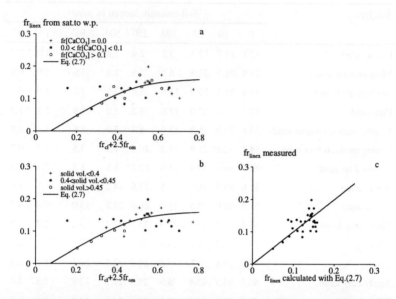

Fig. 2.1: Relation between $fr_{linex,wp}$ and $fr_{cl} + 2.5fr_{om}$ between saturation and wilting point. The curve is calculated with Eq.(2.7). a: for different classes of $CaCO_3$; b: for different classes of solid volume at saturation; c: comparison of measured and calculated values of $fr_{linex,wp}$.

Soil ripening may play an important role in some extreme shrinkage values observed. It might be expected that after some drying and wetting cycles these soils will show a normal shrinkage behaviour. The observed course and magnitude of shrinkage upon drying has important consequences for the field behaviour of clay soils. Some of the clay soils show normal shrinkage from saturation until a suction much higher than 16 bar. This means that, under Dutch climatic conditions, the aggregates of these soil horizons always remain fully saturated. Air is only present in inter-aggregate pores like shrinkage cracks. For other clay soils air enters the soil aggregates at much lower suction. The fraction of linear extensibility (fr_{linex}) is defined as:

$$fr_{linex} = \sqrt[3]{\frac{V_{sat}}{V_{dry}}} - 1 \qquad (2.6)$$

in which:
V_{sat} = volume of a saturated soil aggregate
V_{dry} = volume of the aggregate in a dry state

Taking the aggregate volume at wilting point (V_{wp}) as reference volume gives a relation between $fr_{linex,wp}$ and $fr_{cl} + 2.5fr_{om}$. Data given by Bronswijk and Evers-Vermeer (1990) are shown in Fig. 2.1. The curves are calculated as the smallest root of:

$$(fr_{linex,wp} - 0.18)[fr_{linex,wp} - 0.4(fr_{cl} + 2.5fr_{om}) + 0.02] = 0.0015 \quad (2.7)$$

In ripened soils shrinkage is a reversible process and extends uniformly in all directions (Bronswijk, 1991). It means that a unit volume of soil decreases equally in size in both horizontal directions and in the vertical direction. The relative volume of solids on basis of solid bulk density at saturation per layer equals:

$$V_s = \frac{\rho_{d,sat}}{\rho_{spw}} \quad (2.8)$$

in which:
$\rho_{d,sat}$ = the dry bulk density determined at saturation in kg.m^{-3} soil
ρ_{spw} = solid specific weight in kg.m^{-3} solids.

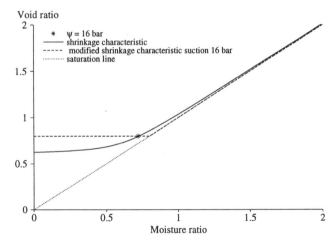

Fig. 2.2: General form of the shrinkage characteristic; void ratio = volume of pores/volume of solids; moisture ratio = volume of water/volume of solids.

Defining moisture ratio and void ratio as (water volume)/(solid volume) and (pores volume)/(solid volume), respectively, gives the general form of the shrinkage characteristic presented in Fig. 2.2. An approximation of shrinkage is given in the water quality model assuming complete saturation of the aggregates till the value of the void ratio at wilting point has been reached, dividing the curve into two linear sections.

The first one follows the saturation line until the void ratio at wilting point has been reached. The second part follows the horizontal line until the moisture ratio at wilting point has been reached. The change in the volume of the aggregates equals, in this schematization, the change in moisture volume in the model layer, until the intersection point of both linear relations. A further decrease of the moisture volume until wilting point then takes place with a constant aggregate volume.

The apparent value of $fr_{linex,ap}$ for any moisture condition between saturation and wilting point is calculated by:

$$fr_{linex,ap} = \sqrt[3]{\frac{V_{sat}}{V_{sat} - (\theta_{sat} - \theta_\psi)}} - 1 \qquad (2.9)$$

The actual value of fr_{linex} is calculated as:

$$fr_{linex} = MIN\,[fr_{linexwp}\,;\,fr_{linexap}] \qquad (2.10)$$

The relation between the dry bulk density of shrinking structure elements and the maximum soil moisture fraction of these elements is given by the expression:

$$\rho = (1 + fr_{linex})^3 \rho_{sat} = \frac{\rho_{sat}}{1 - (\theta_{sat} - \theta_\psi)} \qquad (2.11)$$

in which:

ρ_{sat} = minimum dry bulk density of structure elements at saturation in tonnes.m^{-3}.
ρ_θ = maximum dry bulk density after shrinkage at a moisture fraction θ_ψ.

Roest et al. (1993) give data, originally determined by Kittab (1983), of the relation between soil moisture and dry bulk density of structure elements for a number of Egyptian clay soils, with a clay fraction varying between 0.3 and 0.6. The data are presented in Fig. 2.3. This figure shows also the maximum shrinkage curve, calculated with Eq. (2.11), starting with a dry bulk density of 900 kg.m^{-3} at saturation. The horizontal lines present the maximum bulk densities after shrinkage for soils with a clay fraction of 0.3 and 0.6, respectively. The majority of the observations of Kittab (1983) lie between the calculated boundaries. The line for the relation between dry bulk density and moisture content at saturation for non shrinking soils is also presented.

Considering the volume of solids V_s as being constant in a model layer, gives automatically the following expressions for layer thickness, crack area and crack volume per layer, when defining the layer thickness on the basis of the conditions at saturation:

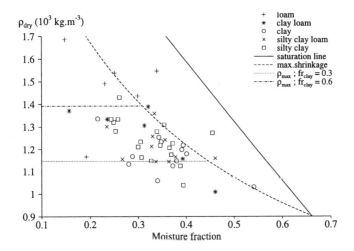

Fig. 2.3: Relation between the moisture content of aggregates and dry bulk density for Egyptian soils, with a clay fraction fr_{cl} varying between 0.3 and 0.6.

Layer thickness:

$$\Delta Z_\theta = (1 - fr_{linex})\Delta Z_{sat} \qquad (2.12)$$

Crack area:

$$A_{cr} = 1 - (1 - fr_{linex})^2 \qquad (2.13)$$

Crack volume per layer:

$$V_{cr} = (1 - fr_{linex})[1 - (1 - fr_{linex})^2]\Delta Z_{sat} \qquad (2.14)$$

The model FLOCR (Oostindie and Bronswijk 1992) is a one-dimensional hydrological model for the dynamic simulation of the flow of water in an unsaturated soil, taking into account swelling and shrinking processes and the cracks related to this phenomenon. The precipitation falls partly in the cracks and partly on the soil matrix. The distribution of the precipitation is proportional to the ratio of crack area and soil matrix area at the soil surface. The FLOCR simulations give a complete water balance, including the transport through the soil cracks, that can be used as input for the calculation of material transport in the water quality model.

2.1.3 Schematization of vertical transport

2.1.3.1 General remarks

Regional hydrological models very often consider the unsaturated domain of the soil as a one layer system. The one-dimensional water balance of the soil accounting for the incoming and outgoing fluxes in the unsaturated zone is based on the equation:

$$\Delta V_\theta = (q_i + q_n^i - q_s - q_t - q_d)\Delta t \qquad (2.15)$$

where:
- ΔV_θ = the change in water storage during a given period of time in m
- q_i = infiltration (precipitation including irrigation) in m.d^{-1}
- q_n^i = net upward flow through the model bottom in m.d^{-1}
- q_s = soil evaporation in m.d^{-1}
- q_t = crop transpiration in m.d^{-1}
- q_d = drainage in m.d^{-1}
- Δt = time increment in d.

Moreover, insufficiently detailed information of soil physical properties is in many cases available in regional studies. Under these conditions simplified water balance models for the unsaturated soils are used. Berghuijs-van Dijk (1985) introduced for this reason a simplified water balance model WATBAL, mainly based on the concept given by Rijtema and Aboukhaled (1975). In WATBAL, the unsaturated zone is divided into three compartments, the crop rootzone, the second layer in which capillary rise of water to the rootzone plays an important role, and a deepest layer in which changes in moisture content are predominantly the result of drainage. The WATBAL model takes into account fluctuating water tables. Recharge of deeper layers already occurs before the rootzone layer is at field capacity. An extension of WATBAL, taking into account soil moisture dependent crack formation and the osmotic effects of salt accumulation in the root zone on crop water uptake is given in the model FAIDS, used in water management policy studies in Egypt (Roest et al. 1993). During field irrigation, water is flowing into soil cracks. Due to hydraulic gradients and high permeability of the cracked topsoil, water is also flowing immediately to the field drains which are in direct connection with the crack volume.

The layer thickness in models like WATBAL and FAIDS is still too large to describe the processes and transport of solutes in the unsaturated soil system. For this reason the water quality models TRANSOL and ANIMO comprise a module BALANCE, which generates waterbalances for the required model layers.

2.1.3.2 The module BALANCE

The module BALANCE transfers the output of these types of models into a multi-layered system for the calculation of the water balances per layer. Based on the amount of data obtained from a hydrological model the moisture content of each layer at the end of the time step can be calculated, starting from the following definitions:
- In its simplest form the moisture fraction in the root zone is assumed to be uniform with depth and in the absence of a water table in the root zone the moisture fraction at the end of the time step can be expressed as:

$$\theta_r(t) = \frac{V_r(t)}{Z_r} \qquad (2.16)$$

where:
θ_r = moisture fraction in the root zone in $m^3.m^{-3}$
V_r = moisture volume in the rootzone in $m^3.m^{-2}$
Z_r = thickness of the root zone in m

- The matrix potential at the top of the subsoil equals the matrix potential of the rootzone. The moisture fractions in the rootzone and at the top of the unsaturated subsoil are described by interpolation of the soil moisture characteristics of both layers. The soil moisture in the subsoil is redistributed over the layers after rewetting of the rootzone by precipitation or irrigation. In this method of operation the effects of hysteresis and preferential flow are also more or less taken into consideration, as rewetting of the subsoil already starts in the model before the rootzone is at field capacity.
- The minimum moisture volume present in the subsoil after maximum extraction through evapotranspiration is assumed to be bounded by the line describing the steady state situation for an upward capillary flux of 0.01 $cm.d^{-1}$ and the soil moisture suction at the boundary of rootzone and subsoil at wilting point. These lines are calculated for ther low suction range by:

$$z = \frac{1}{\alpha} \ln \left[\frac{0.01 + k_o}{0.01 + k_o \exp[-\alpha \psi]} \right] ; \qquad \psi \leq \psi_{max} \qquad (2.17)$$

For the high suction range holds:

$$\frac{dz}{d\psi} = \frac{a\psi^{-1.4}}{0.01 + a\psi^{-1.4}} \approx \varsigma_1 \exp[-\varsigma_2(\psi - \psi_{min})] + \varsigma_3 \qquad (2.18)$$

The numerical solution for the high suction range is here approximated in a few steps with an analytical solution of the differential equation using the empirical expression given in Eq. (2.18). An example is presented in Fig. 2.4 for medium coarse sand.

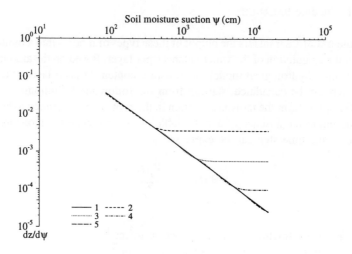

Fig. 2.4: Relation between $dz/d\psi$ and ψ for medium coarse sand.
1: $a = 0.192$ ψ from 85 to 16 000 cm;
2: $\varsigma_1 = 0.026000$, $\varsigma_2 = 0.011700$, $\varsigma_3 = 0.003568$ ψ from 85 to 400 cm;
3: $\varsigma_1 = 0.003800$, $\varsigma_2 = 0.003100$, $\varsigma_3 = 0.000556$ ψ from 400 to 1 500 cm;
4: $\varsigma_1 = 0.000590$, $\varsigma_2 = 0.000875$, $\varsigma_3 = 0.000097$ ψ from 1 500 to 5 000 cm;
5: $\varsigma_1 = 0.000105$, $\varsigma_2 = 0.000300$, $\varsigma_3 = 0.000022$ ψ from 5 000 to 1 6000 cm.

Eq. (2.18) yields after integration:

$$z_{max} = z_{min} + \frac{\varsigma_1}{\varsigma_2}(1 - \exp[-\varsigma_2(\psi_{max} - \psi_{min})]) + \varsigma_3(\psi_{max} - \psi_{min}) \qquad (2.19)$$

With the data presented in Table 2.1 and Table 2.2 the moisture distribution for the minimum water fraction of the subsoil can be calculated. An example of this distribution for different soils is presented in Fig. 2.5. The moisture distribution can be approximated by a set of linear relations. The moisture volume below the rootzone is calculated from the hydrological input:

$$V_s(t) = V_s(t_o) + (\Sigma q^i - \Sigma q^o)\Delta t - \Delta V_r \qquad (2.20)$$

where:
$V_s(t)$ = moisture volume below the root zone at time t in $m^3.m^{-2}$
$V_s(t_o)$ = moisture volume below the root zone at time t_o in $m^3.m^{-2}$
Σq^i = sum of incoming fluxes (precipitation, seepage, infiltration) in $m.d^{-1}$
Σq^o = sum of outgoing fluxes (evapotranspiration, leakage, drainage) in $m.d^{-1}$
Δt = length of time step in d.
ΔV_r = change in moisture volume in the root zone during the time step in $m^3.m^{-2}$

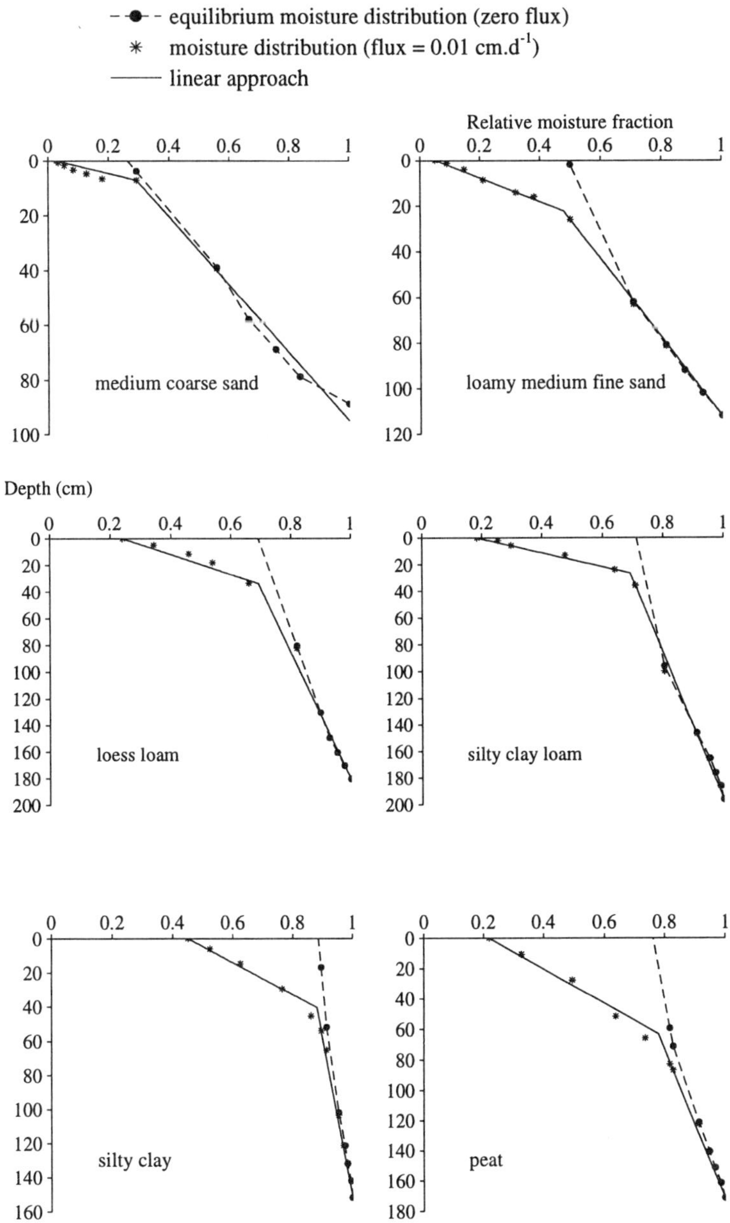

Fig. 2.5: Soil moisture distribution in the unsaturated subsoil as function of depth below the rootzone for different soils with the upper boundary at wilting point and a steady state capillary rise of 0.01 cm.d^{-1}.

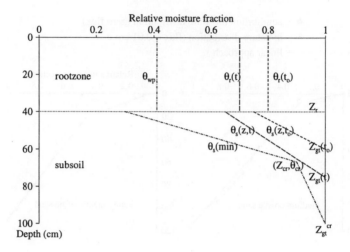

Fig. 2.6: Definition sketch of the distribution of soil moisture below the root zone as used for different situations in the module balance.

The definition sketch presented in Fig. 2.6 is used to calculate the soil moisture distribution and the vertical fluxes between the layers defined in the water quality model. Below the root zone one or two linear relationships of the moisture fraction with depth are considered. These relationships are such that at the phreatic level the moisture fraction equals saturation. At the top of the unsaturated layer the moisture content is in equilibrium with the moisture suction at the bottom of the rootzone and the total moisture volume fits with the water balance for this layer. The depth Z_{cr} below the root zone is considered as a breaking point in the moisture fraction-depth relation.

The moisture volumes $V_s(t)$ in the subsoil are calculated either with:

$$V_{s,1}(t) = \frac{(\theta_{sr}(t) + \theta_{sat})}{2}(Z_{gt}(t) - Z_r) + \theta_{sat}(Z_{deep} - Z_{gt}(t)) \quad (2.21)$$

or:

$$V_{s,2}(t) = \frac{(\theta_{sr}(t) + \theta_{cr})}{2}(Z_{cr} - Z_r) +$$
$$\frac{(\theta_{cr} + \theta_{sat})}{2}(Z_{gt}(t) - Z_{cr}) + \theta_{sat}(Z_{deep} - Z_{gt}(t)) \quad (2.22)$$

where:
Z_{gt} = depth of groundwater table in m
Z_{deep} = depth below deepest observed groundwater table in m
Z_r = depth of the rootzone in m

Z_{cr} = depth of brake point in subsoil moisture distribution in m
θ_r = moisture volume fraction in the rootzone in m^3.m^{-3}
θ_{sr} = subsoil moisture volume fraction at depth Z_r in equilibrium with θ_r in m^3.m^{-3}
θ_{cr} = lowest moisture volume fraction at depth Z_{cr} in m^3.m^{-3}
θ_{sat} = saturated moisture volume fraction at saturation in m^3.m^{-3}

For $V_s(t)$ in Eq. (2.20) holds:

$$V_s(t) = MAX[V_{s,1}(t) ; V_{s,2}(t)] \qquad (2.23)$$

Dividing the soil profile in j layers gives, combined with Eq. (2.21) and Eq. (2.22), a number of expressions for the moisture volume in each layer was derived, depending on the position of the layer in the soil profile. The upper and lower layer boundary of each layer in the module BALANCE are defined as $Z_{j-1/2}$ and $Z_{j+1/2}$. For the rootzone layers ($Z_{j+1/2} < Z_r$) holds:

$$V_{\theta,j} = \theta_r(Z_{j+1/2} - Z_{j-1/2}) \qquad (2.24)$$

When a linear relation for the moisture distribution in the subsoil is present, without a breaking point ($V_s(t) = V_{s,1}(t)$), a number of situations can occur.

- If the upper boundary of the layer is in the rootzone and the lower boundary in the unsaturated subsoil ($Z_{j-1/2} < Z_r < Z_{j+1/2} \leq Z_{gt}$) then the following expression is valid:

$$V_{\theta,j} = \theta_r(Z_r - Z_{j-1/2}) + [\theta_{sr} + \frac{\theta_{sat} - \theta_{sr}}{Z_{gt} - Z_r} \frac{Z_{j+1/2} - Z_r}{2}](Z_{j+1/2} - Z_r) \qquad (2.25)$$

- If both layer boundaries are in the unsaturated subsoil ($Z_r < Z_{j-1/2} < Z_{j+1/2} \leq Z_{gt}$) then the following expression holds:

$$V_{\theta,j} = [\theta_{sr} + \frac{\theta_{sat} - \theta_{sr}}{Z_{gt} - Z_r}(\frac{Z_{j-1/2} + Z_{j+1/2}}{2} - Z_r)](Z_{j+1/2} - Z_{j-1/2}) \qquad (2.26)$$

- If the upper boundary of the layer is in the unsaturated subsoil and the lower one in the saturated subsoil ($Z_r \leq Z_{j-1/2} < Z_{gt} < Z_{j+1/2}$) gives the expression:

$$V_{\theta,j} = [\theta_{sr} + \frac{\theta_{sat} - \theta_{sr}}{Z_{gt} - Z_r}(Z_{j-1/2} + \frac{Z_{gt}}{2} - Z_r)](Z_{gt} - Z_{j-1/2}) + \theta_{sat}(Z_{j+1/2} - Z_{gt}) \qquad (2.27)$$

- If the upper boundary of the layer is in the rootzone and the lower boundary in the saturated subsoil ($Z_{j-1/2} \leq Z_r < Z_{gt} < Z_{j+1/2}$) then the following expression is valid:

$$V_{\theta,j} = \theta_r(Z_r - Z_{j-1/2}) + [\frac{\theta_{sr} + \theta_{sat}}{2}](Z_{gt} - Z_r) + \theta_{sat}(Z_{j+1/2} - Z_{gt}) \qquad (2.28)$$

When two linear relations for the moisture distribution in the subsoil are present, with a breaking point at (Z_{cr}, θ_{cr}), $(V_s(t) = V_{s,2}(t))$, the following situations can be present.

- If the upper boundary of the layer is in the rootzone and the lower boundary in the unsaturated subsoil above the breaking point $(Z_{j-1/2} < Z_r < Z_{j+1/2} \leq Z_{cr})$, gives:

$$V_{\theta,j} = \theta_r(Z_r - Z_{j-1/2}) + [\theta_{sr} + \frac{\theta_{cr} - \theta_{sr}}{Z_{cr} - Z_r} \frac{Z_{j+1/2} - Z_r}{2}](Z_{j+1/2} - Z_r) \qquad (2.29)$$

- If the upper boundary of the layer is in the rootzone and the lower boundary in the unsaturated subsoil below the breaking point $(Z_{j-1/2} < Z_r < Z_{cr} < Z_{j+1/2})$, gives:

$$V_{\theta,j} = \theta_r(Z_r - Z_{j-1/2}) + \frac{\theta_{sr} + \theta_{cr}}{2}(Z_{cr} - Z_r) +$$
$$[\theta_{cr} + \frac{\theta_{sat} - \theta_{cr}}{Z_{gt} - Z_{cr}} \frac{Z_{j+1/2} - Z_r}{2}](Z_{j+1/2} - Z_{cr}) \qquad (2.30)$$

- If the upper boundary of the layer is in the rootzone and the lower boundary in the saturated subsoil below the breaking point $(Z_{j-1/2} < Z_r < Z_{cr} < Z_{gt} < Z_{j+1/2})$, gives:

$$V_{\theta,j} = \theta_r(Z_r - Z_{j-1/2}) + \frac{\theta_{sr} + \theta_{cr}}{2}(Z_{cr} - Z_r) +$$
$$[\frac{\theta_{cr} + \theta_{sat}}{2}](Z_{gt} - Z_{cr}) + \theta_{sat}(Z_{j+1/2} - Z_{gt}) \qquad (2.31)$$

- If both layer boundaries are in the unsaturated subsoil, but above the breaking point $(Z_r < Z_{j-1/2} < Z_{j+1/2} \leq Z_{cr})$, then the following expression holds:

$$V_{\theta,j} = [\theta_{sr} + \frac{\theta_{cr} - \theta_{sr}}{Z_{cr} - Z_r}(\frac{Z_{j-1/2} + Z_{j+1/2}}{2} - Z_r)](Z_{j+1/2} - Z_{j-1/2}) \qquad (2.32)$$

- If both layer boundaries are in the unsaturated subsoil, but below the breaking point $(Z_{cr} < Z_{j-1/2} < Z_{j+1/2} < Z_{gt})$, then the following expression holds:

$$V_{\theta,j} = [\theta_{cr} + \frac{\theta_{sat} - \theta_{cr}}{Z_{gt} - Z_{cr}}(\frac{Z_{j-1/2} + Z_{j+1/2}}{2} - Z_{cr})](Z_{j+1/2} - Z_{j-1/2}) \qquad (2.33)$$

- If both boundaries are in the unsaturated subsoil, but the upper boundary is above the breaking point and the lower one below $(Z_r < Z_{j-1/2} < Z_{cr} < Z_{j+1/2} \leq Z_{gt})$, gives:

$$V_{\theta,j} = [\theta_{sr} + \frac{\theta_{cr} - \theta_{sr}}{Z_{cr} - Z_r}(\frac{Z_{j-1/2}+Z_{cr}}{2} - Z_r)](Z_{cr} - Z_{j-1/2}) +$$
$$[\theta_{cr} + \frac{\theta_{sat} - \theta_{cr}}{Z_{gt} - Z_{cr}}\frac{Z_{j+1/2}-Z_{cr}}{2}](Z_{j+1/2} - Z_{cr}) \quad (2.34)$$

- If the upper boundary is in the unsaturated subsoil below the breaking point and the lower boundary in the saturated subsoil ($Z_{cr} < Z_{j-1/2} < Z_{gt} < Z_{j+1/2}$), gives:

$$V_{\theta,j} = [\theta_{cr} + \frac{\theta_{sat} - \theta_{cr}}{Z_{gt} - Z_{cr}}(\frac{Z_{j-1/2}+Z_{gt}}{2} - Z_{cr})](Z_{gt} - Z_{j-1/2}) +$$
$$\theta_{sat}(Z_{j+1/2} - Z_{gt}) \quad (2.35)$$

- If the upper boundary is in the unsaturated subsoil above the breaking point and the lower boundary in the saturated subsoil ($Z_r < Z_{j-1/2} < Z_{cr} < Z_{gt} < Z_{j+1/2}$), gives:

$$V_{\theta,j} = [\theta_{sr} + \frac{\theta_{cr} - \theta_{sr}}{Z_{cr} - Z_{sr}}(\frac{Z_{j-1/2}+Z_{cr}}{2} - Z_{sr})](Z_{cr} - Z_{j-1/2}) +$$
$$[\frac{\theta_{cr} + \theta_{sat}}{2}](Z_{gt} - Z_{cr}) + \theta_{sat}(Z_{j+1/2} - Z_{gt}) \quad (2.36)$$

- If both layer boundaries are in the saturated subsoil ($Z_{gt} < Z_{j-1/2} < Z_{j+1/2}$):

$$V_{\theta,j} = \theta_{sat}(Z_{j+1/2} - Z_{j-1/2}) \quad (2.37)$$

Z_{gt}^{cr} is defined as the maximum groundwater table depth for a combined water extraction by capillary rise and drainage. A further fall of the groundwater table below this level is only due to drainage. The vertical drainage flux is under these conditions given as:

$$q_d = \frac{(\theta_{sat} - \theta_{cr})(Z_{gt}(t) - Z_{gt}(t_0))}{2(t - t_0)} \quad (2.38)$$

For each distinguished soil layer the water balance is given as:

$$(q_{j-\frac{1}{2}} - q_{j+\frac{1}{2}} - q_d(j) - q_{et}(j))\Delta t + V_{\theta,j}(t) - V_{\theta,j}(t_0) = 0 \quad (2.39)$$

where:
$q_{j-\frac{1}{2}}$ = inflow flux from the previous layer in m.d^{-1}
$q_{j+\frac{1}{2}}$ = outflow to the next layer in m.d^{-1}
$q_d(j)$ = discharge flux to the drainage system in m.d^{-1}
$q_{et}(j)$ = evaporation flux from the layer in m.d^{-1}

Table 2.3: Critical values of soil moisture and depth, as used in the model BALANCE.

Soil type	Critical values of soil moisture and depth				
	θ_{wp} (vol.%)	Z_{cr} (cm)	θ_{cr} (vol.%)	Z_{gt}^{cr} (cm)	θ_{sat} (vol.%)
Coarse sand	1.0	6.3	5.4	57.3	32.1
Medium coarse sand	1.0	7.1	9.9	95.0	33.9
Medium fine sand	1.5	27.3	13.8	116.6	35.8
Fine sand	2.2	27.6	10.7	145.7	37.7
Loamy medium coarse sand	1.2	20.3	13.0	101.1	33.1
Loamy medium fine sand	1.7	22.2	16.4	112.0	34.3
Loamy fine sand	2.3	16.2	13.0	172.8	36.0
Sandy loam	6.1	21.8	22.0	124.5	37.6
Loess loam	10.0	33.7	28.7	180.5	41.5
Fine sandy loam	8.7	21.7	28.2	165.5	40.4
Silt loam	10.8	18.1	32.3	166.6	41.0
Loam	16.1	18.8	31.2	125.2	41.0
Sandy clay loam	7.3	21.6	28.7	211.0	46.6
Silty clay loam	8.5	26.2	31.9	195.7	46.3
Clay loam	11.6	27.7	35.7	219.9	49.0
Light clay	17.5	31.9	32.5	228.1	45.3
Silty clay	25.3	40.0	49.4	151.7	56.1
Basin clay	30.7	40.0	47.9	164.8	56.8
Peat	19.1	63.2	68.0	171.4	87.3

For the first layer $q_{j-\frac{1}{2}}$ equals the precipitation rate plus the irrigation rate minus surface runoff. The outflow at the lower boundary of each layer is calculated at the result of the water balance equation. For the last layer $q_{j+\frac{1}{2}}$ should equal the leakage or seepage flux q_l that has been obtained as an input from the hydrological model used. The relevant critical values of moisture contents and depths for a series of soils are presented in Table 2.3. The soil moisture fractions have been derived from Table 2.2. The values for the critical depth have been derived from the calculated soil moisture distributions as presented in Fig. 2.5.

2.2 Transport in the saturated domain

The direction and the quantity of groundwater flow are influenced by a number of geohydrological factors. In the unsaturated zone and in the saturated zone above drain level

Water Transport in Soils

the flow is mainly vertical. In the soil profile above drain level the effective drainage flux q_d depends on the precipitation q_p, the evaporation flux q_{et} and the change in areic water volume ΔV_θ during a time step Δt. Because only vertical flow needs to be considered, the residence time Γ_d^o above drain level is given by the expression:

$$\Gamma_d^o = \frac{\varepsilon_w Z_d}{q_d} \qquad (2.40)$$

in which:
Z_d = the drain depth in m
q_d = the effective drainage flux m.d^{-1}
ε_w = the effective water-filled pore volume in m^3.m^{-3}

If drainage flow below drain level is considered to be two-dimensional, the main flow is horizontal but in downstream direction the streamlines penetrate deeper into the aquifer. The horizontal flow velocity below drain depth is not constant but increases in the downstream direction. At field level, different drainage situations can be considered.

2.2.1 Perfect drains

2.2.1.1 Theory

In a homogeneous soil, with perfect drains, radial flow near the drains is absent. The flow velocity at distance x from an infiltration point x_i, as shown in Fig 2.7, is for a two-dimensional transect between parallel drains given as:

$$q_x(x) = \varepsilon_w \frac{dx}{dt} = \frac{(x + x_i) q_d}{H_d} \qquad (2.41)$$

in which:
q_x = the horizontal flux in m.d^{-1}
x = the horizontal distance from the point of infiltration in m
x_i = the distance of the point of infiltration from the water divide in m
H_d = the equivalent thickness of drainage layer below drain level in m.

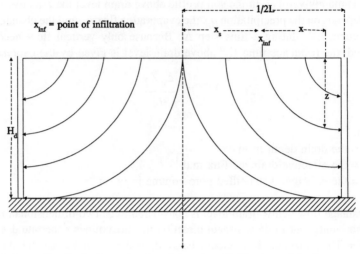

Fig. 2.7: Stream line distribution as function of the point of infiltration and the distance travelled.

Integration of Eq. (2.41) yields:

$$x = x_i\left[\exp[q_d(t - \Gamma_d^o)/\varepsilon_w H_d] - 1\right] \tag{2.42}$$

The depth $z(x)$ of the streamline below drain level at distance x is found as:

$$z(x) = \frac{x}{x + x_i} H_d \tag{2.43}$$

Substitution of Eq. (2.43) into Eq. (2.42) yields an expression for the isochrones as a function of the depth z only:

$$z = H_d\left[1 - \exp[-q_d(t-\Gamma_d^o)/\varepsilon_w H_d]\right] \tag{2.44}$$

The residence time required to reach the perfect drain at $x + x_i = \frac{1}{2}L$ and the corresponding value of $z(\frac{1}{2}L)$ follow from Eq. (2.42) and Eq (2.43) as, respectively:

$$\Gamma_{\frac{1}{2}L} = \Gamma_d^o + \frac{\varepsilon_w H_d}{q_d}\ln\left(\frac{L}{2x_i}\right) \tag{2.45}$$

and

$$z_{\frac{1}{2}L} = \frac{L - 2x_i}{L} H_d \tag{2.46}$$

Combination of the equations (2.45) and (2.46) yields:

$$\Gamma_{\frac{1}{2}L} = \Gamma_d^o + \frac{\varepsilon_w H_d}{q_d} \ln\left(\frac{H_d}{H_d - z_{\frac{1}{2}L}}\right) \quad (2.47)$$

2.2.1.2 Model schematization

One-dimensional leaching models generally represent a vertical soil column. Within the unsaturated zone, chemical substances are transported by vertical flows, whereas in the saturated zone the drainage discharge leaves the vertical column sideways. In this type of model, the distribution of lateral drainage fluxes with depth has been used to simulate the response of the load on the surface water system to the inputs in the groundwater system. In the saturated zone of the model approach a number of horizontal layers have been distinguished, so a pseudo two-dimensional transport is considered, as is shown in Fig. 2.8.

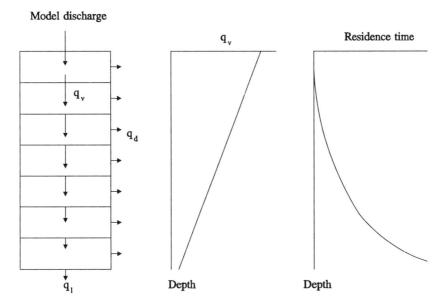

Fig. 2.8: a Model profile with vertical flux q_v, leakage flux q_l and drainage flux q_d; b Vertical flux q_v as function of depth; c Residence time as function of depth.

For the calculation of drain water quality the differences in residence time of the water particles discharging to the drain have to be taken into account. The vertical flux for a single drainage system is given as a linear function with depth in the model schematization, resulting in the expression:

$$q_z(z) = \varepsilon_w \frac{dz}{dt} = q_d(1 - z/H_d) + q_l \qquad (2.48)$$

where:
- q_d = drainage flux obtained from the hydrological model in m.d^{-1}
- q_z = vertical flux according model schematization in m.d^{-1}
- q_l = flux to the aquifer over the lower boundary in m.d^{-1}
- H_d = equivalent thickness of the drainage layer in m
- t = time in d
- z = depth in soil profile in m
- ε_w = porosity in m^3.m^{-3}

Integration of equation (2.48) yields:

$$\Gamma(z) = \Gamma_d^o + \frac{\varepsilon H_d}{q_d} \ln(H_d/(H_d - z)) \qquad (2.49)$$

in which:
- Γ_d^o = residence time above drain level in d
- Γ = residence time below drain level in d

The model schematization results in a similar expression for the relation between residence time and depth in the soil profile as given in 2.2.1.1 for perfect drains.

2.2.2 Line drains in a soil profile with infinite thickness

2.2.2.1 Theory

Ernst (1973) provides a mathematical formulation of a streamline pattern in a saturated soil profile with infinite thickness. Such a hydrological situation can be seen as the most extreme situation for evaluating the influence of the H_d/L ratio. In reality, the drainage flow will occupy less space in the saturated groundwater body and the flow path will be less deep. The streamline pattern is shown graphically in Fig 2.9.

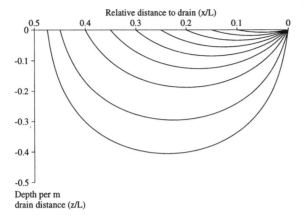

Fig. 2.9: Streamline pattern in a groundwater body with infinite thickness.

The streamlines are described by:

$$\Psi(x,z) = \frac{q_d}{\pi} \arctan\left(\frac{\exp[2\pi \frac{z}{L}]\sin(2\pi \frac{x}{L})}{\exp[2\pi \frac{z}{L}]\cos(2\pi \frac{x}{L}) - 1}\right) \quad (2.50)$$

The deepest point of each streamline bounds the stream zone, which will never be found below that depth. For the deepest point of each streamline holds:

$$\frac{\partial \Psi(x,z)}{\partial x} = 0 \quad (2.51)$$

Partial differentiation of Eq. (2.50) to x yields:

$$\frac{\partial \Psi}{\partial x} = \frac{2q_d}{L} \frac{\exp[2\pi \frac{z}{L}] - \cos(2\pi \frac{x}{L})}{\exp[2\pi \frac{z}{L}] - 2\cos(2\pi \frac{x}{L})} \quad (2.52)$$

Eq. (2.52) equals 0, when the nominator of the expression equals 0, so for the relation between the coordinates (x_d, z_d) of the deepest point of each streamline:

$$x_d = \frac{L}{2\pi} \arccos(\exp[-2\pi \frac{z_d}{L}]) \quad (2.53)$$

The total discharge q_d can be divided into q_d^x flowing to the drain above the depth z_d of each streamline and the flux q_d^z, passing as vertical flux, through this level z_d. The part of the flux to the drain which will never pass through level z_d amounts to:

$$q_d^x = \frac{2x}{L} q_d = \frac{4x_d}{L} q_d = \frac{2}{\pi} q_d \arccos(\exp[2\pi \frac{z_d}{L}]) \qquad (2.54)$$

The drainage water (q_d^z) residing some time deeper than level z_d is given by:

$$q_d^z = q_d - q_d^x = q_d \left(1 - \frac{2}{\pi} \arccos(\exp[2\pi \frac{z_d}{L}])\right) \qquad (2.55)$$

This horizontal flow fraction as a function of the depth per m drain distance (z/L) is graphically presented in Fig 2.10. For practical applications the maximum depth of the flow field is often bounded by the condition $H_d \leq L/4$. In soil profiles with infinite thickness, about 87 percent of the total drain discharge is conveyed above the plane $z/L = 0.25$. In deep soil profiles with a finite thickness, more than 87 percent of the total drain discharge will be transported above this plane.

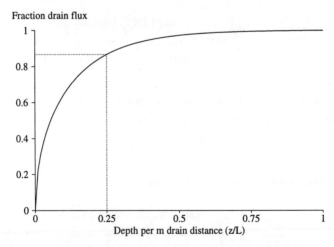

Fig. 2.10: Fraction of total drain discharge which will never be found below z/L.

2.2.2.2 Model schematization

The drainage flux q_d^z is in the model schematization considered as the vertical flux with depth. Indicating z_d by $-z$ gives as a relation between vertical flux and model depth the expression:

$$q_d^z = \varepsilon_w \frac{dz}{dt} = q_d\left(1 - \frac{2}{\pi}\arccos(\exp[-2\pi\frac{z}{L}])\right) \quad (2.56)$$

The residence time $\Gamma(z)$ as a function of the depth z equals:

$$\Gamma(z) = \Gamma_d^o + \int_{z=0}^{z=z} \frac{\varepsilon_w}{q_d} \frac{dz}{1 - \frac{2}{\pi}\arccos(\exp[-2\pi\frac{z}{L}])} = \Gamma_d^o + \int_{z=0}^{z=z} \frac{\varepsilon_w}{q_d} F_z dz \quad (2.57)$$

It appears from Fig. 2.11 that the function of F_z can be approximated by:

$$F_z = \left(1 - \frac{2}{\pi}\arccos(\exp[-2\pi\frac{z}{L}])\right)^{-1} \approx 1.45\exp[6.3632\frac{z}{L}] \quad (2.58)$$

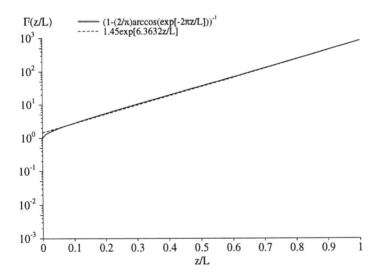

Fig. 2.11: The relation between $F(z/L)$ and z/l.

The residence time as function of z/L can be approximated as:

$$\Gamma^* = \frac{\Gamma(z)}{L} \approx \frac{\Gamma_d^o}{L} + \frac{1.45}{6.3632}(\exp[6.3632z/L] - 1) \quad (2.59)$$

Taking $x = 0$ in the middle between two drains yields:

$$q_d^z = \frac{2x}{L}q_d = q_d\left(1 - \frac{2}{\pi}\arccos(\exp[-2\pi z/L])\right) \quad (2.60)$$

or:

$$\frac{z}{L} = \frac{-1}{2\pi}\ln\cos[\frac{\pi}{2}(1 - \frac{2x}{L})] \qquad (2.61)$$

Substitution of Eq. (2.61) in Eq. (2.59) results in an expression for the residence time in the soil as a function of x/L. Fig. 2.12 gives a presentation of the relation between the residence time in groundwater per m drain distance and the relative distance x/L for a line drain in a profile with infinite thickness and for a perfect drain with a relative drainage layer of $H_d/L = 0.25$. It appears from this figure that the residence time in relation to x/L differs considerably for both systems. For practical applications another approach, taking into account limited values of H_d, has to be considered.

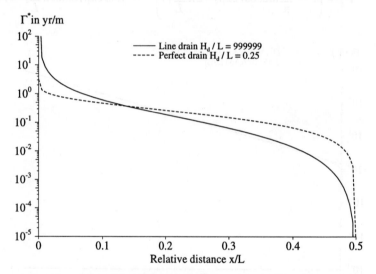

Fig. 2.12: The relation between the residence time in groundwater per m drain distance and the relative distance x/L for a line drain in a soil profile with infinite thickness and a perfect drain with $H_d/L = 0.25$.

2.2.3 Line drains in a soil profile with finite thickness

The drainage flux can be considered as a nearly horizontal flow in the major part of the area in the case of symmetrical drainage between parallel open field drains in a homogeneous soil profile, as presented in Fig 2.13. Near the drains, however, a radial flux is present.

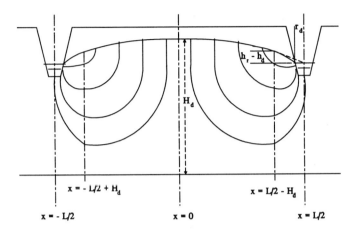

Fig. 2.13: Cross section showing the regions for horizontal flux between $0 \leq 2x \leq L - 2H_d$ and radial flux for the region $L - 2H_d \leq 2x \leq L$.

2.2.3.1 Theory

An approximation for the residence time was obtained considering the drainage flux in the saturated zone as a horizontal one in the region between $0 \leq 2x \leq L - 2H_d$ and as a radial flux between $L - 2H_d \leq 2x \leq L - 2r_d$ (Ernst 1973). The calculation of residence time was performed in two steps. For the region with horizontal flow:

$$\mu \frac{\partial h}{\partial t} = \frac{\partial}{\partial x} kH_d \frac{\partial h}{\partial x} + q_d \tag{2.62}$$

The boundary conditions are:

$$x = 0 \qquad h = h_m \qquad \frac{dh}{dx} = 0$$

$$x = \frac{L}{2} - H_d \qquad h = h_r \tag{2.63}$$

Considering steady state conditions yields:

$$\frac{d}{dx} kH_d \frac{dh}{dx} = -q_d \tag{2.64}$$

Integration of Eq. (2.64) gives:

$$h = -\frac{q_d}{2kH_d} x^2 + \frac{K_1}{kH_d} x + K_2 \tag{2.65}$$

Substitution of the boundary conditions yields:

$$h_x = h_m - \frac{q_d}{2kH_d}x^2 \qquad (2.66)$$

Substituting $x = L/2 - H_d$ gives for h_r the expression:

$$h_r = h_m - \frac{q_d}{8kH_d}(L - 2H_d)^2 \qquad (2.67)$$

Moreover, at $x = L/2 - H_d$:

$$\frac{dh}{dx} = -\frac{q_d}{2kH_d}(L - 2H_d) \qquad (2.68)$$

If the infiltration point is between $0 \leq 2x \leq (L - 2H_d)$, then the time to reach the boundary for radial flux is given by the expression:

$$t = -\frac{\varepsilon_w H_d}{q_d}\ln\left(\frac{2x}{L - 2H_d}\right) \qquad (2.69)$$

An approximation of the cross sectional area (A_r) for radial flux in the region between $L/2 - H_d$ and $L/2 - r_d$ can be derived from Fig. 2.13, where h_d is pressure head in the drain, resulting in:

$$A_r = \frac{\pi}{2}\left(1 + \frac{h_r - h_d}{H_d - (h_r - h_d)}\right)r = \frac{\pi}{2}(1 + \delta_r)r \qquad (2.70)$$

So for the radial flux:

$$\mu\frac{\partial h}{\partial t} = \frac{\partial}{\partial r}\frac{\pi}{2}(1 + \delta_r)rk\frac{\partial h}{\partial r} + q_d \qquad (2.71)$$

Steady state conditions yield:

$$\frac{\pi}{2}(1 + \delta_r)k\frac{d}{dr}\left(r\frac{dh}{dr}\right) = -q_d \qquad (2.72)$$

Integration of Eq. (2.72) gives:

$$\frac{dh}{dr} = \frac{-2q_d}{\pi(1 + \delta_r)k} + \frac{K_1}{r} \qquad (2.73)$$

and

$$h = \frac{-2q_d}{\pi(1+\delta_r)k}r + K_1 \ln r + K_2 \qquad (2.74)$$

The following set of boundary conditions holds for the region with radial flux:

$$x = \frac{L}{2} - H_d \quad r = H_d \quad h_r = h_m - \frac{q_d}{8kH_d}(L - 2H_d)^2$$

$$\frac{dh}{dr} = -\frac{dh}{dx} = \frac{q_d}{2kH_d}(L - 2H_d)$$

$$x = \frac{L}{2} - r_d \quad r = r_d \quad h = h_d \qquad (2.75)$$

This yields for the integration constants when substituting the boundary conditions:

$$K_1 = \frac{q_d}{\pi k(1+\delta_r)}L \qquad (2.76)$$

and

$$K_2 = h_m - \frac{q_d}{8kH_d}(L - 2H_d)^2 + \frac{q_d}{\pi k(1+\delta_r)}(2H_d - L\ln H_d) \qquad (2.77)$$

The expression for the pressure head in the region with radial flow can be written as:

$$h = h_m - \frac{q_d}{8kH_d}(L - 2H_d)^2 + \frac{q_d}{\pi k(1+\delta_r)}(2H_d - 2r + L\ln\frac{r}{H_d}) \qquad (2.78)$$

Differentiation of Eq. (2.78) gives the gradient in the pressure head as:

$$\frac{dh}{dr} = \frac{-q_d}{\pi k(1+\delta_r)}(2 - \frac{L}{r}) \qquad (2.79)$$

For the residence time in the radial flux field holds:

$$\Gamma_r = -\int_{r_1}^{r_2} \frac{\varepsilon_w dr}{k\frac{dh}{dr}} = -\varepsilon_w \int_{r_1}^{r_2} \frac{dr}{\frac{-q_d}{\pi(1+\delta_r)}(2 - \frac{L}{r})} \qquad (2.80)$$

So for $L - 2H_d < 2x < L - 2r_d$ Eq. (2.80) can be rewritten as:

$$\Gamma_r = \frac{-\varepsilon_w \pi(1 + \delta_r)}{4q_d}[L - 2x + L\ln(\frac{2x}{L})] \qquad (2.81)$$

The condition $x = L/2 - H_d$ yields:

$$\Gamma_{max}^r = \frac{-\varepsilon_w \pi(1 + \delta_r)}{4q_d}[2H_d + L\ln(\frac{L - 2H_d}{L})] \qquad (2.82)$$

The residence time in the saturated domain for water infiltrating in the region between $0 < 2x < L - 2H_d$ is given as the sum of Eq.(2.69) and Eq. (2.82) by:

$$\Gamma = \frac{\varepsilon_w H_d}{q_d}\ln(\frac{L - 2H_d}{2x}) - \frac{\varepsilon_w \pi(1 + \delta_r)}{4q_d}[2H_d + L\ln(\frac{L - 2H_d}{L})] \qquad (2.83)$$

Expressing the residence time in days per m drain distance, with $\Gamma^* = \Gamma/L$, $x^* = x/L$, $H_d^* = H_d/L$ and subject to the condition $0 < 2x^* < 1 - 2H_d^*$, gives the expression:

$$\Gamma^* = \frac{\varepsilon_w H_d^*}{q_d}\ln(\frac{1 - 2H_d^*}{2x^*}) - \frac{\varepsilon_w \pi(1 + \delta_r)}{4q_d}[2H_d^* + \ln(1 - 2H_d^*)] \qquad (2.84)$$

and for water draining from the zone with radial flux only $(1 - 4H_d^*/\pi < 2x^* < 1 - 2r_o^*)$, Eq. (2.81) can be rewritten as:

$$\Gamma_r^* = \frac{-\varepsilon_w \pi(1 + \delta_r)}{4q_d}[1 - 2x^* + \ln(2x^*)] \qquad (2.85)$$

Fig. 2.14 shows the effect of taking into account the addition radial cross section related to increasing values of δ_r. The effect of the increased cross section, due to the required pressure head for water transport, on the relative residence time as related to δ_r appears to be small.

2.2.3.2 Model schematization

An approximation for the vertical flux distribution and the residence time in soil profiles with a finite thickness is obtained by combining the linear flux distribution of the perfect drains with the flux distribution of the line drain in a soil profile with infinite thickness of the flow field.

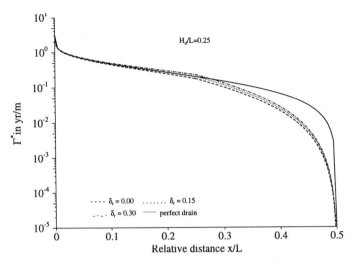

Fig. 2.14: The effect of the slope δ_r of the phreatic water level near the drains as additional cross-section on residence time.

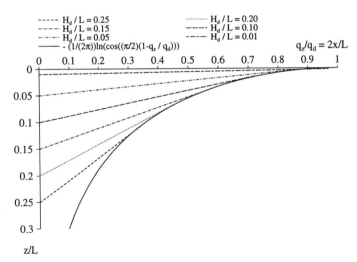

Fig. 2.15: The relation between the relative vertical flux distribution (q_z/q_d) in soil columns with finite thickness and the relative depth (z/L).

This approximation is visualized in Fig. 2.15 for different values of H_d. The curve is given by the expression:

$$\frac{z}{L} = -\frac{1}{2\pi}\ln\cos[\frac{\pi}{2}(1 - \frac{q_z}{q_d})] \tag{2.86}$$

Differentiation of Eq. (2.86) yields the slope of the curve as:

$$\frac{d(z/L)}{d(q_z/q_d)} = \frac{-1}{4}\tan[\frac{\pi}{2}(1 - \frac{q_z}{q_d})] \tag{2.87}$$

For the linear part of the flux distribution as a tangent line to the curve:

$$\frac{z}{L} = \frac{H_d}{L} + a\frac{q_z}{q_d} = \frac{H_d}{L} - \frac{1}{4}\tan[\frac{\pi}{2}(1 - \frac{q_z^*}{q_d})]\frac{q_z}{q_d} \tag{2.88}$$

in which q_z^*/q_d is the value of q_z/q_d at the tangent point of both lines. Combination of Eq. (2.86) and Eq. (2.88) for the tangent point yields the relation between H_d/L and the value of q_z^*/q_d as:

$$\frac{H_d}{L} = \frac{-1}{2\pi}\ln\cos[\frac{\pi}{2}(1 - \frac{q_z^*}{q_d})] + \frac{1}{4}\tan[\frac{\pi}{2}(1 - \frac{q_z^*}{q_d})]\frac{q_z^*}{q_d} \tag{2.89}$$

q_z^*/q_d can be solved from Eq. (2.89) either by the method of Newton-Raphson iteration or by interpolation of a calculated table. This equation can also be approximated for values of H_d/L between 0 and 0.45 with sufficient accuracy by the empirical relation:

$$\frac{q_z^*}{q_d} = \frac{2x_{tan}}{L} = 1.5216\left[0.31 - \frac{H_d}{L} + \sqrt{(0.31 - \frac{H_d}{L})^2 + 0.02448}\right] \tag{2.90}$$

as is shown in Fig. 2.16

The residence time in years per meter drain distance can now be calculated by:

If: $x_{tan}^* \leq x^* < 0.5$:

$$\Gamma^* = \frac{1.45}{6.363204}\frac{\varepsilon_w}{q_d}\left[\exp[-\frac{6.363204}{2\pi}\ln\cos(\frac{\pi}{2}(1 - 2x^*))] - 1\right] \tag{2.91}$$

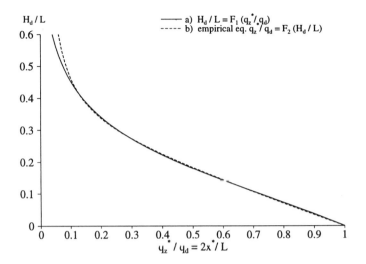

Fig. 2.16: The relation between q_z^*/q_d and H_d/L calculated with
a) Eq. (2.89) and b) Eq. (2.90).

and if: $0 < x^* \leq x_{tan}^*$:

$$\Gamma^* = \frac{\varepsilon_w H_d^*}{q_d} \ln \frac{x_{tan}^*}{x^*} + \frac{1.45}{6.363204} \frac{\varepsilon_w}{q_d} \left[\exp\left[-\frac{6.363204}{2\pi} \ln \cos\left(\frac{\pi}{2}(1 - 2x_{tan}^*)\right)\right] - 1 \right] \quad (2.92)$$

The relative distance of the tangent point per m drain distance is given as:

$$x_{tan}^* = 0.7608 \left[0.31 - H_d^* + \sqrt{(0.31 - H_d^*)^2 + 0.02448} \right] \quad (2.93)$$

Fig. 2.17 gives a presentation of the differences in residence time per m drain distance (Γ^*) and the relative distance (x/L) of the point of infiltration with different values of the relative thickness of the drainage layer (H_d/L). The comparison is given for perfect drains and for line drains in soil profiles with a finite draining depth. For the last situation both the expressions derived for the combination of horizontal and radial flux are presented, as well as the consequences of the model schematization approach. It appears that the model schematization for finite draining soil profiles gives a reasonable approximation for the distribution of the residence time.

Lateral drainage outflow in the model schematization starts at drain depth when tile drains are considered and at the depth of the water level in the drain when dealing with open field drains.

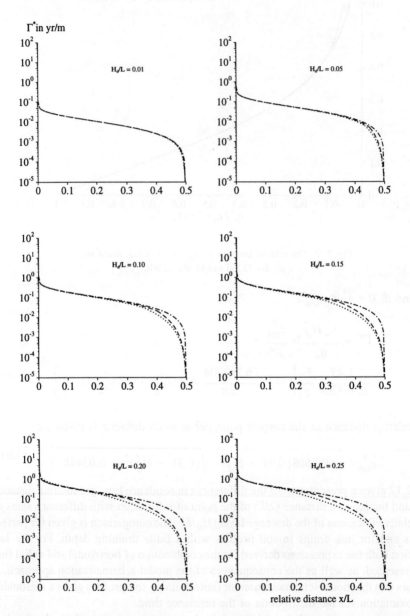

Fig. 2.17: The relation between residence time in groundwater per m drain distance and the relative distance x/L; - - - horizontal + radial flow; · · · · horizontal + infinite field flow; - · - · - perfect drain.

2.3 Model approach for regional transport

In the model approach, the distribution of lateral drainage fluxes with depth has been used to simulate the load to surface water system as a function of the inputs to the groundwater system. The distribution of the lateral fluxes with depth was approximated by means of so-called horizontal model discharge layers. This concept is based on the idea that the summed volumes of the flow systems which contribute to a certain order drainage system can be represented by one horizontal layer. In most cases, the streamline pattern alters from the ideal situation with perfect drains only in the near vicinity of the drains. The discharge flux to a drainage system in a hydrologically homogeneous area with equidistant parallel drains is, according to Ernst (1962), given by:

$$h_m - h_d = q_d(\omega L + \frac{L^2}{8k_h H_d}) = \frac{q_d L^2}{8(k_h H_d)^c} \quad (2.94)$$

where:
q_d = drainage flux in m.d^{-1}
h_m = height of phreatic water level at half drain distance in m
h_d = height of the water level in the drain in m
ω = radial and entry resistance in d.m^{-1}
k_h = horizontal hydraulic conductivity in m.d^{-1}
L = drain distance in m
H_d = equivalent thickness of the drainage layer in m

From Eq. (2.94) follows for the equivalent transmissivity $(k_h H_d)^c$ in m^2.d^{-1}:

$$(k_h H_d)^c = k_h H_d \frac{L}{8\omega k_h H_d + L} \quad (2.95)$$

The flux equation of the whole flow field can be approximated with the equivalent transmissivity for steady state conditions by the expression:

$$\frac{d}{dx}(k_h H_d)^c \frac{dh}{dx} \approx -q_d \quad (2.96)$$

Integration of Eq. (2.96) yields:

$$h(x) \approx h_m - \frac{q_d}{2(k_h H_d)^c} x^2 \quad (2.97)$$

For the average pressure head in the finite element or at the nodal point of a regional groundwater model:

$$\overline{h} = \frac{1}{L/2}\int_{x=0}^{L/2}(h_m - \frac{q_d}{2(k_hH_d)^c}x^2)\mathrm{d}x = h_m - \frac{q_dL^2}{24(k_hH_d)^c} \qquad (2.98)$$

Substitution of Eq. (2.98) in Eq. (2.97) yields for $x = \frac{1}{2}L$:

$$q_d = \frac{8(k_hH_d)^c}{L^2}(h_m - h_d) = \frac{12(k_hH_d)^c}{L^2}(\overline{h} - h_d) \qquad (2.99)$$

For the purpose of water quality simulations, the thickness of the drainage layer has to be limited to a certain depth. In the model the maximum thickness of the equivalent drainage layer H_d^c has been set at $L/4$.

In the saturated zone, the horizontal conductivity is often larger than the vertical conductivity. The general rules with respect to anisotropic conditions of a two-dimensional flow field are:

- hydraulic heads and flow rates are equal to the hydraulic heads and flow rates in an isotropic situation;
- x-coordinate: $x' = x\sqrt{(k_v/k_h)}$;
- z-coordinate: $z' = z$;
- conductivity: $k' = \sqrt{(k_vk_h)}$

For heterogeneous soil profiles, an average anisotropic scale factor has to be considered. The average horizontal and vertical hydraulic conductivity are calculated by:

$$\overline{k}_h = \int_{z=Z_{gwl}}^{z=Z_H} k_h(z)\mathrm{d}z [\int_{z=Z_{gwl}}^{z=Z_H} \mathrm{d}z]^{-1} \qquad \frac{1}{\overline{k}_v} = \int_{z=Z_{gwl}}^{z=Z_H} \frac{1}{k_v(z)}\mathrm{d}z [\int_{z=Z_{gwl}}^{z=Z_H} \mathrm{d}z]^{-1} \qquad (2.100)$$

Applying these rules to the relation between the thickness of the discharge layer H_d and the horizontal drain distance L yields:

$$H_d \leq \frac{L}{4}\sqrt{\frac{\overline{k}_v}{\overline{k}_h}} \qquad (2.101)$$

The formulation for steady discharge to parallel hydrologically uniform drains can also be applied to different types of regional drainage systems, considering each drainage system separately as a homogeneous system. This is schematically presented in Fig. 2.18. By summing up the discharge of the different drain systems the real discharge situation is obtained. The different drain systems are trenches or tile drains, ditches,

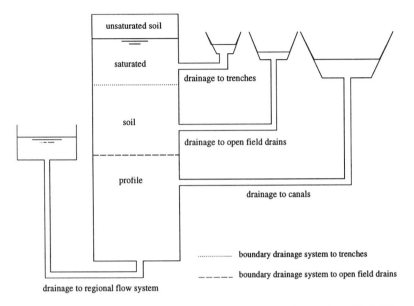

Fig. 2.18: Schematized illustration of the regional drainage concept for a region with 4 different drainage systems. The different drainage systems are field drains, ditches, canals and discharge to the regional flow system.

drain canals or rivers and aquifer recharge. In this schematization, drain canals and rivers also act as ditches and trenches, and ditches act also as trenches.

Fig. 2.19 gives a simplified presentation of the principles of the connection between different columns in a regional transport concept and the water fluxes in the water quality model for a system with a one-dimensional horizontal water flux.

The water balance equation yields:

$$q_{d,i}A_i + k_h W_{i-1,i}(H_{i-1}-H_{d,i-1j_{max}})\frac{h_{i-1}-h_i}{0.5(L_{i-1}+L_i)} =$$

$$A_i \sum_{j=1}^{j_{max}} 12k_{h,i,j}^c H_{d,i,j}\frac{h_i-h_{i,j}}{L_{i,j}^2} + k_h W_{i,i+1}(H_i-H_{d,ij_{max}})\frac{h_i-h_{i+1}}{0.5(L_i+L_{i+1})} \quad (2.102)$$

in which:
i = column number
j = order number of drain system
A_i = column area in m^2
H_i = aquifer depth in m
W_i = width of the column in m
L_i = length of the column in m

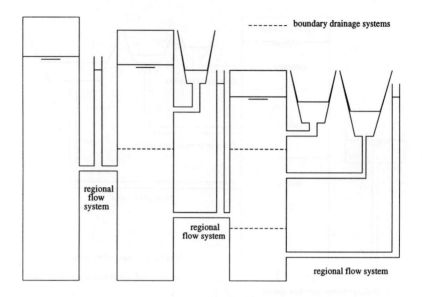

Fig. 2.19: Schematic presentation of coupling columns in a regional transport concept.

h_i = height of the phreatic water level in m
$q_{s,i}$ = vertical flux at the height h_i, positive for downward flux, in m.d^{-1}
$H_{d,i,j}$ = equivalent thickness of the drainage layer of order j in m
$L_{i,j}$ = drain distance of drain system j in m
$h_{i,j}$ = height of the water level in the drain system j in m
$k_{h,i,j}^c$ = equivalent horizontal hydraulic conductivity of drain system j

In regional studies, the water balance per sub-region is somewhat more complicated. Denoting the upstream sub-regions bordering the sub-region i by the number l, and the downstream sub-regions by the number m, gives the parameters:

$L_{l,i}$; $L_{i,m}$ = distance between nodal points l and i, respectively i and m in m
$W_{l,i}$; $W_{i,m}$ = border width between elements l and i, respectively i and m in m

The water balance equation is now given by the expression:

$$q_{d,i} A_i + \sum_{l=1}^{l_{max}} k_h W_{l,i} (H_l - H_{d,l,j_{max}}) \frac{h_l - h_i}{L_{l,i}} =$$

$$A_i \sum_{j=1}^{j_{max}} 12 k_{h,i,j}^c H_{d,i,j} \frac{h_i - h_{i,j}}{L_{i,j}^2} + \sum_{m=1}^{m_{max}} k_h W_{i,m} (H_i - H_{d,i,j_{max}}) \frac{h_i - h_m}{L_{i,m}} \quad (2.103)$$

Water Transport in Soils

Regional discharge does generally give a non-linear relation between measured discharges and groundwater tables. The general shape of such a so-called q/h-relation is given in Fig. 2.20a, from which it appears that there is a clear difference between discharge intensity and depth of the groundwater table.

High discharge intensities by shallow groundwater tables are mainly caused by shallow drains, as open field drains and tile drains, with relatively short residence times, whereas low discharge intensities with deep groundwater tables are the result of drainage canals at larger distance, resulting in large residence times.

In the model schematization the q/h-relation is subdivided in a number of linear relations (Fig. 2.20b), each of them representing a certain type of drain system. From these schematized q/h-relations drainage resistances are derived and they are used as input for the calculation of the drainage quantities per time step and per drain system, as depending on depth of the groundwater table and the surface water level. In most cases three drain systems are sufficient for the description of the discharge to surface waters from the unit elements in regional studies. The discharge to or infiltration from rivers and canals at very large distances is taken into account as leakage and/or seepage over the lower boundary of the model system.

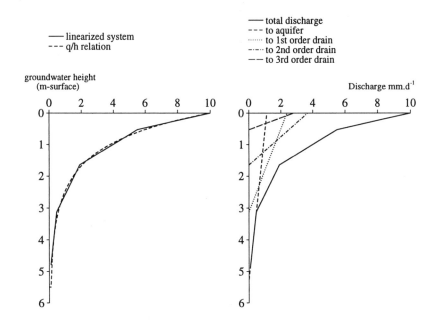

Fig. 2.20: a) Relation between regional discharge and phreatic water level according to Ernst (1978); b) schematized relation with 4 levels of drainage.

In the hydrological schematization for regional water quality studies the following aspects have to be taken into account:

- The thickness of a model discharge layer has to be limited to a certain depth. This depth is determined by the presence of an impervious layer at a certain depth, with the maximum thickness of the drainage layer taken as $H_d \leq 0.25L$ of the drain spacing of the highest order of drains present in a calculation unit.
- In the hydrological simulation generally the average pressure head in the model unit is calculated, which requires a shape factor in the drainage resistance for the calculation of the average discharge of each system.
- For the calculation in regional studies of the layer numbers with a lateral flux to a certain drainage system all drains are considered as perfect drains, so radial resistances are neglected.
- Each higher order drain system also acts as lower order drains. So a drainage canal is also considered as a drain ditch and field drain, while a drain ditch also operates as a field drain.

It must be realized that if no infiltration from the drain into the soil system is present, as is very often the case for field trenches and field tile drains or without seepage from the aquifer, the total discharge is related to the pressure head in the soil (h), the pressure heads in the drains ($h_{d,n}$) and in the aquifer (h_{aq}) by the conditions:

$$\begin{aligned}
h_{d,4} &\leq h \leq 0.0 & n &= 4 \\
h_{d,3} &\leq h \leq h_{d,4} & n &= 3 & q_{d,4} &= 0.0 \\
h_{d,2} &\leq h \leq h_{d,3} & n &= 2 & q_{d,3} &= q_{d,4} = 0.0 \\
h_{aq} &\leq h \leq h_{d,2} & n &= 1 & q_{d,2} &= q_{d,3} = q_{d,4} = 0.0 \quad (2.104)
\end{aligned}$$

Drains continuously containing water have an infiltration function when the water table in the soil becomes lower than the water level in the drain under consideration. When the phreatic water level drops below the aquifer pressure head leakage conditions are changing into seepage conditions. Analogous to the situation with only one drainage system, so-called model discharge layers are identified from which water is discharged to corresponding drainage systems. In this approach for each time step a certain layer with specified thickness is identified for discharge to the distinguished drainage system. The depth of the bottom of the highest order of drainage system present in a unit is considered as the base of the local flow system.

The determination of the layers draining to a certain order of drains is for homogeneous soils and perfect drains approximated from the drainage equation considering horizontal flow only. The differences in residence time of the water discharging to the corresponding systems is then taken into account with the discharging layers proportional to the areic discharge volume rates towards the corresponding drains, or:

$$H_{d,1} : H_{d,2} : H_{d,3} = q_{d,1} : q_{d,2} : q_{d,3} \quad (2.105)$$

and

$$H_{d,1} + H_{d,2} + H_{d,3} \leq \frac{L_1}{4}\sqrt{\frac{\overline{k}_v}{\overline{k}_h}} \qquad (2.106)$$

where:
$H_{d,i}$ = the thickness of the model discharge layer of drainage system i in m
$q_{d,i}$ = the average areic discharge volume rate of drainage system i in $m^3.m^{-2}.d^{-1}$

Fig. 2.21 shows the calculation method for soil profiles with a heterogeneous conductivity distribution with depth. The depth of the bottom of each discharge layer $H_{d,i}$ for soil profiles with a heterogeneous conductivity distribution with depth is obtained by interpolation in the cumulative k-H_d-relation with depth.

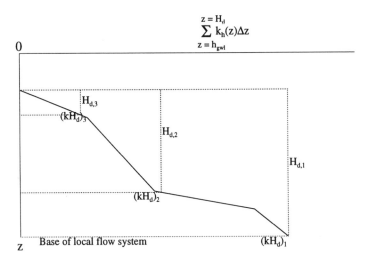

Fig. 2.21: Bottom of model discharge layers as a function of transmissivity in a heterogeneous soil profile.

The lateral flux relation per unit soil depth has a uniform distribution. Lateral fluxes $q^L_{k,i}$ to drainage system k per computation layer i of the model are calculated by, respectively:

$$q^L_{j,i} = q_j \frac{k_{h,i}\Delta z_i}{\sum_{i(z=h(dr,3))}^{i(z=H_{d,j})} k_{h,i}\Delta z_i} \qquad h_{dr,3} < z < H_{d,j} \quad j = 1 \qquad (2.107)$$

and

$$q_{j,i}^L = q_j \frac{k_{h,i}\Delta z_i}{\sum_{i(z=H_{d,j-1})}^{i(z=H_{d,j})} k_{h,i}\Delta z_i} \qquad H_{d,j-1} < z < H_{d,j} \quad j > 1 \qquad (2.108)$$

where:
$k_{h,i}$ = the horizontal conductivity per compartment in m.d^{-1}
Δz_i = layer thickness in m
$i(z=h_{dr,3})$ = number of the compartment in which the groundwater level is situated
$i(z=H_{d,j})$ = number of the bottom compartment of drain system j.

The direction and the quantity of regional groundwater flow are influenced by a number of geo-hydrological factors, depending on the heterogeneity of the aquifer. However, the main flow is considered being horizontal and two-dimensional. It must be realized, however, that in the downstream direction the streamlines penetrate deeper into the aquifer. With the use of a groundwater model all fluxes in each element are known. For the regional water quality calculations the aquifer is subdivided in a relatively small number of horizontal layers each with its own hydrological characteristics, to introduce solute dispersion due to horizontal heterogeneity. This approach also has the advantage that inflow of a solute is not mixed immediately over the complete aquifer thickness.

References

Abrahamse, A.H., Baars G. and Beek E. van, 1982. *Policy analysis of water management for the Netherlands. Model for regional hydrology, agricultural water demands and damage from drought and salinity.* Vol XII. RAND Corporation, Delft Hydraulics Laboratory: Delft, The Netherlands.

Belmans, C., Wesseling, J.G. and Feddes, R.A. 1983. Simulation model of the water balance of a cropped soil: SWATRE. *Journal of Hydrology*, **63**, 271-286.

Berghuijs-van Dijk, J.T. 1985. WATBAL: *a simple water balance model for an unsaturated/saturated soil profiel.* Nota **1670**. Institute Land and Water Management Research: Wageningen, The Netherlands.

Broek, B.J. van den, Elbers, J.A., Huygen, J., Kabat, P., Wesseling, J.G., Dam J.C. van and Feddes, R.A. 1994. *SWAP93, Input instructions manual.* Interne Mededeling **288**. SC–DLO: Wageningen, The Netherlands.

Bronswijk, J.J.B. and Evers-Vermeer, J.J. 1990. Shrinkage of Dutch clay soil aggregates. *Neth. J. Agricultural Science*, **38**, 175-194.

Bronswijk, J.J.B., 1991. *Magnitude, modelling and significance of swelling and shrinkage processes in clay soils.* Ph. D. Thesis. Agricultural University: Wageningen, The Netherlands.

Butijn, J. 1961. *Bodembehandeling in de fruitteelt.* Verslagen Landboukundige Onderzoekingen **66.7**. PUDOC: Wageningen, The Netherlands.
Christensen, H.R. 1944. Capillary conductivity curves for three prairie soils. *Soil Science*, **57**, 381-391
Dam, J.C. van, Huygen, J., Wesseling, J.G., Feddes, R.A., Kabat, P., Walsum, P.E.V. van, Groenendijk, P. and Diepen, C.A. van, 1997. SWAP *version 2 Theory; Simulation of water flow, solute transport and plant growth in the soil-water-atmosphere-plant environment.* Technical Document **45**. SC–DLO: Wageningen, The Netherlands.
Ernst, L.F. 1973. *De bepaling van de transporttijd van het grondwater bij stroming in de verzadigde zone.* Nota **755**. Institute Land and Water Management Research: Wageningen, The Netherlands.
Ernst, L.F. 1962. *Grondwaterstromingen in de verzadigde zone en hun berekening bij aanwezigheid van horizontale open leidingen.* Ph.D Thesis. University Utrecht, Verslagen Landbouwkundige Onderzoekingen **67.15** Pudoc: Wageningen, The Netherlands.
Feddes, R.A., Kowalik, P.J. and Zaradny, H. 1978. *Simulation of field water use and crop yield.* Simulation Monographs, PUDOC: Wageningen, The Netherlands.
Gardner, W.R. and Fireman, M. 1958. Laboratory studies of evaporation from soil columns in the presence of a water table. *Soil Science*, **85**, 244-249.
Genuchten, M.Th. van, 1980. A closed-form equation for predicting the hydraulic conductivity of unsaturated soils. *Soil Science Soc. Am. J.*, **44**, 892-898.
Kittab, H.M.A. 1983. *Seasonal variation in water table and its effect on properties of some Egyptian Soils.* Ph. D. Thesis. Cairo University: Cairo, Egypt.
Oostindie, K. and Bronswijk, J.J.B. 1992. FLOCR - *A simulation model for the calculation of water balance, cracking and surface subsidence of clay soils.* Report **47**. SC–DLO: Wageningen, The Netherlands.
Querner, E.P. and Bakel, P.J.T. van, 1989. *Description of the regional groundwater flow model* SIMGRO. Report **7**. SC–DLO: Wageningen, The Netherlands.
Querner, E.P. 1988. Description of a regional groundwater flow model SIMGRO and some applications. *Agricultural Water Management*, **14**, 209-218.
Richards, L.A. and Moore, D.C. 1952. Influence of capillary conductivity and depth of wetting on moisture retention in soil. *Transactions Am. Geoph. Un.*, **33**, 531-540.
Richards, L.A. and Wilson, B.D. 1936. Capillary conductivity measurements in peat soils. *J. Am. Soc. Agronomy*, **28**, 427-431
Rijtema, P.E., 1965. *An analysis of actual evapotranspiration.* Agricultural Research. Reports **659**. PUDOC: Wageningen, The Netherlands.
Rijtema, P.E. 1969. *Soil moisture forecasting.* Nota **513**. Institute Land and Water Management Research: Wageningen, The Netherlands.
Rijtema, P.E. and Aboukhaled, A. 1975. Crop Water use. In *Research on crop water use, salt affected soils and drainage in the Arabic Republic of Egypt*, pp.5-61. FAO Near East Regional Office: Cairo, Egypt.
Roest, C.W.J., Rijtema, P.E., Abdel Khalek, M.A., Boels, D., Abdel Gawad, S.T. and El Quosy, D.E. 1993. *Formulation of the on-farm water management model* FAIDS. Report **24**, Reuse of Drainage Water Project. DRI: Cairo, Egypt. SC–DLO: Wageningen, The Netherlands.
Rubin, J., Steinhardt R. and Renniger, P. 1964. Soil water relations during rain infiltration II. Moisture content profiles during rains of low intensities. *Soil Sci Soc. Am. Proc.*, **28**, 1-5.

Talsma, T. 1963. *The control of saline groundwater.* Med. Landbouwhogeschool **63.10**. Agricultural University: Wageningen, The Netherlands.

Wilson, B.D. and Richards, S.J. 1938. Capillary conductivity of peat soil at different moisture tensions. *J. Am. Soc. Agronomy,* **30**, 583-588.

Wind, G.P. 1955. A field experiment concerning capillary rise of moisture in a heavy clay soil. *Neth. J. Agricultural Science,* **3**, 60-69.

Wind, G.P. and Hidding A.P. 1961. The soil physical basis of the improvement of clay cover soils. *Neth. J. Agricultural Science,* **9**, 281-292.

Wösten, J.H.M., Veerman, G.J. en Stolte, J. 1994. *Waterretentie- en doorlatendheids karakteristieken van boven- en ondergronden in Nederland: de Staringreeks. Vernieuwde uitgave 1994.* SC–DLO Winand Staring Centre: Wageningen, The Netherlands.

CHAPTER 3

TRANSPORT OF SOLUTES

Models of varying degree of complexity and dimensionality have been developed during the past several decades to quantify the basic physical and chemical processes affecting pollutant transport in the unsaturated zone. Models for variably unsaturated-saturated flow, solute transport, aqueous chemistry and cation exchange were initially developed mostly independently from each other, and there has only recently been a significant effort to couple the different models. A large number of numerical methods may be used to solve the convection-dispersion solute transport equation. Van Genuchten and Šimůnek (1996) give a review of the problems involved when using different numerical approaches to solve the convection-dispersion transport equation. Standard finite difference and finite elements methods provided the early tools for solving solute transport problems and in spite of some limitations are the most popular methods being used at present. Numerical studies have shown that both methods give good results for transport where dispersion is a relatively dominant process. However, both methods can lead to significant numerical oscillations and/or dispersion for convection-dominated transport problems The popularity of numerical methods stems from the fact that the highly non-linear equations can be solved analytically only for a very limited number of cases involving homogeneous soils and relatively constitutive relationships describing the different properties. Time and space discretization using any of these methods leads to a non-linear system of equations. These equations are most often linearized and solved using the iterative computation schemes.

Roest and Rijtema (1983), Berghuijs *et al.* (1985), Kroes and Rijtema (1989) and Roest *et al.* (1993) used local analytical solutions of the dispersion-convection equation. Recently several other methods were suggested which make use of local analytical solutions of the convection-dispersion equation in combination of finite differences (Li *et al.* 1992). The combination of analytical and numerical techniques has one important limitation, since all coefficients such as water content, flow velocity and retardation factors and degradation process parameters must be independent of time during the time increment. This means that combination methods can only solve solute transport problems during steady-state water flow, and hence are not appropriate to transient unsaturated-saturated flow conditions. However, when modelling diffuse pollution in regional studies the transient variably unsaturated-saturated flow in most field conditions during the time increment becomes less important.

The unsaturated topsoil is that part of the soil profile where influences of human activities are predominant. Concern on groundwater contamination has focused attention on the processes that influence chemical fate in soil-water systems. Independent of the mode of introduction of the chemicals, a major concern with respect to groundwater contamination is the passage of these chemicals through the relative thin layers of soil that cover many terrestrial surfaces. This water unsaturated domain of the soil profile, extending from the soil-atmosphere interface to the groundwater table, including the root zone of most plants, is the chemically and biologically most active region of the biosphere. This is also the domain with the largest variations in aeration, soil moisture suction and temperature. In particular the biochemical transformation processes are strongly affected by these environmental conditions. For instance, the nutrient balance of the topsoil, where the influences of agricultural activities are predominant, can be represented as is shown in Fig. 3.1.

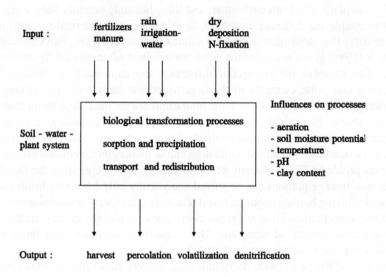

Fig. 3.1: Nutrient balance of the topsoil with respect to N and P.

Inputs of minerals can originate from fertilization with either anorganic fertilizers or animal manures, irrigation water, precipitation and dry atmospheric deposition. Nitrogen gas can also be made available by nitrogen fixation. In the soil-water-plant system the different forms of N and P can be transformed from organic compounds into anorganic compounds and vice versa. In this system sorption processes of NH_4 and mineral PO_4 also take place. Phosphate precipitation has also been considered. Dissolved forms of N and P can be transported to other layers depending on the direction of the water flux. Nutrients leave the topsoil by means of harvest and leaching. N-compounds also can be lost by volatilization as gaseous loss of NH_3 and due to denitrification under anaerobic conditions as gaseous loss of N_2 or N_2O. The dissolved species considered in the C-, N- and P-cycle can be transported with the different water fluxes present in the system as is shown in Fig. 3.2.

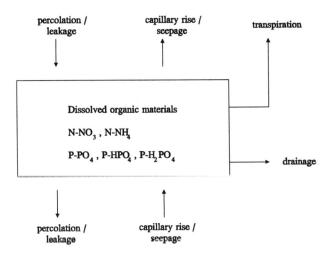

Fig. 3.2: Transport of dissolved compounds to and from a soil layer as influenced by water flow in both the unsaturated and saturated zone.

Transpiration and drainage fluxes extract water and solutes from the soil system. In the unsaturated domain capillary rise is an upward flux of water and solutes to and from layers; percolation is a downward transport to deeper layers. Leakage and seepage are the corresponding transports in the saturated domain. Discharge to drain systems in the saturated domain is taken into account as a lateral outflow. Crops can develop an active uptake of nutrients, which is taken into consideration by the introduction of selectivity coefficients.

3.1 Transport and mass conservation equation

3.1.1 Theory

A mathematical description of transport processes in a soil system must obey the law of conservation of matter and energy. For a compound in a soil system this law can be written in words as:

accumulation = inflow - outflow - sinks + sources

Accumulation is the storage change of a compound and can take place in the liquid phase, as well as in the solid phase. Inflow and outflow of the compound occurs as a solute flux across the boundaries of the system. In water quality models the concentration of the leaching water flux is solved from a general equation which describes the storage change in a soil-water-plant system. By introducing source and sink terms for production, decomposition, plant uptake and lateral drainage to surface water, the following mass conservation equation is formulated:

$$\frac{\partial c^*}{\partial t} = -\frac{\partial J_s}{\partial z} + R_p - R_d - R_u - R_x \tag{3.1}$$

where:
c^* = mass concentration of a compound in the soil in kg.m^{-3}
t = time in d
J_s = vertical solute flux in kg.m^{-2}.d^{-1}
z = depth in the soil in m
R_p = source for production in kg.m^{-3}.d^{-1}
R_d = sink for decomposition in kg.m^{-3}.d^{-1}
R_u = sink for plant uptake in kg.m^{-3}.d^{-1}
R_x = sink for lateral drainage or a source for lateral infiltration in kg.m^{-3}.d^{-1}

The mass concentration of a compound is the sum of concentrations present in the liquid phase and in the solid phases:

$$c^* = \theta c + \rho_d X_e^s + \rho_d X_n^s + \rho_d X^p \tag{3.2}$$

where:
θ = volume fraction of soil water in m^3.m^{-3}
c = solute concentration in kg.m^{-3}
ρ_d = dry bulk density of the soil in kg.m^{-3}
X_e^s = content sorbed to the solid phase in equilibrium with c in kg.kg^{-1}
X_n^s = content sorbed to the solid phase in non-equilibrium with c in kg.kg^{-1}
X^p = content of precipitated compound in kg.kg^{-1}

Traditional descriptions of the transport of solutes in the soil were based on the Richard equation for unsaturated flow and on the Fickian based convection-dispersion transport equation. Convection refers to the transport of a solute at a velocity equivalent to that of the soil water flux. Dispersion is the term used to describe the spreading of a concentration front as a result of spatial variation in aquifer variability, fluid mixing and molecular diffusion. The flux equation for the vertical transport of solutes is a combination of the dispersive and convective flow equation and is given by:

$$J_s = -\theta D_{dd}\frac{\partial c}{\partial z} + qc \tag{3.3}$$

where:
J_s = total solute flux in kg.m^{-2}.d^{-1}
D_{dd} = dispersion (including diffusion) coefficient in m^2.d^{-1}
q = water flux in m.d^{-1}

The general continuity equation for the one-dimensional, vertical non-steady water and solute flow is written as:

$$\frac{\partial \theta c}{\partial t} + \rho_d \left(\frac{\partial X_e^s}{\partial t} + \frac{\partial X_n^s}{\partial t} + \frac{\partial X^p}{\partial t} \right) = \theta D_{dd} \frac{\partial^2 c}{\partial z^2} - q \frac{\partial c}{\partial z} + R_p - R_d - R_u - R_x \quad (3.4)$$

Only a few analytical solutions of this second order partial differential equation are available for a limited set of boundary conditions. The partial differential equation for both water flow and solute behaviour can be solved by numerical approximation methods, using a finite difference method. The equation is then rewritten as an equation with differences instead of partial differentials. A numerical explicit or implicit solution scheme is used to solve sets of difference equations. Examples of models that use this approach are PESTLA (Boesten 1986, Boesten and van der Linden 1991), DAISY (Hansen et al. 1990), LEACHM (Wagenet and Hutson 1992) and WAVE (Vanclooster et al. 1994). The numerical methods preferably use a solution scheme which eliminates or minimizes the numerical dispersion and then reintroduce a physical dispersion.

The accuracy by which the dispersion coefficient D_{dd} can be estimated in field studies is rather small. The value of the dispersion coefficient is primarily determined by the spatial variation of the soil permeability, as caused by preferential flow and soil cracks or by an irregular distribution of the infiltrating water. The dispersion coefficient is generally defined proportionally to the stream velocity, using the expression:

$$D_{dd} = \Lambda \left| \frac{q}{\theta} \right| \quad (3.5)$$

where:
Λ = dispersion length in m

3.1.2 Model schematization

In the water quality models ANIMO and TRANSOL, described here, semi-analytical solutions of the general dispersion-convection equation are used. The semi-analytical approach is well known in modelling point sources in groundwater contamination studies (Davis and Salama 1994) and in modelling the pollutant transport in surface water systems (Todini 1996).

3.1.2.1 Single component transport

The second order partial differential equation is simplified to a first order equation by eliminating the dispersion/diffusion term. The remaining first order differential equation is solved analytically. Dispersion/diffusion is introduced by choosing an appropriate value of the vertical compartment thickness. Rapid and clear solutions combined with the fact that measured values for dispersion/diffusion under field conditions are hardly available are the main arguments for this solution. The soil profile is divided into layers. The depth of the lower boundary of layer number n is given as Z_n. Assuming equal layer thicknesses Δz, the transport term is approximated for the depths Z_{n-1} and Z_n by:

Considering the concentration in the centre of a layer as the average concentration

$$-\frac{\partial J_s}{\partial z} = \frac{\partial}{\partial z}(\theta D_{dd}(\Lambda \frac{\partial c}{\partial z} - qc)) = \frac{\partial}{\partial z}(q(\Lambda \frac{\partial c}{\partial z} - c)) \approx$$

$$\frac{1}{\Delta z}[q(Z_n)(\Lambda \frac{c_{Z,n+\frac{1}{2}} - c_{Z,n-\frac{1}{2}}}{\Delta z} - \frac{c_{Z,n+\frac{1}{2}} + c_{Z,n-\frac{1}{2}}}{2})] -$$

$$\frac{1}{\Delta z}[q(Z_{n-1})(\Lambda \frac{c_{Z,n-\frac{1}{2}} - c_{Z,n-1\frac{1}{2}}}{\Delta z} - \frac{c_{Z,n-\frac{1}{2}} + c_{Z,n-1\frac{1}{2}}}{2})] =$$

$$\frac{q(Z_n)}{\Delta z}(\frac{\Lambda}{\Delta z} - \frac{1}{2})c_{Z,n+\frac{1}{2}} - \frac{q(Z_n)}{\Delta z}(\frac{\Lambda}{\Delta z} - \frac{1}{2})c_{Z,n-\frac{1}{2}} -$$

$$\frac{q(Z_{n-1})}{\Delta z}(\frac{\Lambda}{\Delta z} + \frac{1}{2})c_{Z,n-\frac{1}{2}} + \frac{q(Z_{n-1})}{\Delta z}(\frac{\Lambda}{\Delta z} + \frac{1}{2})c_{Z,n-1\frac{1}{2}} \quad (3.6)$$

of the layer and taking layer thickness as twice the dispersivity ($\Delta z = 2\Lambda$) reduces Eq. (3.6) to an ordinary first order differential equation. The subscripts of the concentration variable have been redefined, since complete mixing within a soil layer is assumed.

Under these conditions for downward flux:

$$\Delta z_n \frac{d\theta_n c_n}{dt} = q_{n-1}c_{n-1} - q_n c_n + S_n(t)\Delta z_n \quad (3.7)$$

For upward flux:

$$\Delta z_n \frac{d\theta_n c_n}{dt} = q_{n+1}c_{n+1} - q_n c_n + S_n(t)\Delta z_n \quad (3.8)$$

The term $S_n(t)$ is a generalized source and sink term given by:

$$S_n(t) = R_p - R_d - R_u - R_x - \rho_{d,n}(\frac{dX_e}{dt} + \frac{dX_n}{dt} + \frac{dX_p}{dt}) \quad (3.9)$$

Transport of Solutes

All decomposition and production processes are described as first or zero order rate kinetics. Uptake of a compound by plants is described proportional to the compound concentration in the liquid phase and the transpiration flux q_t towards plant roots. A multiplication factor has been introduced to account for preferential uptake.

Assuming complete mixing in each identified soil layer, the transport and conservation equation can be written as:

$$\frac{d(\theta_n(t)c_n(t))}{dt} + \rho_{d,n}\frac{dX_{e,n}}{dt} + \rho_{d,n}\frac{dX_{n,n}}{dt} + \rho_{d,n}\frac{dX_{p,n}}{dt} =$$

$$\sum_{l=1}^{l_{max}} \frac{q_l^i(t)c_l^i(t)}{\Delta z_n} - \sum_{k=1}^{k_{max}} \frac{q_k^o(t)c_n(t)}{\Delta z_n} - \frac{q_t \sigma_{pl} c_n(t)}{\Delta z_n} + k_{0,n} + k_{1,n}\theta_n(t)c_n(t) \quad (3.10)$$

where:
- k = number of outgoing fluxes from layer n (-)
- l = number of incoming fluxes into layer n (-)
- n = layer number (-)
- $\sum q^i.c^i$ = total incoming flux of material (kg.m^{-2}.d^{-1})
- $\sum q^o.c$ = total outgoing flux of material (kg.m^{-2}.d^{-1})
- q_t = transpiration flux (m.d^{-1})
- σ_{pl} = selectivity factor for material uptake by plants (-)
- k_0 = zero order production rate coefficient (kg.m^{-3} soil.d^{-1})
- k_1 = first order production rate coefficient (d^{-1})

In general, the following rules are applied:
- The production rate coefficients k_0 and k_1 are positive in case of material production and negative in case of consumption.
- The moisture content θ_n is in aggregated soils replaced by the sum of the moisture content in the macro-pores $\theta_{e,n}$ and the moisture content in the aggregates $\theta_{i,n}$, so: $\theta_n = \theta_{e,n} + \theta_{i,n}$.
- The water balance of each layer is used in the hydrological schematization, considering the drainage flux to surface water as a horizontal outflow from each participating layer below drain depth in the saturated zone.
- The relation between the adsorbed quantity and the concentration in solution is given by an equilibrium expression. The change in the quantity of adsorbed material is given in the transport and conservation equation by:

$$\frac{dX_{e,n}}{dt} = K_d \frac{dc_n(t)}{dt} \quad (3.11)$$

in which the distribution coefficient K_d is linearized ($\bar{K}_d$) if non-linear adsorption is considered (see section 4.2.2).

- The formulation of the effects of time dependent adsorption and precipitation or dissolution of material in the transport and mass conservation equation is discussed in Chapter 4.
- All fluxes are assumed to be constant within a time step, the moisture fraction will change linearly with time and the moisture volume of layer n is given by the expression:

$$V_{\theta,n}(t) = \Delta z_n(\theta_0 + k_\theta t) \tag{3.12}$$

in which:
θ_0 = soil moisture fraction at the beginning of the time step in $m^3.m^{-3}$
k_θ = change in moisture content in $m^3.m^{-3}.d^{-1}$

The total incoming flux of material has to be identified on the basis of all incoming fluxes and their concentrations. Assuming these concentrations are constant within the time step gives the expression:

$$\sum_{l=1}^{l_{max}} q_l^i(t) c_l^i(t) = q_{n+1/2}^i \bar{c}_{n+1} + q_{n-1/2}^i \bar{c}_{n-1} + q_d^i \bar{c}_{id} \tag{3.13}$$

in which:
$q_{n+1/2}^i$ = incoming flux from layer n+1 in $m.d^{-1}$
$q_{n-1/2}^i$ = incoming flux from layer n-1 in $m.d^{-1}$
q_d^i = infiltration flux from the drainage system in $m.d^{-1}$
$\bar{c}$ = average concentration during the time step in $kg.m^{-3}$
$\bar{c}_{id}$ = average concentration of infiltrating water from the drain system in $kg.m^{-3}$

- For the outgoing material transport from layer n the concentration $c(n,t)$ has been considered as a function of time. For the total water flux the upward flow $(q_{n-1/2}^o)$ to layer n-1, downward flow $(q_{n+1/2}^o)$ and lateral flow to the drainage system (q_d^o) have to be considered, so:

$$\sum_{k=1}^{k_{max}} q_k^o(t) c_n(t) = c_n(t)(q_{n+1/2}^o + q_{n-1/2}^o + q_d^o) \tag{3.14}$$

- Material uptake by the plants is considered to be proportional to the transpiration flux and the concentration in layer n. The value of the selectivity factor σ_{pl} can vary from 0, resulting in no uptake by plants, to values exceeding 1. A full discussion on the selectivity factor for plant uptake is presented in Chapter 7.
- The moisture content $\theta_n(t)$ in the first order production term is considered as constant, replacing the variable moisture content by its average value during the time step ($\bar{\theta}_n = \theta_n(t_0) + (k_\theta.t)/2$).

Introducing these relations in the transport and conservation equation gives after rearranging of terms the following general form of the differential equation:

$$\frac{dc_n(t)}{dt} + \frac{\varsigma_1}{\theta_0 + \rho_d K_d + k_\theta t} c_n(t) = \frac{\varsigma_2}{\theta_0 + \rho_d K_d + k_\theta t} \quad (3.15)$$

For the solution of the general transport and mass conservation equation different cases must be distinguished, depending on the sorption equations used and the introduction of instantaneous and/or time dependent precipitation. These conditions determine the values of the constants ς_1 and ς_2. A complete description for the evaluation of the constants ς_1 and ς_2 is given in section 4.6.

Based on certain conditions for the change in moisture content with time k_θ and ς_1 the following solutions of the differential equation are found:

- if $k_\theta \neq 0$ and $\varsigma_1 \neq 0$:

$$c_n(t) = \frac{\varsigma_2}{\varsigma_1} + (c_n(t_0) - \frac{\varsigma_2}{\varsigma_1}) \left(\frac{\theta_0 + \rho_d K_d + k_\theta t}{\theta_0 + \rho_d K_d} \right)^{-\varsigma_1/k_\theta} \quad (3.16)$$

- if $k_\theta = 0$ and $\varsigma_1 \neq 0$:

$$c_n(t) = \frac{\varsigma_2}{\varsigma_1} + (c_n(t_0) - \frac{\varsigma_2}{\varsigma_1}) \exp[-\varsigma_1 t/(\theta_0 + \rho_d K_d)] \quad (3.17)$$

- if $k_\theta \neq 0$ and $\varsigma_1 = 0$:

$$c_n(t) = c_n(t_0) + \frac{\varsigma_2}{k_\theta} \ln(\frac{\theta_0 + \rho_d K_d + k_\theta t}{\theta_0 + \rho_d K_d}) \quad (3.18)$$

- and if $k_\theta = 0$ and $\varsigma_1 = 0$:

$$c_n(t) = c_n(t_0) + \frac{\varsigma_2 t}{\theta_0 + \rho_d K_d} \quad (3.19)$$

The average concentration of each layer during a time step has to be calculated, due to the assumption that the incoming fluxes from other layers have the average concentration of these layers. By integration of Eq. (3.16) through Eq. (3.19) over the time step and dividing by the time step length the following expressions are found:

- if $k_\theta \neq 0$ and $\varsigma_1 \neq 0$ and $k_\theta \neq \varsigma_1$:

$$\bar{c}_n = \frac{\varsigma_2}{\varsigma_1} + \frac{(\theta_0 + \rho_d K_d)(c_n(t) - c_n(t_0)) + k_\theta t(c_n(t) - \frac{\varsigma_2}{\varsigma_1})}{t(k_\theta - \varsigma_1)} \quad (3.20)$$

- if $k_\theta \neq 0$ and $\varsigma_1 \neq 0$ and $k_\theta = \varsigma_1$:

$$\bar{c}_n = \frac{\varsigma_2}{\varsigma_1} + (\theta_0 + \rho_d K_d)\left(\frac{c_n(t_0) - \varsigma_2/\varsigma_1}{k_\theta t}\right) \ln\left(\frac{c_n(t_0) - \varsigma_2/\varsigma_1}{c_n(t) - \varsigma_2/\varsigma_1}\right) \quad (3.21)$$

- if $k_\theta = 0$ and $\varsigma_1 \neq 0$:

$$\bar{c}_n = \frac{\varsigma_2}{\varsigma_1} + \left(\frac{\theta_0 + \rho_d K_d}{\varsigma_1 t}\right)(c_n(t_0) - c_n(t)) \quad (3.22)$$

- if $k_\theta \neq 0$ and $\varsigma_1 = 0$:

$$\bar{c}_n = c_n(t_0) + \frac{\varsigma_2}{k_\theta}[(\theta_0 + \rho_d K_d + k_\theta t)\left(\frac{c_n(t) - c_n(t_0)}{\varsigma_2 t}\right) - 1] \quad (3.23)$$

- and if $k_\theta = 0$ and $\varsigma_1 = 0$:

$$\bar{c}_n = \frac{1}{2}[c_n(t_0) + c_n(t)] \quad (3.24)$$

The calculation procedure has to follow the flow direction, because the concentration from each incoming flux has to be known. It means that the concentration of the inflow through the upper boundary of the top layer (c_r), through the lower boundary of the bottom layer (c_s) and the lateral inflow of the model layers (c_d) must be known. In particular in the unsaturated domain a frequent change in flow direction can be present, due to partial rewetting by rainfall and irrigation and drying by evapotranspiration.

Fig 3.3 gives an example of possible flux directions in the model and the selection procedure for the sequence of the calculations. For each layer the direction of transport at the lower boundary has to be checked. If the direction of flow is upward, a search has to be made at which layer the direction of flow changes again. The procedure starts with the selection of the sequence of the layer numbers in the calculations. This has the advantage that the selection in cases of multi-component transport also only needs to be executed once per time step. For the layers with upward flow the calculations for the concentration and the average concentration are made in reverse order.

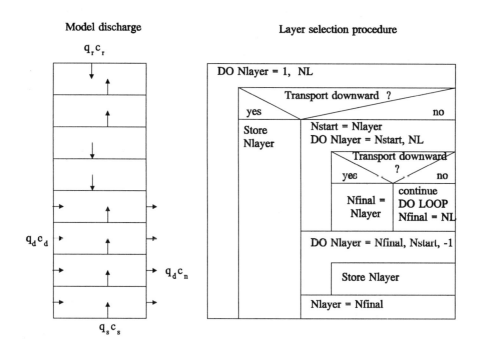

Fig. 3.3: Model transport scheme with layer selection procedure.

3.1.2.2 Multi-component transport

Most modelling efforts involving multi-component transport are focused on the saturated zone, assuming one-dimensional water flow and fixed water flow velocity, temperature and pH. The coupling of two- and three-dimensional models for water flow and solute transport with multi-component chemical equilibrium and non-equilibrium models both in the unsaturated as well as in the saturated zone has become increasingly important for the analysis of policy measures to control water quality.

Yeh and Tripathi (1989) discussed three different methods for mathematically solving multi-component transport problems:

- Mixed differential and algebraic approach: the sets of differential and algebraic equations describing the transport processes and chemical reactions, respectively, are treated simultaneously.
- Direct substitutional approach: the algebraic reactions representing the non-linear chemical reactions are substituted directly into the differential mass balance transport equations.

- Sequential iteration approach: two coupled sets of linear partial differential and algebraic equations are solved sequentially and iteratively.

Based on a study of computer resource requirements, Yeh and Tripathi suggested that only the last method can be applied in two-dimensional multi-component transport problems. For this reason the model development was focused on the last method in combination with an equilibrium multi-component model and semi-analytical solutions for different processes. A complete description of the mathematical solutions used in the multi-component model and in the corresponding transport model for non-equilibrium conditions were given by Groenendijk (1997a, 1997b).

3.1.3 Numerical analysis

3.1.3.1 Numerical and physical dispersion

The dispersion term of Eq. (3.4) comprises the physical apparent dispersion which occurs as a result of a number of natural processes. In some model approaches these natural processes are defined as the sum of molecular diffusion and dispersional mixing. When convective flow is present the dispersion is dominated by the dispersive processes and diffusion can be ignored. The natural dispersion term resulting from physical processes in Eq. (3.4) is replaced by a dispersion term which accounts for the mathematical dispersion. Neglecting sink and source terms and focusing on the concentration in the liquid phase results in the following equation:

$$\frac{\partial c}{\partial t} = \frac{q}{\theta}\frac{\partial c}{\partial z} + D_n \frac{\partial^2 c}{\partial z^2} \qquad (3.25)$$

where:
D_n = the numerical dispersion coefficient.

When D_n equals D_{dd}, the equation gives a good approximation of the convective and dispersive processes as has been described by Eq. (3.4). The convection/dispersion equation was solved by means of a pseudo-analytical method (Groenendijk 1997b).
The computation scheme yields the following sources of numerical dispersion:

- as a result of spatial discretization;
- as a result of temporal discretization;
- and as the result of the assumption of time averaged constant concentration values for incoming fluxes within a time interval.

Groenendijk (1997b) showed that numerical dispersion resulting from the computation algorithm could be quantified under restricted circumstances. An expression for the numerical dispersion coefficient D_n was derived, allowing manipulation of model variables such as time interval and layer thickness to achieve agreement between physical and numerical dispersion. In this analysis, stationary soil moisture flow conditions were assumed and the soil profile was schematized to layers with equal thickness. The spatial term ∂z was discretized to finite increments with thickness Δz. This resulted in a schematization of the convective transport into a flow through series of perfectly mixed soil layers. In the first instance, the dispersion term of Eq. (3.4) was ignored since the computation algorithm introduces a numerical dispersion, which was utilized to describe physical dispersion. Both concentrations in the liquid phase of layer i and of the adjacent upstream layer i-1 are functions of time. This time function $c_{i-1}(t)$ was replaced by the time averaged concentration in the inflowing soil water from an upstream layer to facilitate the solution of the differential equation. This resulted in the differential equation for layer i:

$$\frac{dc_i}{dt} = \frac{q}{\theta \Delta z} \bar{c}_{i-1} - \frac{q}{\theta \Delta z} c_i(t) \tag{3.26}$$

The averaged concentration is determined by calculating the integral and dividing by the length of the time interval considered:

$$\bar{c}_{i-1} = \frac{1}{\Delta t} \int_{t_0}^{t_0 + \Delta t} c_{i-1}(t) \, dt \tag{3.27}$$

The introduction of the time averaged constant value obeys the mass conservation law. Subject to the initial condition $c_i = c_i(t_0)$ at $t = t_0$, the solution of this differential equation reads:

$$c_i = c_i(t_0) \exp\left[-\frac{q}{\theta \Delta z} \Delta t\right] + \bar{c}_{i-1}\left(1 - \exp\left[-\frac{q}{\theta \Delta z} \Delta t\right]\right) \tag{3.28}$$

At small values of Δt, the resulting concentration will be determined almost completely by the initial value of the concentration, The resulting concentration will approach the value of the forcing function $\bar{c}_{i-1}$ at large values of Δt. This solution can be considered as a combination of an explicit and an implicit numerical solution to a finite difference computation scheme. The measure of dependency of the initial value and the value of the forcing function is determined by the exponential function $\exp[-q\Delta t/(\theta \Delta z)]$. The residence time of the soil moisture in the layer considered can be defined as:

$$\Gamma = \frac{\theta \Delta z}{q} \tag{3.29}$$

The term $\Delta t(q/(\theta\Delta z))$ is a dimensionless constant and expresses the number of pore water refreshments within a time increment. The product of the pore water velocity and the layer thickness $q\Delta z/\theta$ can be replaced by:

$$\frac{q}{\theta}\Delta z = \frac{(\Delta z)^2}{\Gamma} \tag{3.30}$$

After some elaborate algebraic manipulations, the resulting expression of the numerical dispersion coefficient becomes:

$$D_n = \frac{(\Delta z)^2}{\Gamma}[\frac{\frac{\Delta t}{\Gamma}}{1 - \exp[-\frac{\Delta t}{\Gamma}]} - \frac{1}{2}(1 + \frac{\Delta t}{\Gamma})] \tag{3.31}$$

The relation between numerical dispersion and number of pore water refreshments have been given for the cascade model TRADE (Roest and Rijtema 1983) and the Mixing Cell model based on a finite difference approximation of the convection/dispersion equation (Van Ommen 1985). The numerical dispersion coefficients are for the Cascade model:

$$D_n = \frac{q}{\theta}\frac{\Delta z}{2} = \frac{1}{2}\frac{(\Delta z)^2}{\Gamma} \tag{3.32}$$

and for the Mixing Cell model:

$$D_n = \frac{q}{\theta}(\frac{\Delta z}{2} - \frac{q}{\theta}\frac{\Delta t}{2}) = \frac{1}{2}\frac{(\Delta z)^2}{\Gamma}(1 - \frac{\Delta t}{\Gamma}) \tag{3.33}$$

The numerical dispersion coefficient can be multiplied by $\Gamma/(\Delta z)^2$ to obtain a dimensionless entity as a function of the number of pore water refreshments.

The results of the analysis are shown in Fig. 3.4. The following conclusions may be drawn from this figure:

- Numerical dispersion can be used to simulate physical dispersion by manipulating the layer thickness when Δt and Γ have been chosen.
- The numerical dispersion derived in this study approximates the numerical dispersion in the Cascade model at small time increments. The characteristic dispersion length Λ, used to quantify dispersion equals, $\Delta z/2$ at $\Delta t/\Gamma \to 0$.
- The numerical dispersion in this study is always larger than in the Cascade model, when equal thicknesses have been chosen in both models.
- Plug flow ($D_n\partial^2 c/\partial z^2 = 0$) cannot be simulated by the model presented in this study or by the Cascade model, but in the Mixing Cell model zero dispersion can be simulated by choosing the time step equal to the residence time.

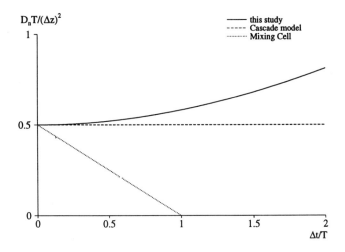

Fig. 3.4: Dimensionless numerical dispersion as a function of the number of pore water refreshments.

3.1.3.2 Stability of the mathematical solution

One of the most well-known procedures for stability estimation of finite difference approaches is the Von Neumann method. The method assumes the propagation of the error E_i to be described by the computation rules for the calculation of the concentration c_i. The error E_i in the concentration can be replaced (Groenendijk 1997b) by the expression:

$$E_i = \Theta^\gamma \exp[i\beta\delta\Delta z] \qquad (3.34)$$

where:
Θ = the amplification factor
β = the frequency of the error
γ = the number of the time step since the beginning of the simulation
δ = the layer number.

The imaginary number i is defined as $i^2 = -1$. Substitution of this expression in the discretized form of the conservation and transport equation (Eq. 3.6) and further elaboration yields the following expression for Θ:

$$\Theta = \exp[-\frac{q\Delta t}{\theta\Delta z}] \qquad (3.35)$$

The mathematical solution is stable when the following condition has been satisfied:

$$\Theta \leq 1 \qquad (3.36)$$

When the ratio $(q\Delta t)/(\theta\Delta z)$ is greater than zero, the amplification factor Θ is always less than one. It is concluded that as long as the computation order proceeds in the direction of water flow the mathematical solution given in this study is always stable.

3.1.4 Model verification with an analytical solution

Jury and Roth (1990) used transfer functions to characterize the outflow solute flux of a soil column as a function of the incoming flux. They provided an analytical solution to determine the concentration in the liquid phase from the convection-dispersion equation in case of the entry of a pulse-type dose at the top of the soil column under steady state water flow. Their solution is only valid for non-sorbing, non-decaying substances in a normalized situation (normalized means that the integral of the total concentration equals 1).

Jury and Roth (1990) indicated how sorption and decay or conversion can be implemented in their solution. Adriaanse (1996) implemented their extended solution to substances with linear sorption, first order decay and an application of an areic dose. This resulted in the following equation for the concentration in the liquid phase:

$$c = \frac{M}{\theta R_a}\exp[-kt]\left[\frac{1}{\sqrt{\pi D_{dd} t/R_a}}\exp\left[-\frac{(z-vt/R_a)^2}{4D_{dd}t/R_a}\right]\right.$$
$$\left. - \frac{v}{2D_{dd}}\exp\left[\frac{vz}{D_{dd}}\right]\mathrm{erfc}\left(\frac{z+vt/R_a}{\sqrt{4D_{dd}t/R_a}}\right)\right] \qquad (3.37)$$

where:
- c = the concentration in the liquid phase in kg m^{-3}
- M = the applied areic mass in kg m^{-2}
- θ = the volume fraction of liquid in m^3 m^{-3}
- R_a = the retardation factor (-)
- k = the conversion rate in d^{-1}
- t = time in d
- v = the apparent velocity in m d^{-1}
- z = the depth in m-surface
- D_{dd} = the apparent dispersion coefficient in m^2 d^{-1}
- erfc = the complementary error function
- exp = the exponential function.

The retardation factor is defined as:

$$R_a = 1 + \frac{\rho_d K_d}{\theta} \qquad (3.38)$$

The numerical solution of the transport equation was verified by Kroes and Rijtema (1996) against the analytical solution given above. The results of the verification are given in Fig. 3.5 at a time step with a complete front in the upper 0.4 m of the soil.

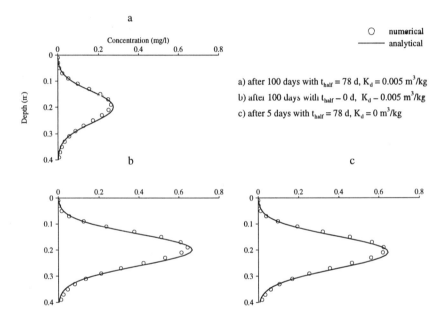

Fig. 3.5: Solute concentration (mg dm^{-3}) as a function of depth simulated with TRANSOL and an analytical solution for 3 different substances: sorption and conversion (a), conversion and no sorption (b) and sorption and no conversion (c).

The two different approaches were tested for three different substances with the following conditions:

- stationary hydrology with a downward apparent velocity of 0.04 m.d^{-1};
- steady state saturated soil profile with a volume fraction liquid of 0.25 m^3.m^{-3};
- applied areic mass of 1 kg.ha^{-1};
- apparent dispersion coefficient of 0.0101 m^2.d^{-1}, which was achieved in the TRANSOL by selecting for each model compartment a thickness of 0.0202 m (Kroes and Rijtema 1996);

The three different substances had the following properties:

- sorption and conversion: half life time of 78 d, combined with a distribution coefficient of 0.005 m^{-3} kg;
- sorption and no conversion: distribution coefficient of 0.005 m^{-3} kg;
- conversion and no sorption: half life time of 78 d.

The simulations of substances with sorption and conversion (Fig. 3.5a) and with only sorption (Fig. 3.5b) were executed with a time step of 1.0 d in the numerical models and was verified after 100 d. Both simulations were regarded as excellent. Differences between the two model approaches are most likely caused by computer inaccuracies. The simulation of a substance without sorption but with conversion was simulated with a time step of 0.02 d and verified after 5 days (Fig. 3.5c). All simulations were regarded as satisfying. Differences between the numerical model approach and the analytical solution are caused by the discretization in space and time.

3.2 Upper soil storage and surface runoff

Generally, materials are added in solid form at the soil surface at any time. Two aspects affect the treatment of additions of fertilizers and other materials. First, the addition is in most cases independent of the weather conditions, so the material remains at the soil surface and will not enter into the first layer. The second aspect is that when it starts raining the quantity of precipitation can be too small to dissolve the material completely. To overcome these difficulties an imaginary storage reservoir has been introduced at the soil surface in which all additions are immediately dissolved. This reservoir can be regarded as layer 0. In the model schematization, the quantity of rain enters this reservoir, and supersedes, without mixing, through the bottom the same quantity of water. This quantity of water infiltrates from this reservoir into the first soil layer, with the average reservoir concentration determined immediately after an addition took place. The total quantity of added material in stock will be depleted after a precipitation volume which equals the reservoir volume. The concentration in the artificial reservoir than equals c_{pr}. The store of materials in this artificial reservoir is evaluated each time interval by means of book keeping. The release rate can be manipulated by the choice of an appropriate thickness (Z_{surf}) of the reservoir.

The release concentration during a time step with rain is calculated by:

- if:

$$\sum_{t=t_0}^{t=n\Delta t} q_{pr}\Delta t \leq Z_{surf} \quad \rightarrow \quad c_i = c_0 \qquad (3.39)$$

- else if:

$$\sum_{t=t_0}^{t=(n-1)\Delta t} q_{pr}\Delta t \leq Z_{surf} \leq \sum_{t=t_0}^{t=n\Delta t} q_{pr}\Delta t$$

$$\rightarrow c_i = \frac{[Z_{surf} - \sum_{t=t_0}^{t=(n-1)\Delta t} q_{pr}\Delta t]c_0 - [Z_{surf} - \sum_{t=t_0}^{t=n\Delta t} q_{pr}\Delta t]c_{pr}}{q_{pr}\Delta t} \quad (3.40)$$

- else

$$\sum_{t=t_0}^{t=n\Delta t} q_{pr}\Delta t \geq Z_{surf} \quad \rightarrow \quad c_i = c_{pr} \quad (3.41)$$

The concentration c_0 after each addition is calculated as:
- if:

$$Z_{surf} - \sum_{t=t_0}^{t=n\Delta t} q_p\Delta t > 0$$

$$\rightarrow c_{0,new} = \frac{[Z_{surf} - \sum_{t=t_0}^{t=n\Delta t} q_p\Delta t]c_{0,old} + [\sum_{t=t_0}^{t=(n)\Delta t} q_p\Delta t]c_p + Q_{add}}{Z_{surf}} \quad (3.42)$$

- else:

$$c_{0,new} = \frac{Z_{surf}c_p + Q_{add}}{Z_{surf}} \quad (3.43)$$

in which:
c_{pr} = concentration in precipitation in kg.m^{-3}
c_i = concentration of infiltrating flux at soil surface in kg.m^{-3}
$c_{0,new}$ = concentration in surface reservoir after addition of solid material in kg.m^{-3}
$c_{0,old}$ = concentration in surface reservoir after previous addition in kg.m^{-3}
q_{pr} = precipitation rate in m.d^{-1}
Δt = time step in d
Z_{surf} = depth of surface reservoir in m
n = number of time steps after an addition (-)
Q_{add} = addition of undissolved material in kg.m^{-2}

Surface runoff may occur in situations where the precipitation rate is higher than the infiltration rate of the soil and in situations where the precipitation excess causes the groundwater level to rise above the soil surface. In both cases the surface runoff to the surface-water system may contain solutes that originate from one of the three sources: precipitation, the surface reservoir and the upper part of the soil profile. The concentration in the precipitation is immediately related to atmospheric pollution and is very often given as wet deposition of atmospheric pollutants. Part of the surface runoff originates from the surface reservoir, resulting in high concentrations in runoff water when a lot of material is present in this reservoir. Diffusion and dispersion of solutes from the soil to the runoff water has been taken into account by supposing a part of the surface runoff flowing through the first model compartment. The surface retardation sub-model and the runoff sub-model are illustrated in Fig. 3.6.

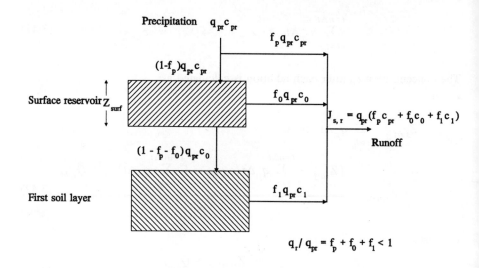

Fig. 3.6: Schematic presentation of the flux distribution through the surface reservoir in combination with surface runoff.

The final surface runoff flux equals the sum of 3 different fluxes:

$$J_{s,r} = q_p[f_p c_p + f_0 c_0 + f_1 c_1] = q_r c_r \qquad (3.44)$$

where:
$J_{s,r}$ = runoff solute flux in kg.m^{-2}.d^{-1}
q_{pr} = precipitation rate in m^3.m^{-2}.d^{-1}
q_r = runoff water flux in m^3.m^{-2}.d^{-1}
c_0 = concentration in the surface reservoir in kg.m^{-3}
c_1 = concentration in the first soil compartment in kg.m^{-3}
c_{pr} = concentration in precipitation in kg.m^{-3}
c_r = the concentration in the runoff water in kg.m^{-3}
f_0 = the fraction of the runoff that passes the surface reservoir (-)

f_1 = the fraction of runoff that passes the first model compartment (-)
f_p = the fraction of runoff that has not been in contact with the soil (-)

The water balance of the surface reservoir is subject to the condition:

$$\frac{q_r}{q_{pr}} = f_p + f_0 + f_1 \leq 1 \qquad (3.45)$$

Shrinkage and soil cracking has important and mainly unwanted consequences for groundwater pollution. An example is the transport of water and solutes as by-pass flow via shrinkage cracks to the subsoil, resulting in rapid pollution of groundwater. When rainfall reaches the surface of a cracked soil, part of the water infiltrates into the soil matrix and part of the water flows into the cracks. This requires adaption of the top boundary conditions of simulation models. In cracked soils, a similar procedure can be used to calculate bypass flow; if rainfall exceeds the maximum infiltration rate of the soil matrix, water flows into the cracks. In addition a certain part of the rainfall falls directly into the cracks. Surface runoff only occurs when cracks are closed, or completely filled with excess precipitation. Matrix infiltration and crack infiltration at a given rainfall intensity can be calculated as follows (Bronswijk 1988, 1991):

$$\begin{array}{llll} q_{pr} \leq I_{max} & I_m = A_m q_{pr} & I_{cr} = q_{pr} \\ q_{pr} > I_{max} & I_m = A_m I_{max} & I_{cr} = A_{cr} I_{max} + A_m(q_{pr} - I_{max}) \end{array} \qquad (3.46)$$

in which:
q_{pr} = rainfall intensity in m.d^{-1}
I_{max} = maximum infiltration rate of soil matrix in m.d^{-1}
I_m = infiltration rate in soil matrix in m.d^{-1}
I_{cr} = infiltration rate in cracks in m.d^{-1}
A_m, A_{cr} = relative areas of soil matrix and cracks respectively

An approximation for the calculation of the relative soil matrix area has been given in section 2.1.2. Most of the water flowing through cracks accumulates at the bottom of the cracks, causing rapid drainage water production. In the present approach, horizontal infiltration into crack walls of water running rapidly downwards along cracks is neglected. For this reason all water infiltrating into cracks is assumed to accumulate at the bottom of the cracks and is added to the moisture content of the corresponding layers. The infiltration rate in the cracks causes an additional incoming flux of solutes in the layers concerned. The solute flux in the cracks is given by:

$$\begin{array}{llll} q_{pr} \leq I_{max} & J_{s,m} = A_m q_{pr} c_0 & J_{s,cr} = q_{pr} c_{pr} \\ q_{pr} > I_{max} & J_{s,m} = A_m I_{max} c_0 & J_{s,cr} = A_{cr} q_{pr} c_{pr} + A_m q_r c_r \end{array} \qquad (3.47)$$

3.3 Emission of pollutants from the soil to surface water

3.3.1 Concentration in drainage water released to perfect drains

Each of the soil layers may be regarded as a perfectly mixed reservoir. Part of the inflow is conveyed to underlying soil layers, the remainder flows horizontally to the water course or tile drain. Assuming steady state conditions, the displacement of a non-reactive solute through the system can be described by a set of linear differential equations. For the first reservoir, the following equation applies:

$$\frac{\theta_{sat} H_d}{N_d} \frac{dc_1}{dt} = q_r^{net} c_{inp} - q_r^{net} c_1 \qquad (3.48)$$

where:
N_d = the number of draining soil layers
c_{inp} = the input concentration in kg.m^{-3}.
q_r^{net} = net recharge in m.d^{-1}

For an arbitrary reservoir i, the change in concentration is described by:

$$\frac{\theta_{sat} H_d}{N_d} \frac{dc_i}{dt} = \frac{N_d-i+1}{N_d} q_r^{net} c_{i-1} - \frac{N_d-i+1}{N_d} q_r^{net} c_i \qquad (3.49)$$

Assuming an initial concentration c_0 which is uniform over the entire depth, the solution to the differential equation yields the concentration course over time in reservoir k:

$$\frac{c_k(t) - c_{inp}}{c_0 - c_{inp}} = \sum_{i=1}^{k} \binom{N_d}{i-1} \binom{N_d-i}{k-i} (-1)^{i+1} \exp\left[-(N_d-i+1)\frac{q_r^{net}}{\theta H_d} t\right] \qquad (3.50)$$

Since the outflow of all reservoirs are assumed equal, the resulting concentration in drainage discharge can be found as the average of all reservoirs. Lengthy but straightforward algebraic summation of the binomial series in Eq. (3.50) yields a simple relation for the concentration in drainage water:

$$\frac{c_d(t) - c_{inp}}{c_0 - c_{inp}} = \frac{1}{N_d} \sum_{k=1}^{N_d} \frac{c_k(t) - c_{inp}}{c_0 - c_{inp}} = \exp\left[-\frac{q_r}{\theta H_d} t\right] \qquad (3.51)$$

This relation is also found if the concentration in the drainage water is modelled by describing the groundwater system as one perfectly stirred reservoir. Overall effects of vertical dispersion which are introduced by defining distinct soil layers can thus be described by using a single reservoir. For the single drainage system, the simulation of solute migration by describing a vertical column with uniform lateral outflow agrees with solutions found by Gelhar and Wilson (1974), Raats (1978) and van Ommen (1986).

Generally, the defined thickness of the soil layers increases with depth and the lateral outflow is described as a function of the depth below the phreatic water table. In that case, the average concentration of discharge water which flows to a third order drainage system is given as:

$$\bar{c}_{dr}(k) = \frac{\sum_{i(z=h_{gwl})}^{i(z=H_{d,k})} \Delta z_i q_{d,i} c_i}{\sum_{i(z=h_{gwl})}^{i(z=H_{d,k})} \Delta z_i q_{d,i}} \qquad k = 3 \qquad (3.52)$$

subject to the conditions:

$\Delta z_i = Z_i - h_{gwl}$ if $i = i(z = h_{gwl})$ and $\Delta z_i = H_{d,3} - Z_{i-1}$ if $i = i(z = H_{d,3})$.

The average drainage concentration for the other drainage systems is given by:

$$\bar{c}_{dr}(k) = \frac{\sum_{i(z=h_{gwl})}^{i(z=H_{d,3})} \Delta z_i q_{d,i} c_i + \sum_{k}^{2} \sum_{i(z=H_{d,k+1})}^{i(z=H_{d,k})} \Delta z_i q_{d,i} c_i}{\sum_{i(z=h_{gwl})}^{i(z=H_{d,3})} \Delta z_i q_{d,i} + \sum_{k}^{2} \sum_{i(z=H_{d,k+1})}^{i(z=H_{d,k})} \Delta z_i q_{d,i}} \qquad k < 3 \qquad (3.53)$$

with $k = 2$ for second order drains and $k = 1$ for first order drains and subject to the conditions:

$\Delta z_i = Z_i - H_{d,k+1}$ if $i = i(z = H_{d,k+1})$ and $\Delta z_i = H_{d,k} - Z_{i-1}$ if $i = i(z = H_{d,k})$.

where:
$i(z=h_{gwl})$ = number of the layer in which the phreatic water level is situated
$i(z=H_{d,k})$ = number of the layer in which the bottom of drain system k is situated
Δz_i = thickness lateral outflow from layer i in m
c_i = concentration in layer i in kg.m^{-3}
$q_{d,i}$ = lateral drainage outflow from layer i in m.d^{-1}

Using these average concentrations, the average concentration $\bar{c}_R$ of the discharge to surface water at the scale of a sub-region is calculated as:

$$\overline{c}_R = \frac{\sum\limits_{i(z=h_{gwl})}^{i(z=H_{d,3})} L_{dr,3}\Delta z_i q_{d,i} c_i + \sum\limits_{k}^{2} \sum\limits_{i(z=H_{d,k+1})}^{i(z=H_{d,k})} L_{dr,k}\Delta z_i q_{d,i} c_i}{\sum\limits_{i(z=h_{gwl})}^{i(z=H_{d,3})} L_{dr,3}\Delta z_i q_{d,i} + \sum\limits_{k}^{2} \sum\limits_{i(z=H_{d,k+1})}^{i(z=H_{d,k})} L_{dr,k}\Delta z_i q_{d,i}} \qquad k < 3 \quad (3.54)$$

in which:

$L_{dr,3}$ = total length of first, second and third order drains in m
$L_{dr,2}$ = total length of first and second order drains in m
$L_{dr,1}$ = total length of first order drains in m

3.3.2 Soil heterogeneity and the concentration in water released to line drains

It can be important to account for the effects of soil heterogeneity near open field drains, the so-called riperian zones, and the corresponding processes in the region with radial flux to the drains on the concentration of incoming nitrate rich seepage water, arriving from the higher upstream situated compartments. In this riperian zone with radial flux to the drain additional denitrification can be present. This situation is schematically presented for the combination of horizontal and radial flux in Fig. 3.7. The concentration of the drainage water is approximated by considering two connected columns. The first one has layers with vertical in- and outflow and lateral outflow from all saturated layers below drain depth to column 2. The vertical inflow and the lateral water flux in column 1 are equal to:

$$q_L^i = (q_d + q_s)(\frac{L - 2H_d}{L}) \qquad (3.55)$$

The second column has a surface area H_d and has vertical in- and outflow and lateral inflow through the boundary at $\frac{1}{2}L - H_d$. The total vertical incoming flux equals $(q_d + q_s).2H_d/L$. The lateral outflow from the saturated layers to the drain system is proportional to the area in direct contact with the drain. When tile drains are present the area of outflow is πr_{dr}, in which r_{dr} is the radius of the drain. In the case of open field drains, the area of outflow equals $h_w(\sin \alpha)^{-1} + W_b/2$ where h_w is the water depth in the open drain, W_b the bottom width of the drain and $\tan \alpha$ the slope of the side wall. The discharge to the drain is controlled by the following outflow conditions:

- If the lower boundary Z_n of the layer n is smaller than the depth of the water level H_{wl} in the drain ($Z_n < H_{wl}$) then: $q_{dr}(n) = 0$
- If the upper boundary Z_{n-1} of layer $n-1$ is smaller than the water level H_{wl} in the drain and the lower boundary Z_n of layer n is smaller than the bottom depth ($H_{wl} + h_w$) of the drain ($Z_{n-1} < H_{wl} < Z_n < H_{wl} + h_w$) then:

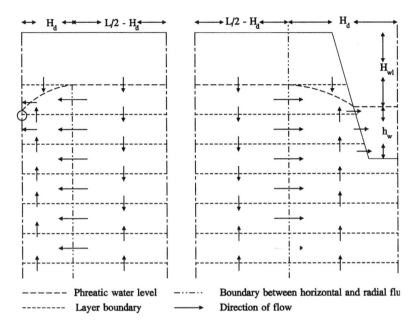

Fig. 3.7: Schematic presentation for the combination of horizontal and radial flux into 2 columns; left: tile drains; right: open field drains.

$$q_{dr}(n) = (q_d + q_s)\frac{2(Z_n - H_{wl})(\sin\alpha)^{-1}}{(2h_w(\sin\alpha)^{-1} + W_b)} \quad (3.56)$$

- If the upper boundary Z_{n-1} of layer n-1 is smaller than the water level H_{wl} in the drain and the lower boundary Z_n of layer n is larger than the bottom depth ($H_{wl} + h_w$) of the drain ($Z_{n-1} < H_{wl} < H_{wl} + h_w < Z_n$) then:

$$q_{dr}(n) = q_d + q_s \quad (3.57)$$

- If the upper boundary Z_{n-1} of layer n-1 is larger than the water level H_{wl} in the drain and the lower boundary Z_n of layer n is smaller than the bottom depth ($H_{wl} + h_w$) of the drain ($H_{wl} < Z_{n-1} < Z_n < H_{wl} + h_w$) then:

$$q_{dr}(n) = (q_d + q_s)\frac{2(Z_n - Z_{n-1})(\sin\alpha)^{-1}}{(2h_w(\sin\alpha)^{-1} + W_b)} \quad (3.58)$$

- If the upper boundary Z_{n-1} of layer n-1 is smaller than or equal to the bottom depth ($H_{wl} + h_w$) and the lower boundary Z_n of layer n is larger than the bottom depth ($H_{wl} + h_w$) of the drain ($Z_{n-1} \leq H_{wl} + h_w < Z_n$) then:

$$q_{dr}(n) = (q_d + q_s)\frac{2(H_{wl} + h_w - Z_{n-1})(\sin\alpha)^{-1} + W_b}{(2h_w(\sin\alpha)^{-1} + W_b)^2} \quad (3.59)$$

- If the upper boundary Z_{n-1} of layer $n-1$ is larger than the bottom depth of the drain ($H_{wl} + h_w$) then: $q_{dr}(n) = 0$.

The transport and conservation equation for the part with radial flux is written as:

$$\frac{d(\theta_n(t)c_n(t))}{dt} + \rho_{d,n}\frac{dX_{e,n}}{dt} + \rho_{d,n}\frac{dX_{n,n}}{dt} + \rho_{d,n}\frac{dX_{p,n}}{dt} =$$
$$\sum_{l=1}^{l_{max}}\frac{q_l^i(t)c_l^i(t)}{\Delta z_n} - \sum_{k=1}^{k_{max}}\frac{q_k^o(t)c_n(t)}{\Delta z_n} - \frac{q_t\sigma_{pl}c_n(t)}{\Delta z_n} + k_{0,n} + k_{1,n}\theta_n(t)c_n(t) \quad (3.60)$$

The total incoming flux of material of each layer is given as:

$$\sum_{l=1}^{l_{max}} q_l^i(t)c_l^i(t) = q_{n+1/2}^i \bar{c}_{n+1} + q_{n-1/2}^i \bar{c}_{n-1} + q_L^i \bar{c}_{n,1} \quad (3.61)$$

in which:
$q_{n+1/2}^i$ = incoming flux from layer $n+1$ in m.d^{-1}
$q_{n-1/2}^i$ = incoming flux from layer $n-1$ in m.d^{-1}
q_L^i = infiltration flux from the system with horizontal flux in m.d^{-1}
$\bar{c}$ = average concentration during the time step in kg.m^{-3}
$\bar{c}_{n,1}$ = average concentration in layer n of the system with horizontal flux in kg.m^{-3}

For the outgoing material transport from layer n the concentration $c(n,t)$ is a function of time. The total outgoing water flux equals the upward flow ($q_{n-1/2}^o$) to layer $n-1$, downward flow ($q_{n+1/2}^o$) and lateral flow to the drainage system ($q_{dr}(n)$), so:

$$\sum_{k=1}^{k_{max}} q_k^o(t)c_n(t) = c_n(t)(q_{n+1/2}^o + q_{n-1/2}^o + q_{dr}(n)) \quad (3.62)$$

Eq. (3.60) is solved according the procedure described in section 3.1.2.1. The average drain water concentration equals:

$$\bar{c}_{dr} = \frac{\sum_{n=1}^{n_{max}}(q_{dr}(n)c_{n,2})}{q_d + q_s} \quad (3.63)$$

3.3.3 Effect of buffer zones on surface water pollution

Buffer zones along open field drains and ditches and rivers are sometimes proposed to reduce the effect of intensive land use on surface water pollution. Next to a direct contamination of the surface water due to these activities, the short residence times near line drains also play a role. In section 2.2.3.2 the relations between z/L, q_z/q_d and $2x/L$ are presented. Denoting the buffer zone width by L_B and the depth of influence of the buffer zone by Z_B gives with Eq. (2.86) and Eq.(2.88) the relation between L_B and Z_B as:

$$\frac{Z_B}{L} = \text{MIN}\left[\frac{-1}{2\pi}\ln\cos\left[\frac{\pi L_B}{2L}\right] \; ; \; \frac{H_d}{L} - \frac{1}{4}\tan\left[\frac{\pi}{2}(1 - \frac{q_z^*}{q_d})(1 - \frac{2L_B}{L})\right]\right] \quad (3.64)$$

The effect of the buffer zone on the pollutant load of surface water caused by the drainage flux is calculated through the introduction of a second column for which the management conditions of the buffer zone hold. For this column the concentrations of the lateral outflow for the required number of layers are also calculated. The average concentration of the drainage water is now calculated, with $Z_B < H_{d,3}$ as:

$$\overline{c}_{dr}(3) = \frac{\sum\limits_{i(z=h(dr,3))}^{i(z=Z_B)} \Delta z_i q_{d,i} c_i^B + \sum\limits_{i(z=Z(B))}^{i(z=H_{d,3})} \Delta z_i q_{d,i} c_i}{\sum\limits_{i(z=h(dr,3))}^{i(z=H_{d,3})} \Delta z_i q_{d,i}} \quad (3.65)$$

The drainage concentration for second ($k = 2$) and first order drains ($k = 1$) equals:

$$\overline{c}_{dr}(k) = \frac{\sum\limits_{i(z=h(dr,3))}^{i(z=Z_B)} \Delta z_i q_{d,i} c_i^B + \sum\limits_{i(z=Z_B)}^{i(z=H_{d,3})} \Delta z_i q_{d,i} c_i + \sum\limits_{k}^{2} \sum\limits_{i(z=H_{d,k+1})}^{i(z=H_{d,k})} \Delta z_i q_{d,i} c_i}{\sum\limits_{i(z=h(dr,3))}^{i(z=H_{d,3})} \Delta z_i q_{d,i} + \sum\limits_{k}^{2} \sum\limits_{i(z=H_{d,k+1})}^{i(z=H_{d,k})} \Delta z_i q_{d,i}} \quad (3.66)$$

subject to the conditions:

$\Delta z_i = Z_i - h_{dr,3}$ if $i = i(z = h_{dr,3})$; $\Delta z_{i,B} = Z_B - Z_{i-1}$ if $i = i(z = Z_B)$;
$\Delta z_i = Z_i - Z_B$ if $i = i(z = Z_B)$; $\Delta z_i = H_{d,3} - Z_{i-1}$ if $i = i(z = H_{d,3})$;
$\Delta z_i = Z_i - H_{d,k+1}$ if $i = i(z = H_{d,k+1})$ and $\Delta z_i = H_{d,k} - Z_{i-1}$ if $i = i(z = H_{d,k})$.

3.4 Model schematization for regional groundwater pollution

The models ANIMO and TRANSOL are used for the nutrient and other pollutant balances within a subregion per calculation unit, respectively. A calculation unit is defined as a unique combination of soil type, hydrological conditions and land use. Fig. 3.8 shows

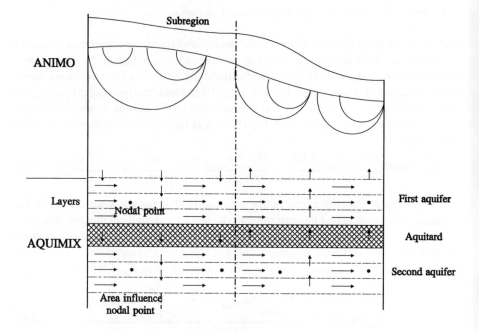

Fig. 3.8: Spatial schematization of the top-system (ANIMO, TRANSOL) and the aquifer system (AQUIMIX).

the spatial schematization of the top-system (ANIMO, TRANSOL) and the deep aquifer system (AQUIMIX). The depth of the boundary between the local top-system and the deep regional groundwater system depends on the depth of the deepest streamline that discharges water to the local drainage systems. This depth depends on the drain distances, the hydraulic conductivity of the different layers in the soil profile and the pattern and magnitude of the regional groundwater flow. This depth is calculated for each sub-region from the output of the hydrological model. ANIMO and TRANSOL calculate the processes in the unsaturated zone and the top layers of the saturated zone for the calculation of the local discharge of nutrients and other pollutants to surface water by drainage systems present in the subregion. The time scale for these calculations is relatively short and varies generally from 1 to 7 days, depending on the problem to be analyzed. The water quality aspects related to the regional groundwater flow between subregions are calculated in the groundwater quality model AQUIMIX. The relation between ANIMO and TRANSOL and the regional model AQUIMIX is determined by leakage and seepage fluxes between the sub-regional top-system and the regional aquifer system.

The model AQUIMIX was initially developed for the transport of N-compounds and phosphate in the aquifer system (van der Bolt et al. 1996). It also considers the reduction in the nitrate concentration in the aquifer under anaerobic conditions through oxidation of organic matter and pyrite. AQUIMIX has been extended for the simulation of regional transport of other pollutants. The following assumptions have been made in the model:

- the model uses the geo-hydrological schematization and element network of the regional groundwater flow model used;
- the smallest horizontal unit is the sphere of influence of a nodal point;
- the smallest vertical depth is user defined;
- the length of a time step is user defined, depending on layer thickness and geohydrological properties of the layers;
- all geochemical reactions are described as first order processes.

The model AQUIMIX calculates per nodal point and per layer the regional solute transport and the geochemical processes in the saturated zone with time steps of 1/2 to 1 year. The calculated concentrations in the top layer of the regional system at the end of the time step are input concentrations for seepage in the models ANIMO and TRANSOL for the next time step of AQUIMIX. The pollutant load to the regional system caused by leakage from the top-system is calculated by summing per type of land use and per subregion the calculated leakage results of ANIMO and TRANSOL for the time step length used in AQUIMIX. Interactions between subregions via the groundwater system are accounted for by linking the models.

References

Adriaanse, P. 1996. *Fate of pesticides in field ditches: The TOXSWA simulation model*. Report **90**. SC–DLO: Wageningen, The Netherlands.
Adriaanse, P., L. Dielen en T. van Geelen, 1994. *Nitraatuitspoeling op een gras/klaver perceel; berekend met simulatiemodellen en getoetst aan gemeten waarden*, eindscriptie IAHL, Velp.
Berghuijs-van Dijk, J.T., Rijtema, P.E. and Roest, C.W.J. 1985. ANIMO: agricultural nitrogen model. Nota **1671**. Inst. for Land and Water Man. Research: Wageningen, The Netherlands.
Boesten, J.J.T.I. 1986. *Behaviour of herbicides in soil: simulation and experimental assessment* Ph.D. Thesis, Agricultural University: Wageningen, The Netherlands, 262 pp.
Boesten, J.J.T.I. and Linden, A.M.A. van der, 1991. Modelling the influence of sorption and transformation on pesticide leaching and persistence. *J. Environmental Quality*, **20**, 425-435.
Bolt, F.J.E. van der, Groenendijk, P. and Oosterom, H.P. 1996. *Nutriënten belasting van grond- en oppervlaktewater in de stroomgebieden van de Beerze, de Reusel en de Rosep. Simulatie van de nutriëntenhuishouding*. Rapport **306.2**. SC–DLO: Wageningen, The Netherlands.
Bronswijk, J.J.B. 1988. Effect of swelling and shrinkage on the calculation of water balance and water transport in clay soils. *Agricultural Water Management*, **14**, 185-193
Bronswijk, J.J.B. 1991. *Magnitude, modelling and significance of swelling and shrinkage processes in clay soils*. Ph. D. Thesis. Agric. Un.: Wageningen, The Netherlands. 145 pp.
Davis, G.B. and Salama R.B., 1996. Modelling point sources of groundwater contamination: a review and case study. In: *Contamination of groundwaters*. (ed. D.C. Adriano), pp. 111-140. Advances in Environmental Science. Science reviews.

Gelhar, L.W. and Wilson, J.L. 1974 Groundwater quality modelling. *Ground Water*, **12**, 399-408.

Groenendijk, P. 1997a. *The calculation of complexation, adsorption, precipitation and dissolution in a soil water system with the geochemical model EPIDIM.* Report **70**. SC–DLO: Wageningen, The Netherlands.

Groenendijk, P. 1997b. *Modelling the influence of sorption and precipitation processes on the availability and leaching of chemical substances in soil.* Report **76**. SC–DLO: Wageningen, The Netherlands.

Hansen, S., Jensen, H.E., Nielsen, N.E. and Svendsen, H. 1990. *Daisy; a soil plant system model.* The Royal Veterinary and Agricultural University, Dept. agric. Sci., Section Soil and Water and Plant Nutrition: Copenhagen, Denmark.

Jury, W.A. and Roth, K. 1990. *Transfer functions and solute movement through soil; theory and applications.* 226 pp. Birkhaeuser: Basel,.

Kroes, J.G. and Rijtema, P. E. 1989. TRANSOL (TRANsport of a SOLute): User's guide. Internal Report **5**. SC–DLO: Wageningen, The Netherlands.

Kroes, J.G. and Rijtema, P.E. 1996. *TRANSOL, a dynamic simulation model for transport and transformation of solutes in soils.* Report **103**. SC–DLO: Wageningen, The Netherlands.

Li, Shu-Guang, F. and McLaughlin, D. 1992. A space-time accurate method for solving solute transport problems. *Water Resources Research.* **28** (9): 2297-2306.

Ommen, H.C. van, 1986. Influence of diffuse sources of contamination on the quality of outflowing groundwater including non-equilibrium adsorption and decomposition. *Journal of Hydrology*, **88**: 79-95.

Raats, P.A.C. 1978. Convective transport of solutes by steady flows. I. General theory. *Agricultural Water Management*, **1**, 201-218.

Roest, C.W.J. and Rijtema, P.E. 1983. *Analysis of a model for transport, adsorption and decomposition of solutes in the soil.* Nota **1404**. Institute Land and Water Management Research: Wageningen, The Netherlands.

Roest, C.W.J., Rijtema, P.E., Abdel Khalek, M.A., Boels, D., Abdel Gawad, S.T. and D.E. El Quosy, 1993. Formulation of the on-farm water management model FAIDS. Report **24**. Reuse of Drainage Water Project. Drainage Research Institute: Cairo, Egypt. SC–DLO: Wageningen, The Netherlands. 118 pp.

Todini, E. 1996. Evaluation of pollutant transport in rivers and coastal waters. In *Regional Approaches to Water Pollution in the Environment*, (ed P.E. Rijtema and V. Elias), pp. 195-225. Kluwer Academics Publishers: Dordrecht, The Netherlands.

Vanclooster, M., Viaene, P., Diels, J. and Christiaens, K. 1994. *WAVE, a mathematical model for simulating water and agrochemicals in the soil and the vadose environment. Reference & user's manual (release 2.0).* Institute for land and water management: Leuven Belgium.

Wagenet, R.J. and Hutson, J.L. 1992. *LEACHM, Leaching Estimation And Chemistry Model. A process-based model of water and solute movement, transformations, plant uptake and chemical reactions in the unsaturated zone. Version 3.* Research Series **92-3**. New York State College of Agriculture and Life Sciences, Cornell University: Ithaca, New York.

Yeh, G.T. and Tripathi, V.S. 1989. A critical evaluation of recent developments in hydro-geochemical transport models of reactive multi-chemical components. *Water Resources - Research*, **25**, 93-108.

CHAPTER 4

PHYSICAL-CHEMICAL PROCESSES

Soils usually consists of three different phases, namely a solid phase, a gas phase and a liquid phase. These phases normally occur simultaneously and are in continuous mutual interaction. Real equilibrium conditions are met rarely in the dynamic soil system. The solid phase, which has a spatial build-up giving rise to a porous system, mainly comprises mixtures of quartz sand, organic compounds and clay minerals. In many cases both last mentioned components only constitute a minor weight fraction of the total solid phase. They nevertheless play a predominant role in the interactions under consideration in the attenuation of pollutant concentration. This must be ascribed to their large specific surface area, which causes an extended contact area between these solids and the other phases. Moreover, they have a relatively reactive nature which may considerably influence the surrounding of adjacent phases. One of the most striking properties in this respect is their ability to adsorb other compounds, especially ions and charged complexes. The soil organic matter also plays a dominant role in different biological decomposition processes.

The presence of solid forms of anorganic minerals in an unsaturated solution may lead to the dissolution of the compounds. The slow dissolution and release of chemical constituents from rocks, clay and other soil particles are considered as weathering reactions. The rate of weathering in soil-water systems and geological formations depends on a number of factors such as size and shape of the particles, the age of the precipitate, composition of the bulk solution, temperature and stream velocity of the percolating water.

Fresh forms of solid material dissolve faster than more stable forms. The fresh form of the compound is generally formed very quickly from strongly over-saturated solutions. Such a fresh precipitate may persist in metastable equilibrium and may convert only slowly through crystallization into a more stable form.

4.1 Chemical equilibria and ion speciation

The ion speciation in a soil solution is calculated with the chemical equilibrium model EPIDIM (Groenendijk 1997a). The model follows the traditional approach in geochemical modelling that emphasizes the role of ion-pairing and complexing in an aqueous solution. To describe the chemistry of a mixture solution a set of independent chemical components has to be chosen in such a way that all the chemical species can be build up from this set. No component can be formed out of other components. In general terms, setting up an inorganic geochemical model involves choosing N unknowns and N governing equations.

Expressions for the concentrations of the species can be formulated according to law of mass action. Schematically the formation of all species S_i from the components A_j can be represented by:

$$a_{i,1}A_1 + \ldots + a_{i,j}A_j + \ldots + a_{i,N}A_N = S_i \tag{4.1}$$

where:
$a_{i,j}$ = stoichiometric coefficient of component j in the formation of species i
N = number of components

The mathematical formulation of the mass law equation describing the formation of species S_i equals:

$$[S_i] = K_i \prod_{j=1}^{N} [A_j]^{a_{i,j}} \tag{4.2}$$

For each component in the aqueous phase a material balance equation is set up. The analytical total concentration of each component can be expressed as the sum of the separate species concentrations according to:

$$[A_j^{aq}]_T = \sum_{i=1}^{m} a_{i,j}[S_i] = \sum_{i=1}^{m} a_{i,j} K_i \prod_{j=1}^{n} [A_j]^{a_{i,j}} \tag{4.3}$$

The chemical system is described by a set of N non-linear algebraic equations with N unknown variables. The system is solved with a Newton-Raphson iteration scheme (Groenendijk 1997a).

The existence of long range electrostatic interactions between ions causes non-ideal behaviour of ions in solutions. In general, the activity of a given ionic species (a_X) is related to the concentration of the uncomplexed (free) ionic species $[X]$ by:

$$a_X = \gamma_X [X] \tag{4.4}$$

The formation constants K_i have been corrected to account for this non-ideal behaviour:

$$K_i = K_i^0 \frac{\prod_{j=1}^{n} \gamma_j^{a_{i,j}}}{\gamma_i} \tag{4.5}$$

where:
γ_i = activity coefficient of species i
γ_j = activity coefficient of component j
K_i = corrected formation coefficient
K_i^0 = uncorrected formation coefficient

Activity coefficients are calculated with the extended Debije-Hückel equation (Stumm and Morgan 1981), which relates the activity coefficient to a weighted mean molarity parameter:

$$- \log \gamma_j = \frac{A_\gamma z_j^2 \sqrt{I}}{1 + dB_\gamma \sqrt{I}} - C_\gamma I \qquad (4.6)$$

with:
γ_j = activity coefficient of component j (-)
z_j = valence of ion j (-)
I = ionic strength in mole.dm^{-3}
A_γ = temperature dependent coefficient, at 25 °C, A = 0.5085 mole$^{-0.5}$.(dm^3)$^{-0.5}$
B_γ = temperature dependent coefficient, at 25 °C, B = 3.281 nm^{-1}. mole$^{-0.5}$.(dm^{-3})$^{-0.5}$
d = diameter of the hydrated ion i in nm
C_γ = ion dependent parameter in mole^{-1}.dm^3

The temperature dependency of A_γ and B_γ is presented in Fig. 4.1.

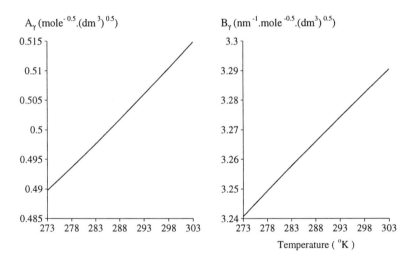

Fig. 4.1: The relation between A_γ and temperature (left) and B_γ and temperature (right).

A possible alternative would be the Davies equation:

$$- \log \gamma_j = \frac{0.5 z_j^2 \sqrt{I}}{1 + \sqrt{I}} - 0.3 I \qquad (4.7)$$

The ionic strength is a measure for the weighted mean molarity of the solution and is calculated according to:

$$I = \frac{1}{2}\sum_{i=1}^{n} z_i^2 [S_i] \qquad (4.8)$$

with

$[S_i]$ = the concentration of species i
z_i = the charge of species i

Ritsema (1993) gives for dilute solutions with ionic strengths up to 0.3 mole.dm^{-3} relations for the calculation of 26 ions. Using the mean salt method, measured mean activity coefficients were converted into single ion activity coefficients. For each ion, the variation in ion activity coefficient with ionic strength was curve-fitted with the first degree extension of the Debye-Hückel equation. The ion dependent parameters in the extended Debije-Hückel equation are presented in Table 4.1. The derived set of relations can be incorporated in EPIDIM, which makes the calculation of ion activities in soil solutions with variable ionic strengths up to 0.3 mole.dm^{-3} simple and accurate.

Table 4.1: Optimized ion-dependent parameters d and C for Eq. (4.6) after Ritsema (1993).

Ion	d (nm)	C (dm^3.mole^{-1})	Ion	d (nm)	C (dm^3.mole^{-1})
K^+	0.4472	- 0.02228	Cl^-	0.4472	- 0.02228
Na^+	0.4680	0.05333	OH^-	0.2210	0.25778
H^+	0.4668	0.23667	I^-	0.4752	0.03722
			F^-	0.3920	0.05544
Fe^{2+}	0.5120	0.17667	Br^-	0.4120	0.01333
Ca^{2+}	0.5328	0.13278	NO_3^-	0.2192	- 0.20222
Mg^{2+}	0.5240	0.22000	HCO_3^-	0.5568	- 0.00622
Mn^{2+}	0.5240	0.14389	$H_2PO_4^-$	0.1820	- 0.22800
Ba^{2+}	0.5380	0.02944			
Co^{2+}	0.5260	0.17778	CO_3^{2-}	0.5490	- 0.01556
Ni^{2+}	0.5260	0.17778	HPO_4^{2-}	0.5100	- 0.26444
Cu^{2+}	0.5080	0.09778	SO_4^{2-}	0.2550	0.01111
Zn^{2+}	0.5740	0.05556			
			PO_4^{3-}	0.5176	0.16444
Al^{3+}	0.5296	0.20667			

4.2 Adsorption

Adsorption is the transfer of compounds from the liquid phase onto the surface of a solid phase. The processes may often be explained by electrical attraction to the solid surface of components with a negative electrical charge. The adsorption results in the formation of a molecular layer of the adsorbate on the surface, but sometimes it is followed by a slow diffusion into the particles of the adsorbent (Bolt et al. 1978, 1982).

Ion exchange is an exchange of ions between a liquid and a solid phase. The exchange process can be explained in the same way as any other chemical process: the chemical energy at equilibrium after the process has been realized is lower than before the process was initiated. Pure adsorption or ion exchange is hardly observed in nature. A mixture of the two processes is most often observed. Adsorption and ion exchange are significant processes in the environmental context and a description of the processes is often included in water quality modelling. Where water is in contact with organic matter or clay a significant transfer of matter by adsorption and desorption may take place.

The adsorption on soil colloids is characterized by a high degree of reversibility: adsorbed compounds can, relatively, easily be exchanged by others. The composition of the adsorption complex depends on the relative concentration in solution and on the relative affinity to the solid phase. The electrostatic bonding of cations on clay minerals and on soil organic matter represents an example of such reversible adsorption. The adsorption is more specific and the adsorption forces are much stronger in the case of chemisorption. This characterizes chemisorption by a low degree of reversibility; exchange is restricted to compounds of a comparable structure, having a comparably high affinity for the absorber.

Data of the cation exchange capacity of soils can be obtained from national soil information systems or can be calculated from clay content and organic matter content. Using an unbuffered solution, the effect of pH should be included. This can be done according to Helling et al. (1964) with the expression:

$$CEC = (0.3 + 0.044\text{pH})M_{cl} + (-0.59 + 0.51\text{pH})M_{oc} \qquad (4.9)$$

where:
CEC = cation exchange capacity in $\text{mol}_c.\text{kg}^{-1}$
M_{cl} = mass fraction of clay
M_{oc} = mass fraction of organic carbon

The strong dependency of the adsorption to organic carbon on pH is probably caused by dissociation of humic and fulvic compounds present in the solid organic matter. The CEC used in the transfer functions for adsorption is defined at pH = 8.2.

4.2.1 Cation exchange

Cation exchange is the process whereby dissolved cations are exchanged by adsorbed cations fixed to a solid. Different cations compete for a fixed amount of exchange sites. It is assumed in the model EPIDIM (Groenendijk 1997a) that the reactions are time-independent and reversible. The capacity of the adsorber is fully occupied and the sub-system obeys the electro-neutrality condition.

A number of models are available to describe the exchange relations, each with its own properties. Three of the most well-known models have been implemented in EPIDIM: exchange relations according to a combination of Kerr and Gapon: a Capon relation for hetero-valent and Kerr for mono-valent exchange, as well as relations according to Gaines-Thomas and Vanselow.

The characterization of an exchange equilibrium requires the composition of the exchange complex and of its equilibrium solution. Notwithstanding the wealth of information available in literature on cation exchange of soil materials, one rarely finds the complete information mentioned above. With, for example, ten measuring points per curve, five electrolyte levels, numerous pairs of cations of interest, and many different types of soils and soil materials, it is easy to see that such information would require too much effort to be executed on a routine basis. Accordingly with the exception of a limited number of systems, the data available tend to be limited in scope.

The different types of selectivity coefficients used most widely are described briefly below. Experimental results are sometimes given in terms of solution equations and in other cases in terms of solution phase activities. All coefficients are expressed as a function of the selectivity coefficient K_N, so with:

$$K_N = \frac{N_X^{1/z_X} a_Y^{1/z_Y}}{N_Y^{1/z_Y} a_X^{1/z_X}} \qquad K_N' = \frac{N_X^{1/z_X} c_Y^{1/z_Y}}{N_Y^{1/z_Y} c_X^{1/z_X}} \qquad (4.10)$$

Homo-valent exchange with $z_X = z_Y = z$, gives the Kerr equation:

$$K_K = \frac{N_X a_Y}{N_Y a_X} = K_N^z \qquad (4.11)$$

Mono-divalent exchange with $z_A = 1$, $z_B = 2$ is described by the Gapon equation as:

$$K_G = \frac{N_B a_A}{N_A \sqrt{a_B}} = (K_N \sqrt{N_B}) \qquad (4.12)$$

Both homo- and hetero-valent exchange is described by a Gapon type expression as:

$$K_{HE} = \frac{N_X \, a_Y^{1/z_Y}}{N_Y \, a_X^{1/z_X}} = K_N \frac{N_X^{(1 - 1/z_X)}}{N_Y^{(1 - 1/z_Y)}} \quad (4.13)$$

A consideration of chemical equilibria with two half reactions is helpful to introduce the hetero-valent exchange equation in a component table. The variables X_Q' and Y_Q' denote the amounts adsorbed expressed in meq.(100 gr)$^{-1}$. This yields the expressions:

$$\frac{1}{z_X} a_X^{z_X+} + Q^- \rightleftarrows X_Q'$$

$$Y_Q' \rightleftarrows \frac{1}{z_Y} a_Y^{z_Y+} + Q^-$$

$$\frac{1}{z_X} a_X^{z_X+} + Y_Q' \rightleftarrows \frac{1}{z_Y} a_Y^{z_Y+} + X_Q' \quad (4.14)$$

From these two half reactions, stability equations can be derived to calculate the concentrations of the adsorbed amounts from the component concentrations:

$$X_Q = \frac{\rho_d}{100\theta} K_{HE}^X (a_X^{z_X+})^{\frac{1}{z_X}} [Q^-]$$

$$Y_Q = \frac{\rho_d}{100\theta} K_{HE}^Y (a_Y^{z_Y+})^{\frac{1}{z_Y}} [Q^-] \quad (4.15)$$

The quotient of K_{HE}^X and K_{HE}^Y equals the exchange coefficient K_{HE}. X_Q and Y_Q are expressed in eq.dm^{-3}. Both the variable Q^- and the soil property factors are eliminated by combining the equations. The equations can be inserted into a component table. In the component table, the reciprocal values of the stoichiometric coefficients are given in the columns of the cations concerned. This leads to a conversion of the concentration units. The contribution of the adsorbed phase in the sum concentrations of the cations will be expressed in mol.dm^{-3}. The sum concentration of the Q^--column is calculated in equivalent charge units. An example of an hetero-valent system with related constants is given in Table 4.2. Ca^{2+} has been taken as the reference ion.

For the Gaines–Thomas equation:

$$K_{GT} = \frac{N_X^{z_Y} \, a_Y^{z_X}}{N_Y^{z_X} \, a_X^{z_Y}} = K_N^{z_X z_Y} \quad (4.16)$$

Table 4.2: Description of the stoichiometry of a system containing four cations exchanging according to the hetero-valent exchange equations.

Species	Components					K_{HE}
	Na^+	Mg^{++}	Ca^{++}	Al^{+++}	Q^-	
Na^+	1	0	0	0	0	0
Mg^{++}	0	1	0	0	0	0
Ca^{++}	0	0	1	0	0	0
Al^{+++}	0	0	0	1	0	0
Adsorbed species						
Na_Q	1	0	0	0	1	$K_{HE}(Na/Ca)$
Mg_Q	0	1/2	0	0	1	$K_{HE}(Mg/Ca)$
Ca_Q	0	0	1/2	0	1	1
Al_Q	0	0	0	1/3	1	$K_{HE}(Al/Ca)$
Sum	Na_T	Mg_T	Ca_T	Al_T	Q_T	

Analogous to the half reactions mentioned earlier in this paragraph, the selectivity equation can be divided into two equations for N_X and N_Y:

$$N_X = (K_N^X)^{z_X} a_X^{z_X^+} [Q^-]^{z_X} \qquad N_Y = (K_N^Y)^{z_Y} a_Y^{z_Y^+} [Q^-]^{z_Y} \qquad (4.17)$$

with:
K_N^X, K_N^Y = stability constants of the half reactions concerned.

The quotient of $(K_N^X)^{z_Y}$ and $(K_N^Y)^{z_X}$ equals K_{GT}. The molar aqueous concentrations of the adsorbed species X_Q and Y_Q are calculated as:

$$X_Q = \frac{1}{z_X} \frac{\rho_d}{1000} CECN_X = \frac{1}{z_X} \frac{\rho_d}{1000} CEC(K_N^X)^{z_X} a_X^{z_X^+} [Q^-]^{z_X}$$

$$Y_Q = \frac{1}{z_Y} \frac{\rho_d}{1000} CECN_Y = \frac{1}{z_Y} \frac{\rho_d}{1000} CEC(K_N^Y)^{z_Y} a_Y^{z_Y^+} [Q^-]^{z_Y} \qquad (4.18)$$

The Vanselow equation assumes the activities of the two exchanger components to be proportional to their respective mole fractions as:

$$K_V = \frac{M_X^{z_Y} a_Y^{z_X}}{M_Y^{z_X} a_X^{z_Y}} = K_N^{z_X z_Y} \frac{z_Y^{z_X}}{z_X^{z_Y}} \left(\sum_{j=1}^{n} \frac{1}{z_{X,j}} \frac{\Gamma_{X,J}}{CEC} \right)^{z_X - z_X} \qquad (4.19)$$

where:
K_V = the Vanselow selectivity coefficient
M_X, M_Y = mole fractions of the adsorbed cations X and Y

For M_X and M_Y the following expressions hold:

$$M_X = (K_M^X)^{z_X} a_X^{z_X^+} [Q^-]^{z_X}$$
$$M_Y = (K_M^Y)^{z_Y} a_Y^{z_Y^+} [Q^-]^{z_Y} \qquad (4.20)$$

where K_M^X and K_M^Y are the stability constants of the mole fractions, based on the Vanselow equation. The quotient of $(K_M^X)^{z_Y}$ and $(K_M^Y)^{z_X}$ equals K_V. The quantity adsorbed X_Q of cation X is expressed in EPIDIM in mol.dm^{-3} as well as in eq.dm^{-3} using the expressions:

- If X_Q is expressed in mole.dm^{-3}:

$$X_Q = \frac{(CEC\rho_d)M_X}{(100\theta)\sum_{j=1}^{n} z_{X,j}M_{X,j}} = \frac{(CEC\rho_d)(K_M^X)^{z_X}a_X^{z_X^+}[Q^-]^{z_X}}{(100\theta)\sum_{j=1}^{n} z_{X,j}(K_M^{X_j})^{z_{X,j}}a_X^{z_{X,j}^+}[Q^-]^{z_{X,j}}} \qquad (4.21)$$

- If X_Q is expressed in eq.dm^{-3}:

$$X_Q = \frac{CEC\rho_d}{100\theta}M_X = \frac{CEC\rho_d}{100\theta}(K_M^X)^{z_X}a_X^{z_X^+}[Q^-]^{z_X} \qquad (4.22)$$

The expressions for X_Q and Y_Q for both the Qaines–Thomas equation and the Vanselow equation are incorporated in a similar way in a component table. An example of a component table in a system with three cations is presented in Table 4.3.

Table 4.3: Description of the stoichiometry of a system containing three cations exchanging according to either the Gaines–Thomas equation or the Vanselow equation.

Species	Components				K_N
	Na^+	Ca^{++}	Al^{+++}	Q^-	
Na^+	1	0	0	0	0
Ca^{++}	0	1	0	0	0
Al^{+++}	0	0	1	0	0
Adsorbed species					
Na_Q	1	0	0	1	K_N (Na/Ca)
Ca_Q	0	1	0	2	1
Al_Q	0	0	1	3	K_N (Al/Ca)
Sum	Na_T	Ca_T	Al_T	Q_T	

4.2.2 Sorption isotherms

A simplified method can be used in those cases where the quantity of the pollutant sorbed to the soil complex is small in comparison with the total soil adsorption complex. In that case changes in the occupation of the soil complex by other cations can be neglected. Instantaneous adsorption is considered when the time scale of the adsorption process by which a solute is transferred from the solution to the adsorption complex of the soil is much smaller than the solute transport time scales. The relation between the adsorbed quantity and the concentration in solution can be given by different equilibrium expressions.

In as much as adsorption in soil comprises many different types of adsorption sites one should expect a decrease of the selectivity with increasing saturation level. It is thus required to describe the adsorption behaviour, particularly at low concentration levels, in terms of a selectivity coefficient as applicable to a limited specified range.

4.2.2.1 Linear adsorption

Many simulation models describing migration of a pollutant in a soil-water system use a linear relation between the adsorbed quantity and the concentration in the liquid phase. If the range between the initial and resulting concentration is relatively small, linearization of a sorption isotherm does not lead to considerable deviations. The linear adsorption isotherm is given by the expression:

$$X^S = K_d c \tag{4.23}$$

where:
X^S = content of substance sorbed in kg.kg^{-1}
K_d = linear sorption coefficient in m^3.kg^{-1}

The disadvantage of this expression is that the amount of adsorbed material can increase indefinitely if the concentration increases. For this reason the maximum concentration c_{max} must be defined by which all the adsorption places X^S_{max} have been used. So the following conditions hold:

$$\begin{aligned} c < c_{max} & \quad X^S = K_d c \\ c \geq c_{max} & \quad X^S = X^S_{max} \end{aligned} \tag{4.24}$$

4.2.2.2 Langmuir adsorption equation

The Langmuir equation was originally derived to describe the simple adsorption of gas molecules on a plane surface having only one kind of elementary space, each of which could hold only one adsorbed molecule. It was assumed that the binding of a molecule on any one elementary space was independent of the binding on the remaining elementary spaces. Although there seems to be no general theoretical justification for the use of the Langmuir equation to describe adsorption of molecular species in soils, it is easily shown that any type of adsorption approaching saturation with increasing aqueous concentrations can be approximated with a Langmuir isotherm.

The Langmuir adsorption isotherm is given by the expression:

$$X^S = X_{max}^S \left(\frac{K_L c}{1 + K_L c} \right) \qquad (4.25)$$

in which:
K_L = the Langmuir distribution coefficient in $m^3.kg$
X_{max}^S = the maximum content of sorbed substance in $kg.kg^{-1}$

The differential adsorption coefficient is approximated by the average value $\overline{K}_d(c)$. This value is assessed by calculating the slope of the chord of the adsorption isotherm:

$$\overline{K}_d(c) = \frac{1}{c(t) - c(t_0)} \int_{t_0}^{t} \left(\frac{dX^S}{dc} \right) dc = \frac{X^S(t) - X^S(t_0)}{c(t) - c(t_0)} \qquad (4.26)$$

Although $\overline{K}_d(c)$ is a function of the concentration, its average value is considered to be constant during the time step.

4.2.2.3 Freundlich adsorption equation

The Freundlich equation is an empirical equation which is used to relate the amount of adsorbed compound with the quantity in solution. The Freundlich isotherm does not exhibit a maximum number of adsorption sites. However, the Freundlich isotherm is more general than the Langmuir isotherm because it allows for taking into account surface heterogeneity of the adsorber. The Freundlich equation is a non-linear adsorption isotherm that expresses the equilibrium conditions as:

$$X^S = K_F c^N \qquad (4.27)$$

where:

K_F = the Freundlich adsorption coefficient in $kg^{1-N}.m^{3N}.kg^{-1}$
c   = the mass concentration in $kg.m^{-3}$
N = Freundlich exponent, usually < 1.0 (-)

K_F and N are compound and soil specific constants.
Eq. (4.27) is subject to the conditions:

$$c < c_{max} \quad X^S = K_F c^n$$
$$c \geq c_{max} \quad X^S = X^S_{max} \quad (4.28)$$

The change in sorbed amount, according to the Freundlich equation, during the time interval has been approximated in the model formulation as:

$$\frac{dX^S}{dt} = \frac{dX^S}{dc}\frac{dc}{dt} \approx K_F \frac{c^n(t) - c^n(t_0)}{c(t) - c(t_0)}\frac{dc}{dt} = \overline{K}_d \frac{dc}{dt} \quad (4.29)$$

4.2.3 Non-equilibrium sorption

Application of the above mentioned equilibrium approach has not worked well in several situations, most notably for many strong adsorbed solutes, many organic chemicals and when used for simulating transport in structured aggregated media. The adsorption rate can be limited by internal diffusion processes, which control the transfer of compounds from the exterior of the adsorbent to the internal surfaces. If internal diffusion processes are rate limiting an adsorption model based on Fick's second diffusion law can be utilized to describe the reaction. A number of chemical-kinetic and diffusion controlled models have been proposed to describe non-equilibrium sorption, varying from relatively simple first order kinetic rate equations, multi-site sorption, to two region transport involving solute exchange between mobile and relatively immobile liquid regions. Van Genuchten and Šimůnek (1996) give an extended discussion of the different time dependent formulations that can be used, taking into account the geometry of the adsorption sites. The different approaches incorporated in TRANSOL are described below.

4.2.3.1 First order rate adsorption

The simplest one-site non-equilibrium formulation is obtained when a chemically controlled first-order linear kinetic rate process is assumed for which:

$$\frac{\partial X^S}{\partial t} = \alpha(X_{eq}^S - X^S) \quad (4.30)$$

where:
α = first order kinetic rate coefficient in d^{-1}
X_{eq}^S = sorbed quantity in equilibrium with the concentration $c(t)$

The difference between the equilibrium situation and the actual adsorbed mass becomes the driving force for mass transfer. The rate equations can be defined as follows:

- Linear adsorption:

$$\frac{\partial X^S}{\partial t} = \alpha(K_D c - X^S) \quad (4.31)$$

- Langmuir equation:

$$\frac{\partial X^S}{\partial t} = \alpha\left(K_L X_{max}^S c - (1 + K_L c) X^S\right) \quad (4.32)$$

- Freundlich equation:

$$\frac{\partial X^S}{\partial t} = \alpha(K_F c^N - X^S) \quad (4.33)$$

Implementation of these equations in computer models, which describe leaching and accumulation of pollutants, can be conducted by linearization. The differential quotient of the time dependent sorption phase is averaged by integrating the expression of $X^S(t)$ between the limits t_0 and t and dividing by the time increment. For the calculation of the quantity X^S sorbed at the end of the time increment, Eq. (4.31), Eq. (4.32) and Eq. (4.33) are integrated, replacing approximation c by $\bar{c}$, the time averaged concentration during the time increment.

This yields:

- Linear adsorption and Freundlich equation:

$$\frac{dX^S}{dt} \approx \frac{X^S(t) - X^S(t_0)}{(t - t_0)} = (X_{eq}^S(\bar{c}) - X^S(t_0)) \frac{(1 - \exp[-\alpha(t - t_0)])}{(t - t_0)} \quad (4.34)$$

- Langmuir equation:

$$\frac{dX^S}{dt} \approx \frac{X^S(t) - X^S(t_0)}{(t-t_0)} = (X_{eq}^S(\bar{c}) - X^S(t_0))\frac{(1 - \exp[-\alpha(1+K_L\bar{c})(t-t_0)])}{(t-t_0)} \quad (4.35)$$

The one-site first order kinetic models, giving generally poor results, can be expanded into a n-site sorption model, assuming that the sorption sites can be divided into a number of fractions to account for the heterogeneity of the soil. The exchange sites are divided in n fractions, of which the sorption to the first fraction is assumed to be instantaneously in equilibrium with the solute concentration. Two- and three-site models are quite successful to describe the transport of pollutants in soils. The effects of so-called aging of sorbed pollutants on the transport of these pollutants can easily be calculated with these multi-site models.

4.2.3.2 Intra-aggregate sorption in structured media

The diffusion of a solute into aggregates can for a single aggregate be given by the general expression:

$$[\theta_i + \rho_s(1-\theta_i)K_d(\bar{c_i})]\frac{\partial c_i}{\partial t} = r^{-m}\frac{\partial}{\partial r}(r^m\theta_i D_s \frac{\partial c_i}{\partial r}) \quad (4.36)$$

where:
θ_i = internal aggregate moisture content in m^3.m^{-3}
c_i = internal concentration in the aggregates in kg.m^{-3}
$\bar{c_i}$ = average internal concentration in the aggregates during time step in kg.m^{-3}
r = radial distance from the surface to the centre of the aggregates in m
m = shape factor for the aggregate structure (-), with
    $m = 0$ for plate aggregates;
    $m = 1$ for cylindrical aggregates;
    $m = 2$ for spherical aggregates
D_s = solute diffusion coefficient in m^2.d^{-1}
ρ_s = specific weight of the soil in kg.m^{-3}

The total number of soil aggregates is divided into N classes with radius R_n. The volume of one soil aggregate $v_{ag,n}$ is given as:

$$v_{ag,n} = \frac{2^{m_n}\pi^{\frac{1}{2}(\frac{3}{m_n}-1)}}{m_n + 1}R_n^{(m_n+1)} \quad (4.37)$$

The total number of aggregates per m³ of soil volume is calculated as:

$$N_{ag,n} = fr_n(1 - \varepsilon_e)\frac{m_n+1}{2^{m_n}\pi^{\frac{1}{2}}\left(\frac{3}{m_n}-1\right)} R_n^{m_n+1} \quad (4.38)$$

where:
fr_n = volume fraction of aggregate class n (-)
ε_e = macro porosity of the soil in m³.m⁻³

The total volume per aggregate class n per m³ soil is given by the expression:

$$V_{ag,n} = fr_n(1 - \varepsilon_e) \quad (4.39)$$

The surface area α_{ag} of an aggregate is given as:

$$\alpha_{ag,n} = 2^{m_n}\pi^{\left(\frac{m_n+1}{2}\right)} R_n^{m} \quad (4.40)$$

The total surface area of the aggregates in each class n per m³ of soil is calculated as:

$$A_{ag,n} = N_{ag,n}\alpha_{ag,n} = fr_n(1 - \varepsilon_e)\frac{m_n+1}{R_n} \quad (4.41)$$

If N equals 1, Eq. (4.41) reduces to:

$$A_{ag} = (1 - \varepsilon_e)\frac{(m+1)}{R} \quad (4.42)$$

Eq. (4.36) can be solved numerically, considering the flow rate of a chemical compound in the aggregates as the flow through a series of compartments and proportional to the concentration gradient in the aggregates (Groenendijk 1997b). A special case is obtained by schematizing the aggregates to only one compartment. In this situation it is assumed that the immobile solution is perfectly mixed and the diffusion process into the aggregates is governed by a first order rate process. Eq. (4.36) is rewritten as:

$$[\theta_i + \rho_s(1-\theta_i)K_d(\overline{c}_{i,n})]\frac{dc_{i,n}}{dt} = 2fr_n(1 - \varepsilon_e)(m_n + 1)\left[\frac{\theta_i D_s}{R_n^2}(c - c_{i,n})\right] \quad (4.43)$$

Denoting the total quantity of a compound present in the aggregates, either adsorbed or in solution, in each aggregate class n by X_n yields:

$$\frac{dX_n(t)}{dt} = \frac{2(m_n + 1)\theta_i D_s}{[\theta_i + \rho_s(1 - \theta_i)K_d(\overline{c}_i, n)]R_n^2}(X_{eq,n}(t) - X_n(t)) =$$
$$k_{ad,n}(X_{eq,n}(t) - X_n(t)) \quad (4.44)$$

where:
X_n = total quantity present in aggregate class n in kg.m^{-3}
$X_{eq,n}$ = the quantity present in aggregate class n at equilibrium when the internal concentration $c_{i,n}$ equals the concentration c_e in the macro-pores in kg.m^{-3}
$k_{ad,n}$ = time step dependent rate coefficient in d^{-1}

The rate coefficient is taken as a constant during the time step and calculated on the basis of the conditions at the beginning of the time step. Substituting for $X_{eq,n}(t)$ the average value $\overline{X}_{eq}$ during a time step $(t - t_0)$ and integrating Eq. (4.44) gives:

$$X_n(t) = \overline{X}_{eq} - (\overline{X}_{eq} - X_n(t_0))\exp[-k_{ad,n}(t - t_0)] \quad (4.45)$$

The differential quotient of the time dependent sorption phase is averaged between the limits t_0 and t by taking $X_n(t) - X_n(t_0)$ and dividing this by the time step $(t - t_0)$:

$$\frac{\overline{dX_n}}{dt} = \frac{X_n(t) - X_n(t_0)}{(t - t_0)} = (\overline{X}_{eq} - X_n(t_0))\frac{(1 - \exp[-k_{ad,n}\Delta t])}{\Delta t} \quad (4.46)$$

4.3 Precipitation and dissolution of minerals

If precipitation is defined as the transformation of two or more dissolved components to a non-dissolved substance, than dissolution and precipitation processes are similar reactions but of opposite direction. Precipitation of a compound occurs when the concentration in the solute exceeds a defined equilibrium or saturation concentration. As soon as the concentration tends to drop below the saturation value dissolution of solid material is present as long as it is present. Precipitation and dissolution can be divided into thermodynamic equilibrium reactions and kinetic reactions The solution-solid equilibrium is characterized, in a general way, by a solubility product for a mineral that dissolves in water according to the reaction:

$$A_{z_b}B_{z_a(s)} \rightleftharpoons z_b A^{z_a^+}{}_{(aq)} + z_a B^{z_b^-}{}_{(aq)} \quad (4.47)$$

If the common standard state convention for aqueous solutions is adopted and the activity of the pure solid phase is set equal to unity, the solubility equation reads:

$$K_{s0} = (A^{z_a^+}{}_{(aq)})^{z_b}{}_{eq} (B^{z_b^-}{}_{(aq)})^{z_a}{}_{eq} \qquad (4.48)$$

The solubility constant K_{s0} is related to molar activities. When the solubility expressions are inserted into mass balance equations, an activity correction has to be taken into account.

Stumm and Morgan (1981) mention two different types of over-saturation ranges, characterizing the precipitation behaviour of electrolytes:

- a super-saturated solution forms a precipitate spontaneously;
- a metastable solution may form no precipitate over a relatively long period.

Often, an active form of the precipitate is formed. Such a fresh precipitate may persist in metastable equilibrium with the solution; it is more soluble than the stable solid phase and may convert only slowly into a more stable phase.

The fresh form of the compound is generally formed very fast from strongly over-saturated solutions. Such a fresh precipitate may persist in metastable equilibrium and may convert only into a more stable form by crystallization. The transformation into more stable forms can be considered as aging of the precipitate. The presence of precipitated forms of anorganic minerals in an unsaturated solution may lead to dissolution of the compounds. Fresh forms dissolve faster than more stable forms. Weathering of minerals is treated as a special type of dissolution.

In order to test whether a solution is over-saturated or under-saturated, the ion activity product (IAP) may be compared with the solubility constant K_{s0}. The IAP is determined from the actual activities in solution:

$$IAP = (A^{z_a^+}{}_{(aq)})^{z_b}{}_{act} (B^{z_b^-}{}_{(aq)})^{z_a}{}_{act} \qquad (4.49)$$

where:
act denotes actual.

The saturation index is calculated as the logarithm of the quotient of IAP and K_{s0}. The state of saturation of a solution with respect to a solid is defined as follows:

oversaturation:	$IAP > K_{s0}$	gives	$\log IAP - \log K_{s0} > 0$
saturation:	$IAP = K_{s0}$	gives	$\log IAP - \log K_{s0} = 0$
undersaturation:	$IAP < K_{s0}$	gives	$\log IAP - \log K_{s0} < 0 \quad (4.50)$

4.3.1 Time dependent multi-component model

A time dependent precipitation reaction is used when the precipitation rate is small in comparison with the time step length. The rate of change in time, expressing the concentration change of an element in the solid phase, is defined to be proportional to the difference between *IAP* and K_{s0}, so:

$$\frac{dA_s}{dt} = \alpha_r[IAP - K_{s0}] = \alpha_r[(A^{z_a+})^{z_b}_{act}(B^{z_b-})^{z_a}_{act} - (A^{z_a+})^{z_b}_{eq}(B^{z_b-})^{z_a}_{eq}] \quad (4.51)$$

Application of the first order rate equation leads to an increase of the solid phase when *IAP* is greater than K_{s0}. In an under-saturated solution, the stock of solid minerals is depleted. In the long term, the system converges to an equilibrium state, with *IAP* being equal to K_{s0}. A dummy variable Z_{pd} is introduced to express the change in the solid phase This variable Z_{pd} is treated as a species in a component table and is calculated for the new activities according to:

$$Z_{pd} = \alpha_r \Delta t[(A^{z_a+})^{z_b}_{act}(B^{z_b-})^{z_a}_{act} - K_{s0}] \quad (4.52)$$

The sum concentrations of the cation A^{z_a+} and the anion B^{z_b-} are given as:

$$A_T = \sum_{i=1}^{m}(a_{ij}K_i\prod_{j=1}^{n}[A_j]^{a_{ij}}) + z_b\alpha_r\Delta t[(A^{z_a+})^{z_b}(B^{z_b-})^{z_a} - K_{s0}]$$

$$B_T = \sum_{i=1}^{m}(a_{ij}K_i\prod_{j=1}^{n}[B_j]^{a_{ij}}) + z_a\alpha_r\Delta t[(A^{z_a+})^{z_b}(B^{z_b-})^{z_a} - K_{s0}] \quad (4.53)$$

The summation term in the equations represents the sum of the aqueous complexes, calculated from the new component concentrations. A precipitation reaction results in a positive value for the concentration of Z_{pd}. A positive contribution of Z_{pd} in the fixed material balance of the aqueous solution causes lower levels of the other balance terms.

When adjusted sum concentrations are defined, it is possible to include the non-equilibrium reactions in a component table. These adjusted sum concentrations are calculated as:

$$A_{adj} = A_T + z_b\alpha_r\Delta t K_{s0} = \sum_{i=1}^{m} a_{ij}K_i\prod_{j=1}^{n}[A_j]^{a_{ij}} + z_b\alpha_r\Delta t(A^{z_a+})^{z_b}(B^{z_b-})^{z_a}$$

$$B_{adj} = B_T + z_a\alpha_r\Delta t K_{s0} = \sum_{i=1}^{m} a_{ij}K_i\prod_{j=1}^{n}[B_j]^{a_{ij}} + z_b\alpha_r\Delta t(A^{z_a+})^{z_a}(B^{z_b-})^{z_a} \quad (4.54)$$

Table 4.4 gives an illustration of the component table of a precipitation and dissolution reaction. After determination of the new apparent equilibrium, the real value of Z_{pd} is calculated by subtracting the fixed part ($\alpha_r\Delta t K_{s0}$) from the dummy variable Z_{pd}^*. The sum concentration of the aqueous system has to be corrected for the quantities precipitated or dissolved. In EPIDIM (Groenendijk 1997a) arrangements have been made to simulate precipitation and dissolution reactions without adjusting the total sum concentrations and the recalculation of the transformed quantities.

Table 4.4: Description of the stoichiometry of a precipitation and dissolution reaction.

Species	Components		log(K)
	A	B	
A	1	0	0
B	0	1	0
Z^*_{pd}	z_b	z_a	$\log(\alpha_r \Delta l)$
SUM	A_{adj}	B_{adj}	-

Instantaneous precipitation of a compound can be considered if the process requires a relatively short time compared to the transport time. The rate itself is in many cases of no concern, so the process is described as an instantaneous reaction.

4.3.2 Transfer of gases

The interaction between the gas and water phase can play an important role in the regulation of the concentration levels of some dissolved compounds. The escape of carbon dioxide from soil moisture generally results in higher pH levels, whereas the volatilization of NH_3 to the atmosphere causes a depletion of NH_4^+ and a lowering of pH. The transfer of gases between the soil atmosphere and the liquid phase can be described as a diffusion controlled process. The flux F of a chemical constituent from the gas phase to the liquid solution can be given as a first order rate equation:

$$F = K_L \left(\frac{c_g}{H} - c_{aq} \right) \tag{4.55}$$

with:
K_L = rate constant in $m^3.d^{-1}$
H = distribution coefficient (-)
c_g = concentration in the soil air in $kg.m^{-3}$
c_{aq} = concentration in the water phase in $kg.m^{-3}$

The distribution coefficient H is obtained from Henry's law constant as:

$$H = (He RT)^{-1} \tag{4.56}$$

in which:
He = Henry's law constant in $Pa^{-1}.m^{-3}.mole$

R = the universal gas constant in Pa.m^3.mole^{-1}.$^\circ$K^{-1}
T = temperature in $^\circ$K

Assuming ideal gas behaviour gives for Henry's law constant:

$$He = \frac{[c_{(aq)}]}{p_g} \qquad (4.57)$$

Combination of Eq. (4.55), Eq. (4.56) and Eq. (4.57) gives:

$$F = K_L(\frac{p_g}{RT} - c_{(aq)}) \qquad (4.58)$$

For CO_2 transport from the soil air to soil water Eq. (4.58) can be approximated as:

$$F = K_L(\frac{p_{CO_2}}{RT} - [CO_{2(aq)}]) \approx K_L(\frac{p_{CO_2}}{RT} - [H_2CO_3^*]) \qquad (4.59)$$

The quantity transferred from the soil air to the water sub-system within a time interval Δt is determined by the product of the flux and the length of the time step, expressed by the dummy variable Z_{CO2}. Taking account of other species such as HCO_3^-, CO_3^{2-}, H^+ and OH^-, the system can be formulated in a component table as is presented in Table 4.5.

Table 4.5: The stoichiometry of a system, describing the transport of CO_2 to the water phase.

Species	Components		log (K)	K_{s0} or c_{eq}
	H^+	HCO_3^-		
H^+	1	0	0	-
OH^-	-1	0	-14.0	-
$H_2CO_3^*$	1	1	6.35	-
HCO_3^-	0	1	0	-
CO_3^{--}	-1	1	-10.33	-
Kinetic species				
Z_{CO2}	1	1	log $(K_L \Delta t)$	p_{CO2}/RT
SUM	H^+_T	$HCO_3^-{}_T$	-	-

Due to the implicit determination of the quantity transferred, the total mass balances have to be updated before the start of the next time step, by subtracting Z_{CO2} from H^+_T and $HCO_3^-{}_T$. The transport of CO_2 to or from the aqueous system causes a shifting of the carbonate equilibria. The chemical system of aqueous species considered in calcareous equilibria and alkalinization is given in Table 4.6.

Table 4.6: Definition of the chemical system of aqueous species considered in calcareous equilibria in aquifers and alkalinization in (semi-)arid regions.

Species	Components								log(K)
	H^+	HCO_3^-	SO_4^{--}	Cl^-	Na^+	K^+	Ca^{++}	Mg^{++}	CEC	
H^+	1	0	0	0	0	0	0	0	0	0.0
OH^-	-1	0	0	0	0	0	0	0	0	-14.0
Na^+	0	0	0	0	1	0	0	0	0	0.0
NaOH	-1	0	0	0	1	0	0	0	0	-14.20
$NaCO_3^-$	-1	1	0	0	1	0	0	0	0	-9.07
$NaHCO_3$	0	1	0	0	1	0	0	0	0	0.24
Na_2CO_3	-1	1	0	0	2	0	0	0	0	-10.32
$NaSO_4^-$	0	0	1	0	1	0	0	0	0	0.70
NaCl	0	0	0	1	1	0	0	0	0	0.0
K^+	0	0	0	0	0	1	0	0	0	0.0
KOH	-1	0	0	0	0	1	0	0	0	-14.50
$KHCO_3$	0	1	0	0	0	1	0	0	0	0.25
K_2CO_3	-1	1	0	0	0	2	0	0	0	-10.35
KSO_4^-	0	0	1	0	0	1	0	0	0	0.85
KCl	0	0	0	1	0	1	0	0	0	-0.70
Ca^{++}	0	0	0	0	0	0	1	0	0	0.0
$CaOH^+$	-1	0	0	0	0	0	1	0	0	-12.70
$CaCO_3$	-1	1	0	0	0	0	1	0	0	-7.13
$CaHCO_3^+$	0	1	0	0	0	0	1	0	0	1.26
$CaSO_4$	0	0	1	0	0	0	1	0	0	2.31
$CaCl^+$	0	0	0	1	0	0	1	0	0	-1.00
$CaCl_2$	0	0	0	2	0	0	1	0	0	0.0
Mg^{++}	0	0	0	0	0	0	0	1	0	0.0
$MgOH^+$	-1	0	0	0	0	0	0	1	0	-11.45
$MgCO_3$	-1	1	0	0	0	0	0	1	0	-6.93
$MgHCO_3^+$	0	1	0	0	0	0	0	1	0	1.16
$MgSO_4$	0	0	1	0	0	0	0	1	0	2.23
$MgCl_2$	0	0	0	2	0	0	0	1	0	-0.03
HCO_3^-	0	1	0	0	0	0	0	0	0	0.0
CO_3^{--}	1	1	0	0	0	0	0	0	0	-10.33
$H_2CO_3^*$	1	1	0	0	0	0	0	0	0	6.36
SO_4^{--}	0	0	1	0	0	0	0	0	0	0.0
Cl^-	0	0	0	1	0	0	0	0	0	0.0
CEC-H	1	0	0	0	0	0	0	0	1	300.0
CEC-Na	0	0	0	0	1	0	0	0	1	1.0
CEC-K	0	0	0	0	0	1	0	0	1	10.0
CEC-Ca	0	0	0	0	0	0	1	0	2	1.0
CEC-Mg	0	0	0	0	0	0	0	1	2	0.9
$CaCO_{3(s)}$	-1	1	0	0	0	0	1	0	0	1.98
$CO_{2(g)}$	1	1	0	0	0	0	0	0	0	-11.34
SUM	H_T	HCO_{3T}	SO_{4T}	Cl_T	Na_T	K_T	Ca_T	Mg_T	CEC_T	-

4.3.3 Schematization in the mass conservation and transport model

For the analysis of multi-component transport a sequential iteration approach is used of the multi-component equilibrium model and the semi-analytical solutions of the mass conservation and transport model

4.3.3.1 Non-equilibrium precipitation and dissolution

A time dependent precipitation reaction is used when the precipitation rate is small in comparison with the time step length considered. The precipitation and dissolution reactions are described as a first order reaction rate proportional to the difference between the actual concentration in the macro-pores and the saturation concentration of the compound, given by the expressions:

$$\frac{dX^P}{dt} = K_{pr}(c - c_{sat}) \qquad c > c_{sat}$$

$$\frac{dX^P}{dt} = K_{di}(c - c_{sat}) \qquad c < c_{sat} \qquad (4.60)$$

where:
K   = the precipitation rate (K_{pr}) or dissolution rate (K_{di}) constant in $m^3.kg^{-1}.d^{-1}$
c_{sat} = saturation concentration in $kg.m^{-3}$
X^P = quantity present as solid material in $kg.kg^{-1}$

Integration of Eq. (4.60) over the time step between t_0 and t gives:

$$X^P(t) = \int_{t_0}^{t} k(c(t) - c_{sat})dt + X^P(t_0) = k(\bar{c} - c_{sat})(t - t_0) + X^P(t_0) \qquad (4.61)$$

where:
$\bar{c}$ = time averaged concentration in the bulk solution in $kg.m^{-3}$

Special conditions are present if the dissolution terminates or the precipitation starts during a time step. In that case the time step has to be split up into two parts under the following conditions for the length of the partial time step τ:

$$\text{if } X^P(t_0) > 0 \quad \text{then } k(\bar{c} - c_{sat})(\tau - t_0) + X^P(t_0) = 0 \qquad (4.62)$$

else:

$$\text{if } t < \tau \; ; \; c(t) < c_{sat}(\tau) \; ; \; X^P(t) = 0 \quad \text{then} \quad \frac{dX^P(t)}{dt} = 0$$

$$\text{else } t \geq \tau \; ; \; c(t) \geq c_{sat}(\tau) \; ; \; X^P(t) \geq 0 \quad \text{then} \quad \frac{dX^P(t)}{dt} > 0 \quad (4.63)$$

Non-linear precipitation behaviour is approximated by a number of parallel first order reactions. Each reaction equation corresponds with the solubility product of a specific compound mineral and its equilibrium concentration.

4.3.3.2 Instantaneous precipitation and dissolution

Instantaneous precipitation of a compound is considered if the process requires a short time compared to the transport time. Under these conditions the process is described as an instantaneous reaction, that is one which is completed immediately. The changes in solid material and in concentrations is given by the conditions:

$$\text{if } X^P(t) = 0 \; ; \; t \leq \tau \; ; \; c(t) \leq c_{eq}(\tau) \quad \text{then} \quad \frac{dX^P(t)}{dt} = 0$$

$$\text{else } X^P(t) > 0 \; ; \; t > \tau \; ; \; c(t) = c_{eq}(\tau) \quad \text{then} \quad \frac{dc(t)}{dt} = 0 \quad (4.64)$$

and

$$\text{if } X^P(t) > 0 \; ; \; t \leq \tau \; ; \; c(t) = c_{eq}(\tau) \quad \text{then} \quad \frac{dc(t)}{dt} = 0$$

$$\text{else } X^P(t) = 0 \; ; \; t > \tau \; ; \; c(t) < c_{eq}(\tau) \quad \text{then} \quad \frac{dX^P(t)}{dt} = 0 \quad (4.65)$$

When applying the model in combination with the instantaneous precipitation reaction, four situations with respect to the concentration course with time within a time increment are relevant to consider. The four possibilities have been explained in Fig. 4.2. In the second and third situations the time interval has to be split up into two parts. The length of the first part τ is calculated from its specific conditions using the conservation equation. The concentration in the solute equals the saturation concentration as long as the store of the solid compound is not exhausted.

4.4 Redox reactions

Redox reactions are modelled in EPIDIM by describing their half reactions: reduction and oxidation. The reductants and oxidants are defined as electron donors and electron

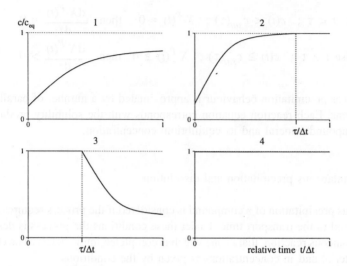

Fig. 4.2: Four possibilities of the concentration course with time within a time step Δt establishing an eventual exceedance of the equilibrium concentration c_{eq}.

acceptors. Because there are no free electrons, every oxidation is accompanied by a reduction and vice versa. In general each reaction can be written as follows:

$$A_{ox} + ae^- = A_{red}$$
$$B_{red} = B_{ox} + be^-$$
$$\overline{bA_{ox} + aB_{red} \rightleftharpoons bA_{red} + aB_{ox}} \qquad (4.66)$$

In Table 4.7 the chemical system considered in the acidification potential of acid sulphate soils is given (Bronswijk and Groenenberg 1994), as an illustration of the method of how to deal with redox-couples in a chemical definition table for the model EPIDIM.

The influence of acid buffering by carbonate equilibria is also taken into account. Sources and sinks have been distinguished for their role they play commonly in soil-water systems. Only an arbitrary set of minerals is mentioned to illustrate their role in the regulation of pH and pe (pe = - log(e⁻)). The transformation of organic matter into carbon-dioxide is accompanied by a donation of protons and electrons. The velocity of this reaction is controlled by the micro-organisms, which mediate the reaction. For $Fe(OH)_3$ the solubility product of the amorph form has been presented. In reality, more than one constant can be determined for iron hydroxide. More stable forms show higher values of log (K). The pyrite (FeS_2) equilibrium has been formulated assuming that the interaction with iron hydroxide can be neglected.

Table 4.7: Definition of the chemical system of aqueous species considered in the research on the acidification potential of acid sulphate soils (Bronswijk and Groenenberg, 1994).

Species	Components											log K
	H^+	Na^+	K^+	Ca^{++}	Mg^{++}	Fe^{++}	Al^{3+}	HCO_3^-	SO_4^{--}	Cl^-	e^-	
H^+	1	0	0	0	0	0	0	0	0	0	0	0
OH^-	-1	0	0	0	0	0	0	0	0	0	0	-14.0
HSO_4^-	1	0	0	0	0	0	0	0	1	0	0	1.98
Na^+	0	1	0	0	0	0	0	0	0	0	0	0
$NaCl$	0	1	0	0	0	0	0	0	0	1	0	0
$NaHCO_3$	0	1	0	0	0	0	0	1	0	0	0	-0.25
$NaSO_4^-$	0	1	0	0	0	0	0	0	1	0	0	0.95
K^+	0	0	1	0	0	0	0	0	0	0	0	0
KSO_4^-	0	0	1	0	0	0	0	0	1	0	0	0.84
Ca^{++}	0	0	0	1	0	0	0	0	0	0	0	0
$CaCl^+$	0	0	0	1	0	0	0	0	0	1	0	-1.0
$CaCl_2$	0	0	0	1	0	0	0	0	0	2	0	0
$CaCO_3$	-1	0	0	1	0	0	0	1	0	0	0	-8.13
$CaHCO_3^+$	0	0	0	1	0	0	0	1	0	0	0	1.26
$CaSO_4$	0	0	0	1	0	0	0	0	1	0	0	2.31
Mg^{++}	0	0	0	0	1	0	0	0	0	0	0	0
$MgCl_2$	0	0	0	0	1	0	0	0	0	2	0	-0.03
$MgCO_3$	-1	0	0	0	1	0	0	1	0	0	0	-6.93
$MgHCO_3^+$	0	0	0	0	1	0	0	1	0	0	0	1.16
$MgSO_4$	0	0	0	0	1	0	0	0	1	0	0	2.25
Fe^{++}	0	0	0	0	0	1	0	0	0	0	0	0
$FeCl_2$	0	0	0	0	0	1	0	0	0	2	0	-0.07
$FeSO_4$	0	0	0	0	0	1	0	0	1	0	0	2.30
Fe^{+++}	0	0	0	0	0	1	0	0	0	0	-1	-13.04
$FeCl^{++}$	0	0	0	0	0	1	0	0	0	1	-1	-11.56
$FeCl_2^+$	0	0	0	0	0	1	0	0	0	2	-1	-10.91
$FeCl_3$	0	0	0	0	0	1	0	0	0	3	-1	-12.27
$FeSO_4^+$	0	0	0	0	0	1	0	0	1	0	-1	-8.9
$Fe(SO_4)_2^-$	0	0	0	0	0	1	0	0	2	0	-1	-7.66
$FeOH^{++}$	-1	0	0	0	0	1	0	0	0	0	-1	-15.24
$Fe_2(OH)_2$	-2	0	0	0	0	2	0	0	0	0	-2	-29.00
Al^{+++}	0	0	0	0	0	0	1	0	0	0	0	0
$AlOH^{++}$	-1	0	0	0	0	0	1	0	0	0	0	-5.02
$Al_2(OH)_2^{4+}$	-2	0	0	0	0	0	2	0	0	0	0	-6.27
$AlSO_4^+$	0	0	0	0	0	0	1	0	1	0	0	3.2
HCO_3^-	0	0	0	0	0	0	0	1	0	0	0	0
CO_3^{--}	-1	0	0	0	0	0	0	1	0	0	0	-10.33
$H_2CO_3^*$	1	0	0	0	0	0	0	1	0	0	0	6.35
SO_4^{--}	0	0	0	0	0	0	0	0	1	0	0	0
Cl^-	0	0	0	0	0	0	0	0	0	1	0	0
Sum	H-OH	Na	K	Ca	Mg	Fe	Al	HCO_3	SO_4	Cl	e	

4.5 Complexation with organic acids

The total concentration of the various dissolved metals M_{dis}^{tot} is determined by the concentration of free ionic species M_{free} and the concentration of complexes M_{comp} with major anions. Major inorganic anions involved in the complexation with heavy metals are Cl^-, SO_4^{--}, OH^- and HCO_3^-. Complexation with organic anions can be described by various models, depending on the description of the organic acid behaviour. Dissolved concentrations of DOC depend on the organic matter content, pH and dissolved Ca concentrations. Values generally range between 10 and 200 mg.dm^{-3}. Empirical models consider the humic/fulvic acids as a mono to n-photic acid. Here fulvic acids stands for the sum of all dissolved fulvic and humic compounds. Monovalent fulvic acids occur mainly at low pH due to dissociation of carboxylic groups, whereas divalent fulvic acids mainly occur at high pH due to dissociation of both carboxylic and phenolic groups. In general dissociation of a di-protonic acid is most often used (Driscoll *et al.* 1994; Bril 1996) according to:

$$H_2FA \leftrightarrows HFA^- + H^+ \qquad [HFA] = K_{a,1}\frac{[H_2FA]}{[H]}$$
$$HFA \leftrightarrows FA^{2-} + H^+ \qquad [FA] = K_{a,2}\frac{[HFA]}{[H]} \qquad (4.67)$$

where:
HFA^- = the monovalent fulvic acid
HFA^{2-} = the divalent fulvic acid
$K_{a,1}$ = the first dissociation constant for fulvic acid in mol.dm^{-3}
$K_{a,2}$ = the second dissociation constant for fulvic acid in mol.dm^{-3}

A complete description requires that FA_{tot} is presented as the sum of protonated and dissociated humic compounds (H_2FA, HFA^- and FA^{2-}), including the complexes of various cations with the dissociated humic compounds. The total concentration of a metal in solution is described as the sum of the ionic free species and the complexes:

$$[M]_{diss} = [M] + [MCl] + [MCl_2] + [MCl_3] + [MSO_4] + [MOH]$$
$$+ [M(OH)_2] + [MHCO_3] + [MCO_3] + [MHFA] + [MFA] \qquad (4.68)$$

The relation between the activity of each of the complexes ML, the free ionic species M and the ligand L is given by:

$$[ML_i] = K_{ML_i}[M][L]^i \qquad (4.69)$$

Introducing f_1 and f_2 as the activity coefficients of monovalent and divalent ionic species or complexes, respectively, yields for the total concentration of a dissolved metal:

$$[M]_{diss} = [M](f_2^{-1} + K_{MCl}[Cl] + K_{MCl_2}f_1^2[Cl]^2 + K_{MCl_3}[Cl]^3 +$$
$$K_{MSO_4}f_2[SO_4] + K_{MOH}[OH] + K_{M(OH)_2}f_1^2[OH]^2 + K_{MHCO_3}[HCO_3]$$
$$+ K_{MCO_3}f_1^2[CO_3] + K_{MHFA}[HFA] + K_{MHFA}f_1^2[FA]) \qquad (4.70)$$

4.6 Model schematization of physical and chemical processes

For the solution of the general transport and conservation equation different cases must be distinguished. If precipitated material is present or formed, precipitation and dissolution can be considered either as instantaneous or as time dependent processes. The relation between the adsorbed quantity and the concentration in solution is given by one of the equilibrium expressions, when the adsorption rate is very fast compared to the model interval time. The change in the quantity of adsorbed material in a layer is given in the transport and conservation equation by the expression:

$$\rho_d \frac{dX_n^S(t)}{dt} = K_d \rho_d \frac{dc_n(t)}{dt} \qquad (4.71)$$

in which the distribution coefficient $K_d = \overline{K}_d$ if non-linear adsorption is considered. The adsorption rate can be limited by the internal diffusion processes that control the transfer of material from the exterior of the adsorbent to the internal surface. The differential quotient of the time dependent sorption phase is averaged between the time boundaries t_0 and t by taking $X_n^S(t) - X_n^S(t_0)$ and dividing by the time step $(t - t_0)$:

$$\frac{dX_{j,n}^S}{dt} = \frac{X_{j,n}^S(t) - X_{j,n}^S(t_0)}{t - t_0} = \left(\overline{X}_{eq,j,n}^S - X_{j,n}^S(t_0)\right) \left(\frac{1 - \exp[-k_{j,n}^S \Delta t]}{\Delta t}\right) \qquad (4.72)$$

where:
$X_{j,n}^S$ = total quantity present at sorption site j or in aggregate class j in kg.kg^{-1}
$\overline{X}_{eq,j,n}^S$ = quantity present at sorption site j or in aggregate class j at equilibrium with the average external concentration $\overline{c}_n$ of the soil solution in kg.kg^{-1}
$k_{j,n}^S$ = sorption rate coefficient in d^{-1}
Δt = $t - t_0$

Multi-site sorption is given as a combined instantaneous and time dependent sorption:

$$\frac{dX_n^S}{dt} = \rho_d \left[\overline{K}_d \frac{dc_n(t)}{dt} + \sum_{j=2}^{j} \left(\overline{X}_{eq,j,n}^S - X_{j,n}^S(t_0)\right) \left(\frac{1 - \exp[-k_{j,n}^S \Delta t]}{\Delta t}\right) \right] \qquad (4.73)$$

4.6.1 Instantaneous precipitation excluded

The rate of the time dependent precipitation and dissolution of material is given by the general expression:

$$\frac{dX_l^P}{dt} = \sum_{l=1}^{l_{max}} [K_l(c_n(t) - c_{eq,l})] \qquad (4.74)$$

subject to the condition:

if $X_l^P = 0$ and $c_n(t) < c_{eq,l}$ then $k_l = 0$

The transport and conservation equation is in that case written as:

$$\Delta z_n \left[\frac{d(\theta_n(t)c_n(t))}{dt} + \rho_d K_d \frac{dc_n(t)}{dt} \right] = \sum_{i=1}^{i_{max}} q_i^i(t)c_i^i(t) - \sum_{j=1}^{j_{max}} q_j^o(t)c_n(t)$$

$$- \sigma_{pl} q_{et} c_n(t) - \Delta z_n \rho_d \left[\sum_{j=2}^{j_{max}} (\overline{X}_{eq,j,n} - X_{j,n}(t_o))\left(\frac{1 - \exp[-k_{adj,n}\Delta t]}{\Delta t}\right) \right]$$

$$+ \Delta z_n [k_{0,n} + \sum_{l=1}^{l_{max}} \rho_d K_l c_{eq,l}] + \Delta z_n [k_{1,n} \theta_n(t) - \sum_{l=1}^{l_{max}} \rho_d K_l] c_n(t) \qquad (4.75)$$

Rearranging of terms in the transport and conservation equation gives the general form of the differential equation presented in paragraph 3.2.1:

$$\frac{dc_n(t)}{dt} + \frac{\varsigma_1}{\theta_0 + \rho_d K_d + k_\theta t} c_n(t) = \frac{\varsigma_2}{\theta_0 + \rho_d K_d + k_\theta t} \qquad (4.76)$$

in which:

$$\varsigma_1 = k_\theta + \frac{1}{\Delta z_n} \left[\sum_{j=1}^{j_{max}} q_j^o + \sigma_{pl} q_{et} \right] - k_{1,n} \overline{\theta}_n + \rho_d \sum_{l=1}^{l_{max}} K_l \qquad (4.77)$$

and

$$\varsigma_2 = \frac{1}{\Delta z_n} \sum_{i=1}^{i_{max}} q_i^i \overline{c}_i^i - \rho_d \sum_{k=1}^{k_{max}} [\overline{X}_{eq,k,n} - X_{j,n}(t_0)]\left(\frac{1 - \exp[-k_{adj,n}\Delta t]}{\Delta t}\right)$$

$$+ k_{0,n} + \sum_{l=1}^{l_{max}} K_l c_{eq,l} \qquad (4.78)$$

4.6.2 Instantaneous precipitation included

Precipitation or dissolution is a chemical reaction with a relatively fast rate. Depending on the time step length, the rate itself is in some instances of no direct concern. In that case, models that describe the equilibrium or the end-point of the reaction for specified boundaries will give good results. In cases where instantaneous precipitation or dissolution is considered, the concentration in the macro-pores remains constant. The solution of the transport and mass conservation equation reduces to:

$$X_n^P(t) = \frac{1}{\Delta z_n \rho_d}[\sum_{i=1}^{i_{max}} q_i \overline{c}_i^i - (\sum_{j=1}^{j_{max}} q_j^o - \sigma_{pl}q_{et})c_{sat,n}] - \frac{c_{sat,n}}{\rho_d}(\theta_n(t) - \theta_n(t_0)) -$$

$$\sum_{k=1}^{k_{max}} [X_{eq,j,n} - X_{j,n}(t_0)](\frac{1-\exp[-k_{adj,n}\Delta t]}{\Delta t}) + \frac{1}{\rho_d}[k_{0,n} + k_{1,n}\theta_n c_{sat,n}] + X_n^P(t_0) \quad (4.79)$$

where:
c_{sat} = the equilibrium or saturation concentration in kg.m^{-3}

4.7 Some special applications

4.7.1 Organic micro-pollutants

Water insolubility and the tendency to adsorb on the solid phase are phenomena which have been shown to be closely related, particularly in soils with a high organic matter content. Sorption processes for most compounds are the most important factor determining the transport velocity in the soil system. Basically this involves an estimation of the Freundlich constants K_F and N. The equation for the Freundlich sorption isotherm has the limitation that Freundlich coefficients of a specific substance that have been measured on different types of soil cannot be correctly compared, because the unit of K_F is a function of the exponent N. To overcome this problem Boesten (1993) suggested the following form of the Freundlich equation:

$$X^S = K_d c_{ref}(\frac{c}{c_{ref}})^N \quad (4.80)$$

where:
K_d = the distribution coefficient in m^3.kg
c_{ref} = the reference mass concentration at which K_d has been measured in kg.m^{-3}

In this case K_d becomes the tangent of the linear isotherm at c_{ref} and X_{ref}^S.

Combination of Eq. (4.27) and Eq.(4.80) gives:

$$K_F = K_d c_{ref}^{1-N} \qquad (4.81)$$

Adsorption of organic compounds in soil is often assumed to be linearly related to the organic carbon content of the soil:

$$K_F = 1000 M_{oc} K_{oc} c_{ref}^{1-N} \qquad (4.82)$$

where:
M_{oc} = mass fraction of organic carbon in kg.kg^{-1}
K_{oc} = adsorption coefficient to organic carbon in dm^3.kg^{-1}

The parameters on the right hand side of this equation are input to the model. For c_{ref} a default value of 0.001 is used. N varies from 0.7 to 1.0 for organic pollutants.

For the modelling of aging of persistent organic micro-pollutants a multi-site adsorption model has to be considered. Boesten (1986) and Boesten et al. (1989) had to introduce 3 classes of sorption sites to explain the leaching results of field experiments with cyanazine and metribuzin in a loamy sand soil by model simulation. They introduced the sorption characteristics presented in Table 4.8 to be used in the simulation model.

Table 4.8: Adsorption characteristics for two herbicides in a loamy sand soil.

Soil properties	Cyanazine	Metribuzin
Freundlich exponent (N)	0.91	0.89
Class 1 site		
$K_{F,1}$ (m$^{3/N}$.kg$^{-1/N}$)	24.10^{-5}	11. 10^{-5}
k_1 (d^{-1})	150	180
Class 2 site		
$K_{F,2}$ (m$^{3/N}$.kg$^{-1/N}$)	10.10^{-5}	4.10^{-5}
k_2 (d^{-1})	0.5	0.4
Class 3 site		
$K_{F,3}$ (m$^{3/N}$.kg$^{-1/N}$)	13.10^{-5}	4. 10^{-5}
k_3 (d^{-1})	0.02	0.01

Experimental data on adsorption of different organic pollutants to soils and sediments were collected and generalized in such a way that in many cases predictions can be made with respect to the sorptive behaviour of substances of which only a few physical-chemical parameters, such as the octanol/water partition coefficient and the water solubility, are known (Briggs 1973, Karickhoff et al. 1979, Friesel 1987, Lagas and Hammers 1987). Kooper *et al.* (1987) presented the following relations based on data collected from literature for 127 organic compounds:

$$\log K_{oc} = 0.989 \log K_{ow} - 0.346 \tag{4.83}$$

and

$$\log K_{oc} = -0.729 \log S + 0.231 \tag{4.84}$$

where:
K_{ow} = the octanol/water partition coefficient (-)
S = the water solubility in mole dm^{-3}

4.7.2 Heavy metals

It is well known that in soil, heavy metal ions are adsorbed highly selectively when present at trace levels in solution. Originally the interest was centred, for these ions, on their role as micro-nutrients, where deficiencies could cause significant yield depressions (e.g. Zn, Co, Cu). Lately the presence of heavy metals in toxic amounts has become a matter of concern.

The equilibrium processes included in soil models are adsorption and complexation. Below two possible options are described to include adsorption and complexation, i.e.:

- non-linear equilibrium adsorption combined with all possible complexation reactions with inorganic and organic anions.
- linear equilibrium partitioning of heavy metals over the solid phase and dissolved organic carbon (DOC) and the soil solution, neglecting complexation with inorganic anions;

The first option, which was used in a recent study on calculating critical loads for Pb, Cd and Cu for European forest soils (Reinds *et al.* 1995), may lead to more reliable critical loads when appropriate data are available, but it requires the use of the chemical equilibrium model EPIDIM. A complete description requires that FA_{tot} is described as the sum of protonated and dissociated humic compounds (H_2FA, HFA^- and FA^{--}), including the complexes of various cations with the dissociated humic compounds. This does require a set of complexation constants.

Table 4.9 gives equilibrium constants for the complexation of different cations with the major inorganic anions as well as with monovalent and divalent fulvic acids, as presented by de Vries and Bakker (1996), based on data collected from various literature sources.

Table 4.9: Logarithmic values of formation constants of complexes at 25 °C and ionic strength = 0, after de Vries and Bakker (1996).

Species	Log K								
	H	Al	Ca	Mg	Pb	Cd	Cu	Zn	Ni	Cr
OH	14.0	9.0	1.3	2.55	6.29	3.9	6.3	5.04	4.14	10.0
2 OH	--	10.7	--	--	10.88	7.7	14.32	11.1	9.0	18.3
Cl	--	--	0.4	1.98	1.6	1.98	0.43	0.43	0.4	--
2 Cl	--	--	--	--	1.8	2.60	0.16	0.45	0.96	--
3 Cl	--	--	--	--	1.7	2.4	--	0.50	0.2	--
SO_4	--	3.2	2.31	2.23	2.75	2.46	2.31	2.33	2.29	3.2
HCO_3	9.36	--	1.13	1.07	2.9	2.1	2.7	2.1	2.14	4.5
CO_3	10.33	--	3.15	3.24	7.0	4.1	6.73	5.3	5.8	13.5
HFA	4.4	7	3.7	3.5	5.5	4.1	5.7	4.3	4.3	7
FA	9.4	14	6.1	6.0	10.0	6.8	10.7	7.2	7.3	14

Adsorption of heavy metals to solid adsorption complexes is described with the Freundlich equation, using for the adsorption coefficient the relation (Bril 1996):

$$\log K_F = \varsigma_0 + \varsigma_1 pH + \varsigma_2 \log CEC + \varsigma_3 \log M_{oc} + \varsigma_4 \log M_{cl} - 0.5 N \log a_{Ca} \quad (4.85)$$

with:
CEC = the cation exchange capacity determined at pH 8.2 in $mol_c.kg^{-1}$
M_{oc} = mass fraction of organic carbon in $kg.kg^{-1}$
M_{cl} = mass fraction of clay in $kg.kg^{-1}$
a_{Ca} = activity of Ca in solution in $mol.m^{-3}$
N = Freundlich exponent (-)

Values for the constants ς_n in Eq. (4.85) are presented in Table 4.10.

Table 4.10: Values for the Freundlich exponent (N) and the coefficients ς_n in the transfer function with the Freundlich adsorption coefficient K_F (after de Vries and Bakker 1996).

Metal	N	ς_0	ς_1	ς_2	ς_3	ς_4	R^2
Pb	0.55	-3.48	0.60	0.62	0.46	-	0.78
Cd	0.82	-3.63	0.50	1.00	-	-0.24	0.96
Cu	0.55	-3.21	0.70	0.52	0.46	-0.14	0.94
Zn	0.75	-3.42	0.75	1.30	-	-	0.88

The second option may be favoured for regional applications, because of its simplicity and the lower data demand. Neglecting complexation of heavy metals with inorganic anions and limiting complexation with organic anions to a lumped adsorption expression with *DOC* implies that only information on the partition coefficient of heavy metal M between *DOC* and soil solution is required. An indication of K_p^{DOC} was derived from the concentration ratio of uncomplexed and total heavy metal M concentration in solution under the following conditions (de Vries and Bakker 1996):

- Adsorption of a heavy metal M to *DOC* is equal to the complexation with dissociated monovalent fulvic acid;
- The dissociation of fulvic acid can be described as the dissociation of a reactive mono-protic acid
- The total concentration of fulvic acids [FA]tot, equals the sum of protonated [HFA] and dissociated [HA] fulvic acids,
- The total concentration of fulvic compounds, denoted as fulvic acids [FA]$_{tot}$ is linearly related to the *DOC* concentration.

These conditions lead to the following expression:

$$K_p^{DOC} = 10^{-3} K_c^M m \frac{K_a}{K_a + [H]} \qquad (4.86)$$

where
m = concentration of acidic functional groups per kg *DOC* in mol$_c$.kg^{-1}
K_p^{DOC} = partition coefficient of heavy metal between *DOC* and soil solution in m^3.kg^{-1}
K_c^M = the complexation constant for heavy metal (mol$_c^{-1}$.dm^3)
K_a = the dissociation constant for organic acid (mol$_c$.dm^{-3})

The value of m can be set at 5.5 mol$_c$.(kg *DOC*)$^{-1}$ for soils (Henriksen and Seip 1980, Bril 1996, de Vries and Bakker 1996). De Vries and Bakker (1996) give values for pK_a and log K_c^M for different heavy metals as presented in Table 4.11.

Table 4.11: Values of pK_a (- log K_a) and log K_c^M describing the dissociation and complexation of organic acids with heavy metals by a mono-protic acid after de Vries and Bakker (1996).

Heavy metal	pK_a	log K_c^M	Heavy metal	pK_a	log K_c^M
Pb	9.4	10.5	Zn	4.4	4.3
Cd	4.4	4.1	Ni	4.4	4.4
Cu	9.4	10.7	Cr	9.4	12.7

Deviating values of m and pK_a in soils may be present due to the formation of high concentrations of acetic acid (CH_3COOH, $m = 16.7$, $pK_a = 4.76$), proprionic acid (CH_3CH_2COOH, $m = 13.5$, $pK_a = 4.88$) and 1-butanoic acid ($CH_3CH_2CH_2COOH$, $m = 11.4$, $pK_a = 4.82$) under anaerobic conditions with land treatment of waste water and infiltration of leachate from sanitary landfills.

Assuming equilibrium partitioning without complexation with inorganic anions, the total content of a heavy metal sorbed to solids in both agricultural and non-agricultural soils can be calculated according to:

$$X_M^S = \frac{K_d}{1 + K_p^{DOC} c_{DOC}} c_M^{tot} = K_d^{tot} c_M^{tot} \qquad (4.87)$$

with:

c_{DOC} = concentration of dissolved organic carbon in $kg.m^{-3}$
c_M^{tot} = total concentration of dissolved free and heavy metal complexes in $kg.m^{-3}$
K_d = partition coefficient between the adsorbed and the dissolved metal in $m^3.kg^{-1}$.
X_M^S = the quantity of heavy metal sorbed to the solid phase of the soil in $kg.kg^{-1}$

De Vries and Bakker (1996) give on the basis of a linear regression analysis an expression relating K_d^{tot} with generally available soil parameters as:

$$\log K_d^{tot} = \varsigma_0 + \varsigma_1 pH_{CaCl_2} + \varsigma_2 \log M_{om} + \varsigma_3 \log M_{cl} + \varsigma_4 \log [Fe]_{ox} \qquad (4.88)$$

where:
$[Fe]_{ox}$ = the oxalate extractable Fe content in $mol.kg^{-1}$
ς_n = constant; values of ς_n are presented in Table 4.12

Combination of Eq. (4.87) and Eq. (4.88) yields for K_d:

$$\log K_d = \varsigma_0 + \varsigma_1 pH_{CaCl_2} + \varsigma_2 \log M_{om} + \varsigma_3 \log M_{cl} + \varsigma_4 \log [Fe]_{ox} + \log(1 + K_p^{DOC} c_{DOC}) \qquad (4.89)$$

Reinds et al. (1995) showed for forest soils, with organic matter contents between 0.2 and 10 %, that c_{DOC} can be approximated by:

$$\log c_{DOC} = -0.611 + 0.38 \log M_{om} \qquad (4.90)$$

Eq. (4.90) is a very crude approximation and extrapolation may lead to an overestimation of c_{DOC}. For organic soils with an organic matter fraction of 0.6, Eq. (4.90) gives a value of c_{DOC} of about 0.2 $kg.m^{-3}$. Vermeulen and Hendriks (1996) give a maximum value for peat soils of 0.12 $kg.m^{-3}$ and Van Breemen et al. (1986) report concentrations up to 0.14 $kg.m^{-3}$.

As far as c_{DOC} is calculated as a function of M_{om} it is realistic to introduce a maximum value for c_{DOC} of 0.14 kg.m^{-3}, using the condition:

$$\log c_{DOC} = \text{MIN}[-0.611 + 0.38\log M_{om}; -0.854] \quad (4.91)$$

Table 4.12: Values of the coefficients ς_n used in Eq. (4.89) for the relationship between the partition coefficient K_d^{tot} and soil properties after de Vries and Bakker (1996).

heavy metal	ς_0	ς_1	ς_2	ς_3	ς_4	R^2
Pb	-0.53	0.24	-0.43	-	0.49	0.71
Pb	-0.95	0.35	-	-	-	0.60
Cd	-4.85	0.48	-0.71	-	-	0.70
Cu	-1.09	0.23	-	-	0.63	0.63
Cu	-2.62	0.36	-	-	-	0.49
Zn	-2.06	0.45	-	0.60	-	0.85
Ni	-0.86	0.25	-	0.57	-	0.74
Cr	0.72	0.15	-	-	0.50	0.69
Cr	-0.36	0.21	-	-	-	0.54

An intermediate approach is to use a linear partition coefficient for the relation between the metal concentration in solution and complexed with DOC, combined with a non-linear adsorption equation. When one wants to apply a non-linear equilibrium adsorption equation, combined with a simple partitioning approach for complexation, this content can be calculated according to:

$$X_M^S = K_F \left(\frac{c_M^{tot}}{1 + K_p^{DOC} c_{DOC}} \right)^N = K_F^{tot} (c_M^{tot})^N \quad (4.92)$$

For application of the multi-component transport model the required EPIDIM component table can be composed, using the given tables and equations.

4.7.3 Phosphate sorption and precipitation

Application of animal manure slurries to mineral soils has resulted in increased leaching of phosphate to surface water systems. In many cases the sorption capacity of iron and aluminium minerals have been utilized to fix the phosphate to such a degree that leaching of dissolved phosphate to surface waters can be expected. Transformation processes which influence phosphate mobility and solution concentrations in soils have been observed to proceed as a series of kinetic reactions. Model analyses have shown that even when the disposal of manure is reduced to very low levels, it may last more than a number of decades before the release of accumulated phosphate is diminished to acceptable levels. Desorption of phosphates, even when far-reaching measures are implemented, may cause an exceedance of water quality standards for long periods. Although considerable efforts have been made to model phosphate behaviour in soils (van der Zee 1988, van der Zee and van Riemsdijk 1991), the current theory defining the exposure integral as a variable characterizing long-term sorption/precipitation has not yet delivered operational tools for prediction of leaching after fertilization reductions.

The Langmuir isotherm is derived from the assumption of a homogeneous monolayer of adsorbate on the adsorbent. The Langmuir equation is used to describe instantaneous sorption of phosphates to soils (Enfield et al. 1981, van Noordwijk et al. 1990), as:

$$X^S = X^S_{max}\frac{K_L c}{1 + K_L c} \tag{4.93}$$

The amount of chemical sorbed to the soil matrix never exceeds the maximum sorption capacity X^S_{max}. The formulations given above can be incorporated into the conservation equation by elaborating the differential quotient $\partial X^S/\partial t$:

$$\rho_d \frac{\partial X^S}{\partial t} = \rho_d (\frac{dX^S}{dc})\frac{\partial c}{\partial t} = \rho_d K_d(c)\frac{\partial c}{\partial t} \tag{4.94}$$

where:
$K_d(c)$ = the differential sorption coefficient.

The differential adsorption coefficient is approximated by the average value $\overline{K}_d(c)$. This value is assessed by calculating the slope of the chord of the adsorption isotherm:

$$\overline{K}_d(c) = \frac{1}{c(t) - c(t_0)}\int_{t_0}^{t}(\frac{dX^S}{dc})dc = \frac{X^S(t) - X^S(t_0)}{c(t) - c(t_0)} \tag{4.95}$$

Although $\overline{K}_d(c)$ is a function of the concentration, its average value is considered to be constant during the time step. Schoumans (1995) established the following set of parameters of the Langmuir isotherm which describes the fast phosphate sorption for a wide range of sandy soils: K_L = 1129 (m^3 kg^{-1} P) and X^S_{max} = 5.167 10^{-6} ρ_d[Al+Fe] (kg.m$_s^{-3}$ P) where [Al+Fe] is the aluminium + iron content of the soil in mmol.kg^{-1}. If equilibrium is not achieved rapidly enough compared to advective transport the sorption process must be addressed by kinetic models (Pignatello 1989, Selim and Amacher, 1988). In such cases, sorption appears to be limited by a type of chemical reaction rate coefficient or physical mass transfer resistance. The general formulation of a first order rate sorption (chemical non-equilibrium) model is:

$$\rho_d \frac{\partial X^S}{\partial t} = \rho_d k_s (f_s(c) - X^S) \quad (4.96)$$

where:
k_s (d^{-1}) = a rate constant
$f_s(c)$ = the solid phase solute concentration based on equilibrium adsorption isotherms.

The difference between the equilibrium concentration which is reached in the steady state situation and the actual solid phase concentration becomes the driving force for mass transfer. In the case of kinetic phosphate sorption, the rate equation is defined as follows:

$$\rho_d \frac{\partial X^S}{\partial t} = \rho_d k_s (k_F c^N - X^S) \quad (4.97)$$

When appropriate parameters are chosen, the formulations given in the preceding section can be utilized to simulate the phosphate diffusion/precipitation model presented by van der Zee and Bolt (1991) and van der Zee and van Riemsdijk (1991). Mansell et al. (1977) applied a rate dependent model to describe non-equilibrium behaviour of phosphate sorption in sandy soils. The ANIMO-model describes the rate dependent phosphate sorption to soil constituents by considering three separate sorption sites:

$$\rho_d \frac{\partial X^S}{\partial t} = \rho_d \sum_{i=1}^{3} \frac{\partial X_i^S}{\partial t} = \rho_d \sum_{i=1}^{3} k_{s,i} (K_{F,i} c^{N_i} - X_i^S) \quad (4.98)$$

In a validation study for a wide range of Dutch sandy soils, Schoumans (1995) assessed the parameters in Eq. (4.98), as presented in Table 4.13.

Table 4.13: Parameters describing the rate dependent phosphate sorption for a wide range of Dutch sandy soils (after Schoumans, 1995).

Sorption class i	$k_{s,i}$ (d^{-1})	$K_{F,i}$ (kg.m$_s^{-3}$).(kg.m$_w^{-3}$)$^{-N}$	N_i (-)
1	1.1755	11.87 10^{-6} ρ_d [Al+Fe]	0.5357
2	0.0334	4.667 10^{-6} ρ_d [Al+Fe]	0.1995
3	0.001438	9.711 10^{-6} ρ_d [Al+Fe]	0.2604

[a] dry bulk density in kg.m^{-3} [b] aluminium and iron content in mmol.kg^{-1}

Numerical elaboration of the rate dependent sorption equation requires an expression for the differential quotient. This expression is obtained by taking the value of the time averaged concentration instead of the concentration $c(t)$. Integrating the differential equation between limits t_0 and $t_0+\Delta t$ and dividing by the time increment Δt yields:

$$\frac{1}{\Delta t} \int_{t_0}^{t_0+\Delta t} \frac{dX^S}{dt} dt = \frac{X^S(t_0+\Delta t) - X^S(t_0)}{\Delta t} \tag{4.99}$$

where:

$X^S(t_0)$ = the amount of chemical bounded to non-equilibrium sorption sites at the beginning of the time interval.

In order to calculate the amount of chemical assigned to X^S at the end of the time interval, the function which describes the exchange between solution and time dependent sorption phase $f_s(c)$ is linearized. The function $f_s(c)$ is approximated by $f_s(\bar{c})$ where $\bar{c}$ is the average concentration during the time interval, so:

$$\frac{\partial X^S}{\partial t} = k_s(f_s(\bar{c}) - X^S) \tag{4.100}$$

Although the average concentration is unknown at the start of the computations, $\bar{c}$ can be considered as a constant. The following solution can be derived subject to the condition $X^S = X^S(t_0)$ at the beginning of the time interval.

$$X^S(t_0+\Delta t) = f_s(\bar{c}) + (X^S(t_0) - f_s(\bar{c}))\exp[-k_s\Delta t] \tag{4.101}$$

In order to obtain an expression for Eq. (4.99), Eq. (4.101) can be rewritten as:

$$\frac{X^S(t_0+\Delta t) - X^S(t_0)}{\Delta t} = (f_s(\bar{c}) - X^S(t_0))\frac{(1 - \exp[-k_s\Delta t])}{\Delta t} \tag{4.102}$$

For concentration ranges below the equilibrium level which result from phosphate precipitation processes, the following conservation-transport equation holds:

$$\frac{d\theta(t)c(t)}{dt} + \rho_d \bar{K}_d(c)\frac{dc(t)}{dt} = \frac{1}{\Delta z}[q_{i-1/2}c_{i-1} - q_{i+1/2}c(t) - \sigma_{pl}q_tc(t)]$$
$$- k_1\theta(t)c(t) + k_0 + \rho_d\sum_{i=1}^{3}(X_i^S(t_0) - K_{F,i}\bar{c}^{N_i})\frac{(1 - \exp[-k_{s,i}\Delta t])}{\Delta t} \tag{4.103}$$

Phosphate precipitation takes place when the concentration of the bulk solution tends to exceed a defined equilibrium concentration c_{eq}. This precipitation reaction is modelled according to two mechanisms as:

- an instantaneous reaction: the reaction occurs immediately and is complete when the solute concentration exceeds the equilibrium concentration c_{eq}. The precipitated minerals dissolve immediately when the concentration of the water phase drops below the buffer concentration. When the store of the precipitated minerals has been exhausted, the term $\partial X_p/\partial t$ equals zero;

- a first order reaction: the precipitation and dissolution reaction rate depends proportionally on the difference between the actual phosphate concentration in the soil moisture phase and a defined equilibrium concentration:

$$\rho_d \frac{\partial X^P}{\partial t} = k_{di}(c - c_{eq}) \qquad c < c_{eq} \qquad (4.104)$$

and

$$\rho_d \frac{\partial X^P}{\partial t} = k_{pr}(c - c_{eq}) \qquad c > c_{eq} \qquad (4.105)$$

where:
k_{pr} = precipitation rate constant in $m_w^3 \cdot m_s^{-3} \cdot d^{-1}$
k_{di} = dissolution rate constant in $m_w^3 \cdot m_s^{-3} \cdot d^{-1}$
c_{eq} = equilibrium concentration, dependent on pH and Ca^{++} concentration.

Enfield *et al.* (1981) have studied the phosphate behaviour in calcareous soils assuming first order kinetics to describe the formation and dissolution of phosphate minerals. They present a relation between pH and the rate constant for different phosphate minerals. The general shape of the pH dependent relation is:

$$k_{pr} = 10^{\varsigma_1 pH - \varsigma_2} \qquad (4.106)$$

where:
ς_1 and ς_2 are regression coefficients obtained by curve fitting from experimental data.

In most applications of the ANIMO model for Dutch sandy soils, the parametrization of the model has been restricted to the instantaneous precipitation formulation. For establishing the equilibrium concentration, the following relation between pH and c_{eq} has been utilized:

$$c_{eq} = 0.135 \cdot 3^{5-pH} \approx 10^{-0.447\, pH + 1.516} \qquad (4.107)$$

Non-linear behaviour of phosphate precipitation can be approximated by a number of parallel linear first order reactions. The mechanism of parallel precipitation reactions is explained in Fig. 4.3. Each reaction is assigned to the solubility product of a specific phosphate mineral. The equilibrium concentration of each reaction corresponds to the solubility product of the mineral concerned. The angles between the cutting lines in Fig. 4.3 indicate the rate constants of the related reaction.

When applying the model in combination with the instantaneous precipitation reaction, four situations with respect to the concentration course with time within a time increment are considered. In the second and third situations the time increment has to be split up into two parts. The length of the first part τ is calculated from its specific conditions using the conservation equation. As a result of the reaction, the concentration $c(t)$ as well as the average concentration $\bar{c}$ equal the equilibrium concentration c_{eq}. The conservation equation equals:

$$\rho_d \frac{dX^P(t)}{dt} = \frac{1}{\Delta z}[q_{i-1/2}c_{i-1} - q_{i+1/2}c(t) - \sigma_{pl}q_t c(t)] - k_1 \theta(t)c(t)$$
$$+ k_0 + \rho_d \sum_{i=1}^{3}(X_i^S(t_0) - K_{F,i}\bar{c}^{N_i})\frac{(1 - \exp[-k_{s,i}\Delta t])}{\Delta t} \quad (4.108)$$

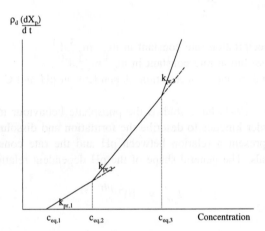

Fig. 4.3: Phosphate precipitation rate as a function of soil moisture concentration, characterized by a number of equilibrium concentration levels c_{eq} and corresponding rate constants k_{pr}.

1) $t_0 \leq$ time $\leq t$: $c < c_{eq}$, $X_p = 0$, $\partial X_p/\partial t = 0$;
2) $t_0 \leq$ time $\leq \tau$: $c < c_{eq}$, $X_p = 0$, $\partial X_p/\partial t = 0$;
   $\tau \leq$ time $\leq t$: $c = c_{eq}$, $X_p \geq 0$, $\partial c/\partial t = 0$;
3) $t_0 \leq$ time $< \tau$: $c = c_{eq}$, $X_p > 0$, $\partial c/\partial t = 0$;
   $\tau \leq$ time $\leq t$: $c \leq c_{eq}$, $X_p \geq 0$, $\partial X_p/\partial t = 0$;
4) $t_0 \leq$ time $< t$: $c = c_{eq}$, $X_p > 0$, $\partial c/\partial t = 0$.

References

Boesten, J.J.T.I. 1986. *Behaviour of herbicides in soil: simulation and experimental assessment* Ph.D. Thesis, Agricultural University: Wageningen, The Netherlands, 262 pp.

Boesten, J.J.T.I. 1994. Simulation of bentazone leaching in sandy loam soil from Mellby (Sweden) with the PESTLA model. *J. Environ.l Science and Health,* A **29** (6), 1231-1253.

Boesten, J.J.T.I., Pas, L.J.T. van der, and Smelt, J.H. 1989. Field test of a mathematical model for non-equilibrium transport of pesticides in soil. *Pesticide Science,* **25**, 187-203.

Bolt, G.H.(ed), 1982. *Soil chemistry B. Physico-chemical models*, Developments in soil science **5B**, Elsevier Scientific Publishing Company: Amsterdam,.The Netherlands.

Bolt, G.H. and Bruggenwert, M.G.M. (ed.), 1978. *Soil Chemistry A. Basic elements.* Developments in soil science 5A. Elsevier scientific publishing company: Amsterdam.

Bolt, G.H. and Bruggenwert, M.G.H. 1978. Composition of the soil. In *Soil Chemistry A. Basic elements.* (ed. G.H. Bolt and M.G.M. Bruggenwert) Developments in soil science 5A. pp. 1-12. Elsevier Scientific Publishing Company: Amsterdam, The Netherlands.

Breemen, N. van, Visser, W.F.J. and Pape, T (ed), 1988. Biochemistry of an oak-woodland ecosystem in the Netherlands affected by acid deposition. *Agricultural Research Reports*, **930**. PUDOC: Wageningen, The Netherlands.

Briggs, G.G. 1973. A simple relationship between adsorption of organic chemicals and their octanol/water partition coefficient. *Proc. 7th British Insecticide & Fungicide Conf.*, **1**, 83 86.

Brill, J. 1996. *The behaviour of dissolved humic substances in neutral soils.* Report DLO Institute Soil Fertility Research: Haren, The Netherlands.

Bronswijk, J.J.B. and Groenenberg, J.E. 1994. SMASS: A simulation model for acid sulphate soils: I. Basic principles. In *Selected papers Saigon Symposium Acid Sulphate Soils*, (ed D. Dent and M.E.F. van Mensvoort), ILRI Publ. **52**, Institute for Land Reclamation and Improvement: Wageningen, The Netherlands.

Bruggenwert, M.G.M. and Kamphorst, A. 1982. Survey of experimental information on cation exchange in soil systems. In *Soil chemistry B. Physico-chemical models*, (ed. G.H. Bolt) Developments in soil science **5B**, pp.141-203. Elsevier Scientific Publishing Company: Amsterdam, The Netherlands.

Driscoll, C.T. Lethinen, M.D. and Sullivan, T.J. 1994. Modelling the acid-base chemistry of organic solutes in Adirondack, New York, lakes. *Water Resources Research*, **30**, 297-306.

Enfield, C.G., Phan, T., Walters, D.M. and Ellis Jr, R. 1981. Kinetic model for phosphate transport and transformation in calcareous soils: I. Kinetics of transformation. *Soil Science. Soc. Am. Journal*, **45**, 1059-1064.

Friesel, P. 1987. Vulnerability of groundwater in relation to subsurface behaviour of organic pollutants. In *Vulnerability of soil and groundwater to pollutants*, (ed. W van Duijvenbooden and H.G. van Waegeningh), Proceedings and Informations, **38**, 729-740. TNO Committee on Hydrological Research: Den Haag, The Netherlands.

Genuchten, M. Th van and Šimůnek, J. 1996 Evaluation of pollutant transport in the unsaturated zone. In *Regional approaches to water pollution in the environment*, (ed. P.E. Rijtema and V. Eliáš) NATO ASI Series 2: Environment, **20**, 139-172. Kluwer Academic Publishers: Dordrecht, The Netherlands.

Groenendijk, P. 1997a. *The calculation of complexation, adsorption, precipitation and dissolution in a soil water system with the geochemical model EPIDIM.* Report **70** SC–DLO: Wageningen, The Netherlands.

Groenendijk, P. 1997b. *Modelling the influence of sorption and precipitation processes on the availability and leaching of chemical substances in soil.* Report **76**. SC–DLO: Wageningen, The Netherlands. (in press)

Helling, C.S., Chesters, G. and Corey, R.B. 1964. Contribution of organic matter and clay to soil cation-exchange capacity as affected by the pH of the saturating solution. *Soil Science Soc. Am. Proc.*, **28**, 517-520.

Henriksen, A. and Seip, H.M. 1980. Strong and weak acids in surface waters in southern Scotland *Water Research*, **14**, 809.

Karickhoff, S.W., Brown, O.S. and Scott, T.H. 1979. Sorption of hydrophobic pollutants on natural sediments. *Water Research*, **13**, 241-248.

Kooper, W.F., Meijden A.M. van der, and Driessen, A.T.P. 1987. Soil quality and the chemical physical equilibrium between soil and groundwater. In *Vulnerability of soil and groundwater to pollutants*, (ed. W van Duijvenbooden and H.G. van Waegeningh), Proc. and Informations, **38**, 1037-1048. TNO Committee on Hydrological Research: Den Haag, The Netherlands.

Lagas, P. and Hammers, W.E. 1987. Vulnerability of various soils to leaching of chlorophenols. In *Vulnerability of soil and groundwater to pollutants,* (ed. W van Duijvenbooden and H.G. van Waegeningh), Proc. and Informations, **38**, 775-785. TNO Committee on Hydrological Research: Den Haag, The Netherlands.

Mansell, R.S., Selim, H.M., Kanchanasut, P., Davidson, J.M. and Fiskell, J.G.A. 1977. Experimental and simulated transport of phosphorus through sandy soils. Water Resources. Research, **13**, 189-194.

Noordwijk, P. van, Willigen, P. de, Ehlert, P.A.I. and Chardon, W.J. 1990. A simple model of P uptake by crops as a possible basis for P fertilizer recommendations. *Neth. J. Agr. Sci.*, **38**, 317-332.

Pignatello, J.J. 1989. Sorption dynamics of organic compounds in soils and sediments. In *Reaction and movement of organic chemicals in soils*, (ed. B.L. Shawney and K. Brown), SSSA Special Publication, **22**, pp 45-80, SSSA and ASA: Madison, WI. .

Reinds, G.J., Bril, J., Vries, W. de, Groenenberg, J.E. and Breeuwsma, A. 1995. *Critical loads and excess loads of cadmium, copper and lead for European forest soils.* Report **96**. SC-DLO: Wageningen, The Netherlands.

Ritsema, C.J. 1993. Estimation of activity coefficients of individual ions in solutions with ionic strengths up to 0.3 mol.dm^{-3}. *J. of Soil Science,* **44**, 307-315.

Schoumans, O.F., 1995. *Beschrijving en validatie van de procesformulering van de abiotische fosfaatreacties in kalkloze zandgronden*, Rapport **381**, SC-DLO: Wageningen, The Netherlands.

Selim, H.M. and Amacher, M.C. 1988. A second-order kinetic approach for modelling solute retention and transport in soils. *Water Resource. Research*, **24**, 2061-2075.

Stumm, W. and Morgan, J.J. 1981. *Aquatic chemistry. An introduction emphasizing chemical equilibria in natural waters.* John Wiley and Sons, INC: New York.

Vermeulen, J. and Hendriks, R.F.A. *Bepaling van afbraaksnelheden van organische stof in laagveen; Ademhalingsmetingen aan ongestoorde veenmonsters in het laboratorium.* Rapport **288**. DLO Winand Staring Centre: Wageningen, The Netherlands.

Vries, W. de and Bakker, D.J. 1996. Manual for calculating critical loads of heavy metals for soils and surface waters; Preliminary guidelines for environmental quality criteria, calculation methods and input data. Report **114** SC-DLO: Wageningen, The Netherlands.

Zee, S.E.A.T.M. van der, 1988. *Transport of reactive contaminants in hetergeneous soil systems.* Ph.D. Thesis, Agricultural University: Wageningen, The Netherlands.

Zee, S.E.A.T.M. van der, and Bolt, G.H. 1991. Deterministic and stochastic modelling of reactive solute transport. *Journal of Contaminant Hydrology,* **7**, 75-93.

Zee, S.E.A.T.M. van der, and Riemsdijk, W.H. van, 1991. Model for the reaction kinetics of phosphate with oxides and soil. In *Interactions at the Soil Colloid - Soil Solution Interface*, (ed G.H. Bolt, M.F. de Boodt and M.H.B. Hayes), NATO ASI Series E Applied Sciences, **190**, 205-239, Kluwer Academic Publishers: Dordrecht, The Netherlands.

CHAPTER 5

BIO-CHEMICAL PROCESSES

Many compounds when present in the soil are subject to conversion reactions of a biological and especially microbiological nature. This is so for specific materials like organic chemicals, but also for organic materials of natural origin, which are added to the soil, e.g. as manure, falling foliage and remaining rests of harvests. Conversion of such organic materials into smaller compounds is strongly enhanced by the activity of micro-organisms. These micro-organisms use the organic materials both for energy supply and assimilation.

A special group of degradation reactions involves decay chains in which solutes are subject to sequential or consecutive degradation reactions. Problems of solute transport involving sequential degradation reactions frequently occur in soil and groundwater systems.

Biological conversion reactions constitute the main processes for disappearance of organic compounds from the soil system. The first step in the decomposition process, with big molecules like cellulose, hemicellulose, pectin and lignin, is the decomposition of these big molecules into smaller compounds. Micro-organisms use exo-enzymes, operating outside the organism, to perform this task. The oxidation starts with the formation of peroxides, primary alcohols and mono-carboxylic acids. Generally spoken, the smaller the compounds formed the better their solubility is. The smaller molecules can be taken up by the micro-organisms to be decomposed finally to CO_2.

In particular for oil components, halogenated hydrocarbons and pesticides, the intermediate products must be considered separately because these products may play an important role in groundwater pollution, with regard to taste and toxic effects. Halogenated hydrocarbons and pesticides may form intermediate products that remain toxic and are decomposed only at a very low rate, so they are likely to be persistent in groundwater

Solute reactions and transformations can be highly dynamic and non-linear in time and space, especially for pesticides and nutrient products. For example, among the nitrogen transformation processes that may need to be considered are mineralization of organic material, nitrification, denitrification and nitrogen uptake by plants. For microbially induced organic and inorganic transformations, the degradation process should also consider the growth and maintenance of soil biomass

When the interest in biodegradation is compound specific, as for instance is the case when dealing with soil pollution through organic fluids as oil, halogenated hydrocarbons or leaching of pesticides, it is not necessary to consider the complete organic matter balance of the soil. In that case the organic matter in the soil is considered as being constant with time. The model TRANSOL can be used when only the transport and fate of the specific compounds and their metabolites will be studied and not the complete soil organic matter balance.

The changes in the complete organic matter balance of the soil as a result of, for instance, application of organic manures, waste water irrigation or decomposing plant material, are accounted for in the model ANIMO, when nutrient leaching is considered.

5.1 The model TRANSOL

The large number of organic fluids hazardous to groundwater known today can be divided into two main groups based on fluid mechanics, i.e. fluids miscible with water and fluids immiscible with water. Of course there are transitional forms between both groups, which shall not, however, be treated here as separate groups. The transport of fluids miscible with water can be handled in the same way as the transport of compounds dissolved in water. The migration of fluids immiscible with water is considerably more complicated.

Although, two phase fluid mechanics is beyond the scope of the present study, attention to the fate of immiscible fluids has to be given in the framework of soil pollution. The transport of the immiscible phase stops as soon as a residual saturation is reached. The residual saturation is defined as being the minimum fluid content in the soil, required for fluid movement or alternatively the threshold below which it is no longer able to move. Its physical meaning can be compared more or less with the definition of field capacity for soil moisture.

Polluted soils in which fluids immiscible with water are found at residual saturation remain penetrable for infiltrating water in the unsaturated zone and for groundwater inflow in the saturated zone without serious limitations. Following the fluctuations of the groundwater table, a significant part of the immiscible fluid at residual saturation can remain immersed in the saturated zone. These fluids are not absolutely insoluble in water. They exhibit a more or less weak solubility, which can be practically meaningless under certain conditions only. The immiscible fluid droplets dispersed in the pores result in a good contact with the infiltrating water into which soluble compounds diffuse. With the very low flow-velocities which usually occur, the saturation concentration is reached after only a short flow-path.

Immiscible fluid pollution in the soil at or below residual saturation can be considered in the same way as the dissolution of precipitates of inorganic compounds. The only difference is that the fluid droplets are also subjected to biodegradation. With the biodegradation of the fluid droplets compounds are formed which have a larger solubility. Oxidation starts with the formation of peroxides, primary alcohols and then mono-carboxylic acids. The final stage of the decomposition is the formation of carbon dioxide, water and biomass.

5.1.1 Kinetic models describing biological decomposition

In calculation methods the loss by chemical and biological degradation is very often described by a first order loss equation. From experiments it is, however, known that organic compounds do not always exhibit first order biodegradation kinetics. Possible other kinetics include zero order, logistic and logarithmic kinetics.

Van Genuchten and Šimůnek (1996) give a review of some of the equations used to represent the kinetics of biodegradation. Most of the expressions have a theoretical basis, but they are commonly used only in an empirical fashion by fitting the equations to observed data. Zero and first order kinetic equations remain the most popular expression for describing the biodegradation of organic compounds, mostly because of the simplicity and ease in which these equations can be incorporated in solute transport models.

One special group of degradation reactions involves decay chains in which solutes are subject to sequential or consecutive decay reactions. Problems of solute transport involving sequential first-order decay reactions frequently occur in soil and groundwater systems. Examples are the migration of simultaneous movement of interacting nitrogen species, organic phosphate transport and the transport of pesticides and their metabolites.

In many cases it is hard to decide from experimental data which rate model is best applicable. Usually the simplified power function will be taken as a point of departure, if a biological function has to be taken into account at all. If this is not the case a first order reaction function is used to describe the decomposition. If during decomposition intermediate products are formed that have to be considered for further modelling, the same functions, but with opposite signs, can be used as production functions. This is the case when, for instance, solid organic compounds are transformed into soluble intermediate organic compounds. For dissolved organic compounds the following expression holds:

$$\frac{dc(t)}{dt} = -k_1 c(t) \qquad (5.1)$$

A similar expression is used when the decomposition of solid organic materials or fluids at residual saturation in soils is described as:

$$\frac{dQ(t)}{dt} = -k_1 Q(t) \qquad (5.2)$$

in which:
k_1 = first order decomposition rate coefficient in d^{-1}
c = concentration of dissolved organic material in $kg.m^{-3}$
Q = the quantity of solid organic material present in $kg.m^{-3}$

5.1.2 Model schematization in TRANSOL

As the quantity of persistent organic compounds in the soil is generally too low to affect the growth of the biodegrading population, a first order model with a compound specific decomposition rate is suitable for this situation. Another reason to choose the first order model is a more practical one as most information on the biodegradation of these synthetic compounds is as first order degradation rate or as half-life time. The special group of degradation reactions involving decay chains in which solutes are subject to sequential or consecutive decay reactions require special attention. If during decomposition intermediate products are formed which require further modelling, the same functions, but with opposite sign, are used as production functions.

The increase in dissolved organic compounds caused by decomposition of solid pesticides and immiscible fluids at residual saturation is in the model described as a zero order production function. The production rate is given as:

$$k_0(m) = fr_m \chi_{m-1}^m Z \rho_d X_{m-1}^P(t_0) \frac{1 - \exp[-k_{1(m-1)}(t - t_0)]}{t - t_0} \qquad (5.3)$$

in which:
fr_m = the formation fraction of the daughter compound (-)
k_0 = the zero order production rate of the daughter compound in $kg.m^{-2}.d^{-1}$
X_{m-1}^P = the quantity of the solid parent compound in $kg.kg^{-1}$
χ_{m-1}^m = ratio of the mole mass of the metabolite over the mole mass of the parent
Z = depth in m

The production of new intermediate compounds from dissolved parent compounds is also calculated as a zero order production using the expression:

$$k_0^d(m) = fr_m \chi_{m-1}^m k_1(m-1) \overline{\theta c}_{m-1} \qquad (5.4)$$

Decomposition of instantaneously adsorbed organic compounds is described as:

$$k_0^S(m) = fr_m \chi_{m-1}^m k_1^S(m-1) \rho_d \overline{X_{m-1}^S} = fr_m \chi_{m-1}^m k_1^S(m-1) K_D \overline{c}_{m-1} \qquad (5.5)$$

Combination of both expressions for zero order production of intermediate compounds originating from dissolved and adsorbed compounds gives:

$$k_0(m) = k_0^d(m) + k_0^S(m) = fr_m \chi_{m-1}^m (k_1(m-1)\theta + k_1^S(m-1)K_D) \overline{c}_{m-1} \qquad (5.6)$$

The formation of intermediate compounds can be considered both simultaneously, i.e. more than one metabolite is formed from the parent material during the first degradation step, as well as sequentially as a successive production. The distribution of metabolite compounds is calculated with the transport and conservation equation, using the metabolite specific first order decomposition rate k_1 as well as the adsorption coefficient $\bar{K}_d$. When in a sequence of decomposition different compounds have to be considered then the calculation procedure must be repeated for each compound distinguished.

5.2 Organic matter models

Organic compounds are generally decomposed relatively quickly as long as enough molecular oxygen is available. Where this is not the case, the degradation takes place with resource to oxygen of nitrates and sulphates. At sufficient oxygen supply these reactions terminate in the production of CO_2 and H_2O. Decomposition of organic matter is mainly an oxidation process. The micro-organisms involved in the decomposition process are mostly aerobic or facultative aerobic, which means that they can live under both aerobic and anaerobic conditions. The anaerobic decomposition process is 100 to 1 000 times smaller than the process under aerobic conditions (Hämäläinen 1991).

Dead plant parts and all other organic materials added to the soil are given as additions of fresh organic materials. Living plant roots excrete soluble organic materials into the soil solution, but also dead root cells during growth. These products become available for decomposition and partake in the carbon and nutrient cycles too. When this material starts to decompose, it is partially oxidized to CO_2 and H_2O and partially transformed into biomass. The ratio between formed soil biomass and total amount of material transformed is given as the assimilation efficiency. At least a part of these transformations take place via the stage of dissolved organic material. The first step in the decomposition process when big molecules, such as cellulose, hemicellulose, pectin and lignin are involved is a splitting up of these molecules into smaller parts. Micro-organisms use exo-enzymes, operating outside the biomass cells, to perform this task. Generally spoken the smaller the compounds formed, the higher their solubility is. These smaller molecules are taken up by the micro-organism cell for further transformation.

Two main approaches are followed in modelling the mineralization of organic materials, both based on the principle of first order kinetics. One approach is to partition substrates into various components each with its own characteristic constant decomposition rate. The other approach is to treat organic materials with a characteristic mineralization rate, which is described either as a concentration dependent function or as a time dependent function.

5.2.1 Multi-component models

Natural organic materials show a reduction in the decomposition rate with time caused by the heterogenetic composition of the organic material considered. Easily decomposing material such as carbohydrates and proteins will be used first, resulting in a relative increase of more resistant compounds of the residual material.

Minderman (1968) showed on the basis of a theoretical organic material composed of sugar, hemicellulose, cellulose, lignin, waxes and resins, assuming first order decomposition with a compound dependent decomposition rate, that the overall decomposition rate decreased with time.

Non-linear behaviour of decomposition can be approximated by a number of parallel linear first order reactions. Each reaction equation is than related to a specific type of organic matter, with its own specific concentration and rate coefficient.

Multi-component models can be used to describe the partitioning of incoming and existing organic materials into several components and the transformation of these predefined organic compounds into CO_2. Other multi-component models describe transformations of one or more organic components into other organic compounds and/or into CO_2. Because of the transformations from one compound into others, this type of model requires many short time intervals, while at the end of an time interval the quantities of the various state variables have to be adjusted for the transformations that took place during the interval.

If during decomposition intermediate products are formed, which are of interest for further modelling, the same functions are used with opposite sign as production functions, taking into account a formation factor.

Jenkinson and Rayner (1977) considered in their model 5 different fractions of organic compounds. Two of them are connected to fresh organic material, i.e. a rapid decomposable plant material (DPM) and a slowly decomposing material (RPM). The soil organic material is divided into three different fractions. The first fraction is considered as biomass, with a low C/N ratio and a relatively high decomposition rate. The second fraction is the active humus compound of physically stabilized organic material with a medium C/N ratio. The third fraction is chemically stabilized organic material with a high value of the C/N ratio and a very low decomposition rate. Part of each fraction of the soil organic material in the Jenkinson and Rayner model returns during decomposition as a result of assimilation into the three soil organic compounds.

The term decomposition has in the Jenkinson and Rayner model a different definition than in the other models discussed before. In this model decomposition is defined as the turnover of each fraction in physically stabilized organic material, chemically stabilized organic material and CO_2. Each turnover process is described as a first order reaction. Both fresh organic material and the turnover products are given as a pulse in their model at the beginning of each time step.

5.2.2 Models with a time dependent decomposition rate factor

For a good description of the decomposition of organic material over a period of years it is very often necessary to consider a time dependent rate factor. Kolenbrander (1969) described the residual organic material as a function of time as:

Bio-chemical Processes

$$Q(t) = Q_0 \exp[-(\varsigma_1 + \frac{\varsigma_2}{t+1})t] \quad (5.7)$$

in which:
- $Q(t)$ = quantity of organic material present at time t in kg.m^{-2}
- Q_0 = quantity of organic material present at time $t = 0$ in kg.m^{-2}
- t = time in years
- ς_1 = material constant, depending on the material under consideration in a^{-1}
- ς_2 = material constant, depending on the material under consideration (-)

Janssen (1986) derived a time dependent decomposition rate of peat and other organic materials using the data published by Kolenbrander (1969, 1974). Janssen introduced a so-called apparent age for different organic materials, plotting the data on log-log paper, resulting in the expression:

$$k_t = k_i(\tau_a + t)^{-1.6} \quad (5.8)$$

in which:
- τ_a = apparent age of the organic material in years
- k_i = initial decomposition rate at time t equalling (1 - τ_a) in a$^{0.6}$

Janssen concluded from the experimental data that k_i was equal to 2.82 a$^{0.6}$. Substitution of Eq. (5.8) into Eq. (5.2) and integration yields for the quantity of organic material present at time t:

$$Q(t) = Q_0 \exp[4.7\left((\tau_a + t)^{-0.6} - \tau_a^{-0.6}\right)] \quad (5.9)$$

The model of Janssen suggests that a certain organic material transfers into another type of material, as related to the decomposition rate. The essence of this model is the use of one general relationship between log(k) and log (t) for all organic materials. In the long run this agrees with the conclusions of Allison (1973), that from different organic materials after a period of decomposition, products are formed with a similar structure. Although the model of Janssen has the advantage of being a one parameter time dependent model it appears from the formulation in Eq. (5.9) that after a long period of decomposition, with t approaching infinity, the remaining quantity of inert organic material depends on the apparent age τ_a. Values of τ_a of 1, 4, 8 and 14 years give a non-degradable fraction of 0.0091, 0.129, 0.259 and 0.381, respectively.

Yang (1996) tested both models with time dependent relative mineralization rates. They gave poor results with observations, mainly because the quantities of organic matter remaining after some years were substantially higher in the data of the experiments cited by Yang than in the experiments originally used for the formulation of the models.

Values for the start decomposition rates and apparent ages of some organic materials following the Janssen model were derived by Hendriks (1991) and they are given in Table 5.1.

Table 5.1: Start decomposition rates k_i and apparent age of different organic materials after Hendriks (1991).

Organic material	Start decomposition rate k_i ($a^{0.6}$)	Apparent age τ_a (a)
foliage	2.87	0.99
straw	1.47	1.50
litter deciduous trees	0.75	2.28
farmyard manure	0.67	2.45
litter spruce trees	0.41	3.34
sawdust	0.36	3.65
peat nr 4	0.31	3.97
peat nr 3	0.18	5.58
peat nr 2	0.10	7.94
peat nr 1	0.044	13.55

5.3 Biomass production and decomposition

Living biomass has a certain lifetime and renews itself at relatively high rates. The biomass concentration in the soil depends on the availability of decomposable material. The total living biomass present depends on the formation rate and the death rate of the biomass. Dead biomass will be transformed in its turn into biomass, CO_2 and H_2O.

Decomposition of organic compounds by biological and in particular by microbiological reactions is related to two processes, i.e. maintenance respiration and cell synthesis. Dead organic soil material and living biomass will generally be considered as soil organic material. The formation of biomass is calculated by the introduction of the assimilation factor fr_{as}. The ratio between respiration and assimilation depends on the conditions and the type of organisms (Hendriks 1991). The ratio between assimilation and assimilation + respiration is indicated as the efficiency of the biosynthesis. The value of the efficiency increases with decreasing C/N ratio, with an average value of 0.33, but a variation between 0 and 0.70 has been reported. It is assumed that the total biomass concentration is proportional to the decomposition rate of organic matter. When the decomposition rate decreases, the death rate of the biomass is larger than the growth rate.

The carbon efficiency determines the ratio between organic carbon incorporated in living biomass and the organic carbon evolving CO_2. This is an important parameter, determining the amount of assimilated carbon and indirectly, through the biomass C/N ratio, the amount of assimilated N. It must be realized that not all the assimilated carbon is available for growth. A part is used as maintenance energy (Babiuk and Paul 1970, Verstraete 1977, Smith et al. 1986, Smith et al. 1989).

The soil microbial biomass amount is the result of a continuous process of growing and dying micro-organisms. Biomass turnover data as collected from literature by Dendooven (1990) are given in Table 5.2. Death rate and growth rate are both dependent on the amount of biomass and indirectly on substrate availability (van Veen et al. 1985). Addition of organic material increases the growth rate, but finally also the death rate.

The apparent microbial biomass death rates range between 0.41 and 5.51 a^{-1} and are a result of competition, antagonistic interaction and predator-prey relations. Independently from the method used for the determination of living biomass, microbial biomass quantities are often determined between 1 and 5 % of the soil organic matter (Jenkinson 1977, Jenkinson and Ladd 1981, Deconinck 1984, Paul 1984, Brookes et al. 1985, Bonde et al. 1988, Andersen and Domsch 1989, Widmer 1989).

Table 5.2: Biomass carbon turnover rates as collected from literature by Dendooven (1990).

Specification	Turnover (a^{-1})	Reference
In soil	5.512	McGill et al. (1981)
In soil	0.402	Jenkinson and Ladd (1981)
Pure culture	350.4	Dendooven (1990)
Biomass protected	1.825	van Veen et al. (1985)
Biomass	0.256	Jenkinson and Rayner (1977)
Biomass, 12 weeks	0.409	(Paul 1984)
Active non-biomass	1.387	(Paul 1984)
Biomass	0.464	(Paul 1984)

Some data of average decay rates, as collected from literature by Dendooven (1990), are presented in Table 5.3. However, on the basis of the published data the conclusion remains uncertain as to whether these data give the real degradation of dead biomass, or the net result of assimilation and respiration.

Dead microbial material is generally considered to contain different fractions, each with a specific decay rate. Fungal material should mineralize less rapidly than bacterial owing to the greater amount of resistant structural material in fungal cells. Cytoplasmic portions decompose more rapidly than cell walls. It does mean that the major part of the dead biomass decomposes with a high decomposition rate. The data in Table 5.3 show a huge variation in mean decomposition rates. Considering the rates presented by Jenkinson and Rayner (1977) and by van Veen et al. (1985) for protected biomass as good mean values for decomposition of biomass in soils gives a decomposition rate of 1.756 a^{-1}.

Table 5.3: Decay rates for soil biomass as given in literature (see Dendooven, 1990).

Comments	Rate (a^{-1})	Reference
Biomass	0.500	Amato and Ladd (1980)
Biomass	1.690	Jenkinson and Rayner (1977)
Glucose amended soil	1.095	Behra and Wagner (1974)
Extrapolated from literature	14.600	Barber and Lynch (1977)
Protected	1.825	van Veen et al. (1985)
Unprotected	273.750	van Veen et al. (1985)

5.4 Schematization in the nutrient model ANIMO

In many cases it is hard to decide from experimental data which rate model is best applicable. In particular when dealing with regional studies with land use systems, in which a continuous supply of fresh organic materials originatinating from different sources is present, complex time dependent or concentration dependent functions are not suitable. Usually the simplified power function will be taken as a point of departure, if a biological function has to be taken into account at all. If this is not the case a first order reaction function is used to describe the decomposition. Biological production and decomposition of material are described in ANIMO with either a first order reaction equation or a zero order one.

For dissolved organic compounds the following expression holds:

$$\frac{dc(t)}{dt} = -k_1 c(t) \quad \rightarrow \quad c(t) = c(t_0) \exp[-k_1(t - t_0)] \quad (5.10)$$

A similar expression is used for the decomposition of solid organic materials:

$$Q(t) = Q(t_0) \exp[-k_1(t - t_0)] \quad (5.11)$$

in which:
k_1 = first order decomposition rate in d^{-1}
c = the concentration of organic material in solution in $kg.m^{-3}$
Q = the quantity of organic material present in $kg.m^{-2}$

The decomposition of stepwise added materials is described by a first order rate equation. The additions are given at the start of a time step, yielding the expression:

$$Q(t) = (Q(t_0) + \Delta Q(t_0))\exp[-k_1(t - t_0)] \tag{5.12}$$

in which:
ΔQ = addition at the beginning of the time step in kg.m^{-2}
t_0 = time at the beginning of the time step in d

If during decomposition intermediate products are formed that have to be considered for further modelling, the same functions, but with opposite signs, are used as production functions. This is the case when for instance solid organic compounds are transformed into soluble intermediate organic compounds, or for the continuous production of dead root material and exudates during the growth of crops. This production is taken into account per time step as a zero order production, resulting in the expression:

$$\frac{dQ}{dt} = k_0 - k_1 Q \rightarrow Q(t) = \frac{k_0}{k_1} + (Q(t_0) - \frac{k_0}{k_1})\exp[-k_1(t - t_0)] \tag{5.13}$$

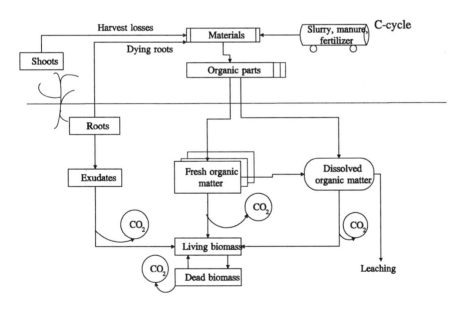

Fig. 5.1: Simplified presentation of the carbon cycle in soil.

Fig. 5.1 gives a simplified organic matter or carbon cycle in the soil, as used in ANIMO. Different kinds of organic materials, partaking in the soil carbon, nitrogen and phosphate cycle, can be introduced in ANIMO. These are:

- Fresh organic materials, as dead plant parts formed by crop residues after harvesting, dead roots and litter of trees, but also the main part of the organic material in animal slurries will be considered as solid organic material. These materials are very often added stepwise, at certain clearly defined points of time. However, sometimes they must be considered as zero order additions during the time period under consideration, as is for instance the case with animal droppings and grazing losses on grassland.
- Root exudates are organic products excreted by living roots and dead root cells discarded by the plant. These products come available continuously when living roots are present. So for plant root material a distinction has been made between the continuously produced part and the part that comes available at harvest.
- Soil organic material is material formed from part of the available fresh material and root exudates. It consists of dead soil organic material and of living biomass.
- Dissolved organic material. As mentioned before, at least a part of the organic material transformations passes the stage of solubilization. For each added material a fraction of the addition can be defined as dissolved organic material.

5.4.1 Decomposition of fresh organic materials

Fresh organic materials can be added to the soil system and can be mixed in the top layers by ploughing. These materials can vary strongly in composition, so each kind of fresh material is considered to exist of a few, say n_f, fractions with a fraction number f_n each with its own specific rate. If different fresh organic materials are present it is convenient to be able to use the same equation for all fractions of the different materials. Therefore, a fixed number of possible decomposition rates of material fractions n_m and possible fractions n_f are defined, of which each kind of fresh organic material contains a few. This is done with the introduction of a matrix $FR(m_n, f_n)$ consisting of numbers smaller than 1, that defines if and for which part a fraction f_n is present in material number m_n. This gives a maximum freedom of material definition when a new material is introduced.

The data given by Kolenbrander (1969) have been used to define these materials in terms of decomposition rates. These data are given in Table 5.4, in combination with the constants used in Eq. (5.7), and the apparent age τ_a as defined by Janssen (1986). A comparison of the three models is given in Fig. 5.2 for different types of peat and in Fig 5.3 for other organic materials. It appears from these figures that the models give similar results. The main advantage of a one parameter model as proposed by Janssen is that only one constant has to be determined on the basis of decomposition experiments of relatively short duration. However, the models given by Kolenbrander and by Janssen have in natural systems with a continuous supply of fresh organic materials, such as litter from trees, dying roots and root exudates, harvest residues and animal manures, the disadvantage that the apparent age of the fresh material and the year of addition to the system have to be considered in the calculations. The fraction distribution of an organic material, following the ANIMO concept, changes with time, but the properties of the composing compounds remain constant.

Table 5.4: Constants ς_1 and ς_2 of Eq. (5.7), the apparent age of organic materials (Hendriks, 1991) and fractions of different organic materials as defined by their decomposition rates, at an average soil temperatrure of 10 °C.

Organic material	constants Eq.(5.7)		Apparent age	Decomposition rate (a^{-1}) and fractions				
	ς_1	ς_2		2.15	0.80	0.25	0.02	0.005
foliage	0.25	2.75	0.99	0.70	0.21	0.06	0.02	0.01
straw	0.15	1.55	1.50	0.52	0.30	0.11	0.05	0.02
litter deciduous trees	0.15	0.95	2.28	0.20	0.36	0.30	0.10	0.03
farmyard manure	0.10	0.90	2.45	0.19	0.37	0.26	0.13	0.05
litter spruce trees	0.14	0.45	3.34	0.07	0.34	0.34	0.17	0.08
saw dust	0.14	0.30	3.65	0.03	0.28	0.38	0.21	0.10
peat 4	0.11	0.32	3.97	0.02	0.23	0.39	0.26	0.10
peat 3	0.07	0.14	5.58	0.00	0.08	0.38	0.37	0.16
peat 2	0.05	0.10	7.94	0.00	0.03	0.29	0.47	0.21
peat 1	0.02	0.06	13.55	0.00	0.00	0.10	0.60	0.30

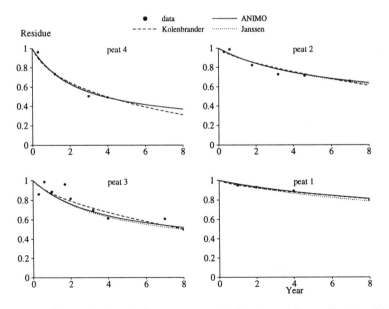

Fig. 5.2: Decomposition of different types of peat calculated with three different models.

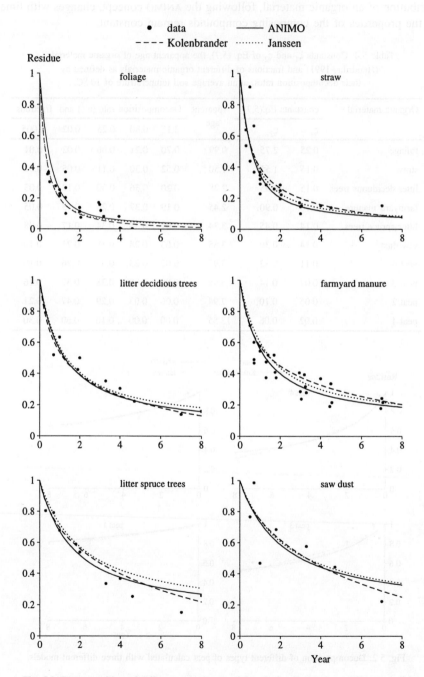

Fig. 5.3: Decomposition of different organic materials calculated with three different models.

To derive a relation between the Janssen concept and the ANIMO approach, a hypothetical organic material has been introduced with a fraction distribution that gives the corresponding distribution for the materials given in Table 5.4 as a function of the apparent age. Table 5.5 presents the fraction distribution of the hypothetical material.

Table 5.5: Fraction distribution of the hypothetical organic material, with an apparent age 0, as the basis for the fraction distribution of organic materials as a function of the apparent age.

fraction number (n)	1	2	3	4	5
decomposition rate $(k_{1,n})$ a^{-1}	2.15	0.80	0.25	0.02	0.005
fraction $(fr_n(0))$	0.9150	0.0678	0.0124	0.0034	0.0014

The fraction distribution of each organic material as a function of the apparent age is calculated using the equation:

$$fr_n(\tau_a) = \frac{fr_n(0)\exp[-k_1(n)\tau_a]}{\sum_{n=1}^{5}(fr_n(0)\exp[-k_1(n)\tau_a])} \quad (5.14)$$

with:
τ_a = the apparent age of an organic material given by Janssen in y.

The relation between the value of the fractions of the various materials and the apparent age after Janssen is presented in Fig. 5.4, showing a well defined relation between the value of the different fractions and the apparent age. So, Eq. (5.14) is used to calculate the fraction distribution in the ANIMO concept of organic materials of which only the apparent age is known. Janssen (1995) gives for different organic manures only the apparent age of these materials. The calculated fraction distribution for the ANIMO model, giving a similar course of the material decomposition, is presented in Table 5.6.

Table 5.6: The apparent age of organic manures (Janssen, 1995) and fractions of different organic materials as defined by their decomposition rates.

Organic material	Apparent age	Decomposition rate (a^{-1}) and fractions				
		2.15	0.80	0.25	0.02	0.005
compost	3.70	0.03	0.27	0.37	0.23	0.10
sewage sludge	1.35	0.58	0.26	0.10	0.04	0.02
cattle slurry	1.35	0.58	0.26	0.10	0.04	0.02
other slurries	1.35	0.58	0.26	0.10	0.04	0.02
poultry manure	1.45	0.54	0.28	0.12	0.04	0.02
pig manure	2.45	0.18	0.38	0.26	0.13	0.05

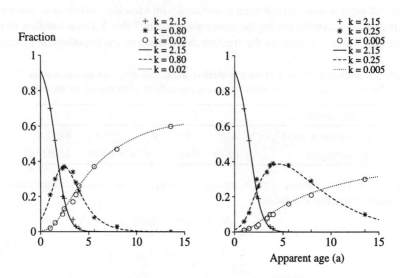

Fig. 5.4: The relation between the apparent age of the organic material after Janssen (1986) and the fraction distribution for the various decomposition rates. The points are the data from Table 5.4; the curves are calculated with Eq. (5.14).

5.4.2 Additions of fresh materials

In ANIMO use has been made of the multi-component model with first order decomposition rates. In the case of instantaneously added material, the total decomposition of fresh organic material, including the part added, is given as:

$$\sum_{f_n=1}^{n_f} Q(t,f_n) = \sum_{f_n=1}^{n_f} \left[(Q(t_0,f_n) + \Delta Q(t_0,f_n)) \exp[-k_1(f_n)(t-t_0)] \right] \quad (5.15)$$

in which:
$Q(t_0)$ = organic material present at the start of a new time step in kg.m^{-2}
$\Delta Q(t_0)$ = material added at the start of a new time step in kg.m^{-2}

5.4.3 Additions of root materials

An extensive discussion of root production and losses of root material is given in Chapter 7. For the addition of root materials in the model ANIMO two different systems are introduced. The root mass present at harvest is treated as an instantaneous addition after harvest. The continuous loss of root material by dying root hairs and the excretion of exudates is treated as a continuous supply depending on the growth rate and the quantity of living roots present at any time. The change of dry matter in living roots is given by:

$$\frac{dP_r}{dt} = P_r^{gr} - k_r P_r \tag{5.16}$$

Considering the value of P_r^{gr} during a time step as a constant gives as expression for the dry matter present in living roots at the end of each time interval:

$$P_r(t) = \frac{1}{k_r}P_r^{gr} + (P_r(t_0) - \frac{1}{k_r}P_r^{gr})\exp[-k_r(t - t_0)] \tag{5.17}$$

in which:
P_r = dry weight of living roots in kg.m^{-2}
P_r^{gr} = gross root production rate during the time interval in kg.m^{-2}.d^{-1}
k_r = rate coefficient for root mass consumption in d^{-1}

The average exudate production rate Q_{ex}^{pr} per time step is calculated as:

$$Q_{ex}^{pr} = \frac{\int_{t=t_0}^{t} k_r P_r dt}{t - t_0} = P_r^{gr} + (P_r(t_0) - \frac{1}{k_r}P_r^{gr})\frac{(1 - \exp[-k_r(t - t_0)])}{t - t_0} \tag{5.18}$$

It is assumed that a linear relation is present between the depth z in the root zone and the root mass distribution, given by the expression:

$$P_r(t,z) = 2P_r(t)(1 - \frac{z}{Z_r}) \tag{5.19}$$

in which :
Z_r = depth of rooting in m
z = depth in root zone, subject to the condition $0 \leq z \leq Z_r$, in m

The exudate production $Q_{ex}^{pr}(z)$, in kg. m^{-2}, is given by a similar expression as:

$$Q_{ex}^{pr}(z) = 2Q_{ex}^{pr}(1 - \frac{z}{Z_r}) \tag{5.20}$$

The decomposition of exudates is described as a first order reaction with a rate constant k_{ex} which yields for the change in the amount of exudates present:

$$\frac{dQ_{ex}}{dt} = Q_{ex}^{pr} - k_{ex}Q_{ex} \tag{5.21}$$

Integrating Eq (5.21) gives the amount of exudates present:

$$Q_{ex}(t) = \frac{Q_{ex}^{pr}}{k_{ex}} - (\frac{Q_{ex}^{pr}}{k_{ex}} - Q_{ex}(t_0))\exp[-k_{ex}(t - t_0)] \tag{5.22}$$

5.4.4 Production and decomposition of dissolved organic material

The production rate of dissolved organic material follows from the calculations in the previous paragraph. It will be clear that the production rate from slowly decomposing fractions will be less than that of faster decomposing fractions of fresh organic materials. Once in solution, all the fractions are assumed to decompose with the same rate. So for the organic material itself it is not necessary to consider these fractions separately in dissolved form. The zero order production rate of dissolved organic material is calculated for each fraction using the expression:

$$k_{0,om}^{dis} = \sum_{n=1}^{n_{max}} \left(fr_{om}^{s}(f_n) \frac{Q(f_n,t_0) - Q(f_n,t)}{t - t_0} \right) = \sum_{n=1}^{n_{max}} \left(fr_{om}^{s}(f_n) Q(f_n,t_0) \frac{1 - \exp[-k_1(f_n)(t - t_0)]}{t - t_0} \right) \tag{5.23}$$

in which:
$k_{0,om}^{dis}$ = zero order production rate coefficient of dissolved material in kg.m^{-2}.d^{-1}
fr_{om}^{s} = fraction of solid organic material that comes into solution (-)

5.4.5 Biomass production and consumption

Normally, the rates of microbiological growth, in solution cultures, vary between 0.3 and 2 h^{-1}. This is 10 to 20 times faster than the net growth rates measured in soils. It does mean that after addition of organic materials to the soil the biomass increases by a factor of 2 to 10 within 24 hours, so biomass concentration in the topsoil is generally high enough to respond immediately within the time step under consideration on changes in the decomposition rate due to addition of fresh organic material or increase in temperature. The change in living biomass is, therefore, given by the expression:

Bio-chemical Processes

$$\frac{dQ_{biom}}{dt} =$$

$$fr_{as}\sum_{f_n=1}^{n_f}\left[\left[(1 - fr^{s}_{om}(f_n))k_1(f_n)Q(f_n,t_0)\exp[-k_1(f_n)(t - t_0)] + k^{dis}_{1,om}\overline{Vc}_{om}\right]\right.$$

$$\left. + fr_{as}k_{ex}\left[\frac{Q^{pr}_{ex}}{k_{ex}} - \left(\frac{Q^{pr}_{ex}}{k_{ex}} - Q_{ex}(t_0)\right)\exp[-k_{ex}(t - t_0)]\right]\right] - k_{bio}Q_{biom} \quad (5.24)$$

in which:
Q_{biom} = concentration of biomass in kg.m^{-2}
V = areic moisture volume in m^3.m^{-2}
fr_{as} = assimilation or biomass formation coefficient (-)
k_{bio} = apparent death rate of biomass $(1-fr_{as}).k_d$ in d^{-1}
k_d = real death rate in d^{-1}

Integration of Eq. (5.24) yields as a general solution:

$$Q_{biom}(t) = Q^{som}_{biom}(t) + Q^{ex}_{biom}(t) + Q^{dis}_{biom}(t) =$$

$$\sum_{f_n=1}^{n_f}[\frac{fr_{as}(1-fr^{s}_{om}(f_n))k_1(f_n)}{k_{bio}-k_1(f_n)}Q(f_n,t_0)(\exp[-k_1(f_n)(t-t_0)]-\exp[-k_{bio}(t-t_0)])]$$

$$+\frac{fr_{as}Q^{pr}_{ex}}{k_{bio}}(1-\exp[-k_{bio}(t-t_0)]) +$$

$$\frac{fr_{as}(Q^{pr}_{ex}-k_{ex}Q_{ex}(t_0))}{k_{bio}-k_{ex}}(\exp[-k_{bio}(t-t_0)]-\exp[-k_{ex}(t-t_0)]) +$$

$$\frac{fr_{as}}{k_{bio}}k^{dis}_{1,om}\overline{Vc}_{om}(1-\exp[-k_{bio}(t-t_0)])+Q_{biom}(t_0)\exp[-k_{bio}(t-t_0)] \quad (5.25)$$

in which:
Q^{som}_{biom} = biomass production related to solid organic materials
Q^{ex}_{biom} = biomass production related to exudates
Q^{dis}_{biom} = biomass production related to dissolved organic matter

If k_{bio} equals k_{ex}, the exudate term in Eq. (5.25) becomes:

$$Q^{ex}_{biom}(t)=\frac{fr_{as}Q^{pr}_{ex}}{k_{bio}}(1-\exp[-k_{bio}t])+fr_{as}(Q^{pr}_{ex}-k_{ex}Q_{ex}(t_o))t\exp[-k_{bio}t] \quad (5.26)$$

If k_{bio} equals $k_1(f_n)$, then the increase in biomass related to the decomposition of solid organic material becomes after integration:

$$Q_{biom}^{som}(t) = \sum_{f_n=1}^{n-1} [\frac{fr_{as}(1-fr_{om}^s(f_n)k_1(f_n))}{k_{bio}-k_1(f_n)}Q_{f_n,t_0}(\exp[-k_1(f_n)(t-t_0)] - \exp[-k_{bio}(t-t_0)])]$$
$$+ fr_{as}(1 - fr_{om}^s(n)k_1(n))Q_{n,t_o}(t - t_0)\exp[-k_{bio}(t - t_0)] +$$
$$\sum_{f_n=n+1}^{n_f} [\frac{fr_{as}(1-fr_{om}^s(f_n)k_1(f_n))}{k_{bio}-k_1(f_n)}Q_{f_n,t_0}(\exp[-k_1(f_n)(t-t_0)] - \exp[-k_{bio}(t-t_0)])] \quad (5.27)$$

Considering steady state conditions with a constant yearly addition of fresh material for each of the different organic materials presented in Table 5.4 gives for the calculation of the steady state fraction distribution $fr^*(f_n)$ of the soil organic material immediately after the addition the expression:

$$fr^*(f_n) = \frac{\frac{\exp[k(f_n)]}{\exp[k(f_n)] - 1}fr(f_n)}{\sum_{f_n=1}^{n_f} \frac{\exp[k(f_n)]}{\exp[k(f_n)] - 1}fr(f_n)} \quad (5.28)$$

The minimum living biomass expressed as a fraction of the total soil organic matter just before the addition follows from:

$$fr_{biom} = \sum_{f_n=1}^{n_f} \left[\frac{fr_{as}k_1(f_n)}{k_{bio} - k_1(f_n)} fr^*(f_n) \frac{\exp[-k_1(f_n)] - \exp[-k_{bio}]}{1 - \exp[-k_{bio}]} \right] \quad (5.29)$$

Table 5.7: Fraction of living biomass under steady state conditions for different fresh materials with an assumed assimilation factor of 0.3 and different biomass death rates.

Fresh material	Fraction of living biomass, as dependent on death rate k_{bio} in a^{-1}				
	3.0	2.5	2.0	1.5	1.0
foliage	0.0152	0.0194	0.0258	0.0367	0.0591
straw	0.0091	0.0098	0.0149	0.0298	0.0329
litter deciduous trees	0.0070	0.0087	0.0111	0.0152	0.0234
farmyard manure	0.0055	0.0069	0.0087	0.0119	0.0183
litter spruce trees	0.0042	0.0052	0.0064	0.0087	0.0133
sawdust	0.0033	0.0041	0.0050	0.0068	0.0103

The minimum relative value fr_{biom} of living biomass has been calculated for steady state conditions of soil organic material originating from different types of fresh organic material and different values of k_{bio}, assuming an assimilation coefficient of 0.3. The results are presented in Table 5.7, indicating that for $k_{bio} \approx 2.0$ realistic values for the minimum fraction of living biomass under steady-state conditions are obtained.

The chemical composition of living and dead biomass is considered to be equal, so it is not necessary for nutrient balances to take into account the assimilation of new biomass from dead biomass material but to consider only the net decomposition of biomass material. The biomass quantity by unlimited growth (Q_{biom}^{nl}) from other organic matter sources is calculated using Eq. (5.24), taking $k_{bio} = 0$, yielding after integration:

$$Q_{biom}^{nl}(t) = Q_{biom}(t_0) +$$

$$fr_{as} \sum_{f_n=1}^{n_f} [(1 - fr_{om}^s(f_n))Q(f_n,t_0)(1 - \exp[-k_1(f_n)(t - t_0)])] + fr_{as} Q_{ex}^{pr} t +$$

$$fr_{as}[\frac{Q_{ex}^{pr}}{k_{ex}} - Q_{ex}(t_0)](1 - \exp[-k_{ex}(t - t_0)]) + fr_{as} k_{1,om}^{dis} \overline{Vc}_{om}(t - t_0) \quad (5.30)$$

The change in dead biomass with time is now given by the expression:

$$\frac{dQ_{dbiom}}{dt} = \frac{Q_{biom}^{nl}(t) - Q_{biom}(t)}{t - t_0} - k_{dbio} Q_{dbiom} = Q_{dbiom}^{pr} - k_{dbio} Q_{dbiom} \quad (5.31)$$

Integration yields the quantity of dead biomass present:

$$Q_{dbiom}(t) = \frac{Q_{dbiom}^{pr}}{k_{dbio}} + \left[Q_{dbiom}(t_0) - \frac{Q_{dbiom}^{pr}}{k_{dbio}}\right] \exp[-k_{dbio}(t - t_0)] \quad (5.32)$$

5.4.6 Nutrient availability in organic materials

Fresh organic materials, dissolved organic compounds and root exudates decompose by first order processes, through which organic N and P is transformed into inorganic nutrients, which become available for uptake by plants and can be transported to groundwater by leaching. The decomposition rate, the nutrient fraction and the organic material fraction of the different materials used in ANIMO are given in Table 5.8.

Root exudates consist of many different materials, such as: carbohydrates, amino acids, organic acids and others. Most research has been done on carbohydrates and amino acids, these being the main components. Barber and Gunn (in Russel 1977) found a ratio of 9 for these components in the exudates of young barley plants. Taking the average C/N ratio of carbohydrates at 100 (% N = 0.68) and that of amino acids at 3.5 (% N = 19) gives 0.025 as the mean N-content of the exudates.

Table 5.8: Decomposition rates, nutrient fractions and organic material fractions of different materials used in ANIMO.

Decomposition rate (a^{-1})	2.15	0.8	0.25	0.02	0.005
N-fraction	var.	0.05	0.035	0.035	0.015
P-fraction	var.	0.006	0.0015	0.0002	0.00015
Material	Organic material fractions				
Compost	0.03	0.27	0.37	0.23	0.10
Sewage sludge	0.58	0.26	0.10	0.04	0.02
Cattle slurry	0.58	0.26	0.10	0.04	0.02
Other slurries	0.58	0.26	0.10	0.04	0.02
Poultry manure	0.54	0.28	0.12	0.04	0.02
Pig manure	0.18	0.38	0.26	0.13	0.05
Cattle manure	0.19	0.37	0.26	0.13	0.05
Foliage plants	0.70	0.21	0.06	0.02	0.01
Litter deciduous trees	0.20	0.36	0.30	0.10	0.04
Litter spruce trees	0.07	0.34	0.34	0.17	0.08
Stubble + roots	0.52	0.30	0.11	0.05	0.02
Eutrophic peat	0.0	0.0	0.20	0.75	0.05
Mesotr. peat	0.0	0.0	0.04	0.52	0.44

Part of the nutrients, however, is immobilized through the biomass-synthesis in the living biomass. Depending on the C/N and C/P ratios of the original organic materials and the same ratios in the biomass a net mineralization or immobilization can be present. The rate of immobilization depends on the growth rate of the biomass and both its C/N and C/P ratio. The C/N and C/P ratio for bacteria and actinomycetes are 5 and 33, respectively (Alexander 1977). For fungi these ratios are respectively 10 and 67. Under optimum soil conditions both groups of biomass are more or less equally present, giving an average value of 8 and 50 for the C/N and C/P ratio, respectively. If the efficiency for bio-synthesis is taken as 0.33, then an equilibrium situation will be present when the decomposing dead organic material has a C/N and C/P ratio of 24 and 150, respectively. If the ratios of the decomposing materials are higher a net immobilization will be present at the expense of available inorganic nutrients.

Decomposition of organic materials is mainly an oxidation process. The biomass involved consists of mainly aerobe or facultatively anaerobe bacteria. The last group can live without oxygen, and these organisms can use nitrate and sulphate as oxygen donors.

5.4.7 Nitrogen processes

Fig. 5.5 gives a schematic presentation of the nitrogen cycle, as used in ANIMO.

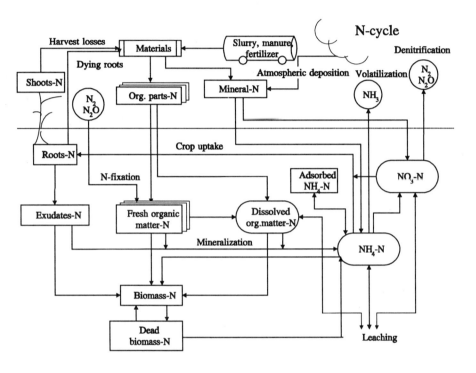

Fig. 5.5: Schematic presentation of the nitrogen cycle in soils.

The processes that control the nitrogen mineralization and immobilization in relation to the processes in the organic matter cycle are:

- The net total mineralization rate of NH_4 during the time step, $k_{0,NH4}$, follows from the formation/decomposition balance of the different organic materials, taking into account their diverse N contents.
- The decomposition (oxidation) rate of NH_4, $k_{1,NH4}$, has a certain basic value, following from literature data, which is reduced for partial anaerobic conditions. The formation rate of NO_3 is derived from the decomposition rate of NH_4.
- The decomposition rate of NO_3 is determined by the part of the total decomposition of organic material that takes place under anaerobic conditions. The decomposition rates under strict anaerobic conditions are much lower and in topsoils complete anaerobic decomposition can mostly be neglected.

Other processes in the nutrient cycles which are not directly parallel with those in the carbon cycle are described briefly.

- The oxidation of ammonium to nitrate is called nitrification. The process takes place in two steps, performed by different groups of micro-organisms:

$$2NH_4^+ + 3O_2 \rightarrow 2NO_2^- + 4H^+ + 2H_2O$$
$$2NO_2^- + O_2 \rightarrow 2NO_3^- \qquad (5.33)$$

Under normal circumstances the second step is much faster than the first, so no accumulation of nitrite (NO_2) will occur.
- Volatilization is the process of gaseous losses of nitrogen from the top soil to the free atmosphere after the formation of ammonia gas (NH_3) from NH_4.
- In addition to animal slurry and inorganic nitrogen fertilizers, another source of nitrogen is available for crops as a result of nitrogen fixation by free living soil bacteria and symbiosis with leguminous plants.

5.4.7.1 Net mineralization

Total net decomposition of organic material per time step is calculated as:

$$J_{om} = \sum_{f_n=1}^{n_f} [Q(f_n,t_0) + \Delta Q(f_n,t_0) - Q(f_n,t)] + V(t_0)c_{om}(t_0) - V(t)c_{om}(t)$$
$$+ \left(k_{0,om}^{dis} + q^i \bar{c}_{om}^i - q^o \bar{c}_{om}^o\right)(t-t_0) + Q_{ex}(t_0) - Q_{ex}(t) + Q_{ex}^{pr}(t-t_0) +$$
$$Q_{biom}(t_0) - Q_{biom}(t) + Q_{dbiom}(t_0) - Q_{dbiom}(t) \qquad (5.34)$$

in which:
c_{om} = concentration of dissolved organic material in $kg.m^{-3}$
c_{om}^i = inflow concentration of dissolved organic material into the layer in $kg.m^{-3}$
$q^o.c_{om}^o$ = outflow concentration of dissolved organic material from the layer in $kg.m^{-3}$
J_{om} = net decomposition of organic material during the time step in $kg.m^{-2}$
V = moisture volume in m
q^i = inflow of water into the layer in $m.d^{-1}$
q^o = outflow of water from the layer in $m.d^{-1}$
$q^i.c_{om}^i$ = total inflow of dissolved organic material into the layer in $kg.m^{-2}.d^{-1}$
$q^o.c_{om}^o$ = total outflow of dissolved organic material from the layer in $kg.m^{-2}.d^{-1}$

Net mineralization is the result of the release of mineral N in the form of NH_4 due to decomposition of organic material and the immobilization of NH_4 as a result of the formation of new biomass. Depending upon the difference between both processes net

Bio-chemical Processes   155

mineralization can be both positive or negative. Net mineralization or ammonification during the time step, $M_N^{\Delta t}$ in kg.m^{-2}, then follows from:

$$M_N^{\Delta t} = \sum_{f_n=1}^{n_f} [fr_N(f_n)[Q(f_n,t_0) + \Delta Q(f_n,t_0) - Q(f_n,t)]] +$$

$$fr_{N,om}^{dis}[V(t_0)c_{om}(t_0) - V(t)c_{om}(t) + (k_{0,om}^{dis} + q^i \overline{c}_{om}^i - q^o \overline{c}_{om}^o)(t - t_0)]$$

$$+ fr_N^{ex}[Q_{ex}(t_0) - Q_{ex}(t) + (Q_{ex}^{pr} + q^i c_{ex}^i - q^o c_{ex}^o)(t - t_0)]$$

$$+ fr_N^{biom}[Q_{biom}(t_0) - Q_{biom}(t) + Q_{dbiom}(t_0) - Q_{dbiom}(t)] \quad (5.35)$$

in which:
fr_N = nitrogen fraction in organic material
$fr_{N,om}^{dis}$ = nitrogen fraction of dissolved organic material
fr_N^{ex} = nitrogen fraction in exudates
fr_N^{biom} = nitrogen fraction in biomass

If $M_N^{\Delta t} < 0$, both NO$_3$ and NH$_4$ will be immobilized. In that case, for the maximum immobilization:

$$M_N^{\Delta t*} = M_N^{\Delta t} + V(t_0)c_{NO_3}(t_0) + [\rho_d K_{d,NH_4} + V(t_0)]c_{NH_4}(t_0)$$

$$+ [q^i \overline{c}_{NO_3}^i - q^o \overline{c}_{NO_3}^o + q^i \overline{c}_{NH_4}^i - q^o \overline{c}_{NH_4}^o](t - t_0) \quad (5.36)$$

with
c_{NO3} = NO$_3$-N concentration in kg.m^{-3}
c_{NH4} = NH$_4$-N concentration in kg.m^{-3}

If $M_N^{\Delta t*} \geq 0$, then the immobilization is controlled by the decomposition rate of the organic materials present, else if $M_N^{\Delta t*} < 0$, the decomposition rate of all organic materials will be reduced by multiplication with a reduction factor, due to lack of nitrate. For the reduction factor:

$$\xi_N = \text{MIN}[1 \; ; \; 1 - \frac{M_N^{\Delta t*}}{fr_N^{biom}[Q_{biom}(t_0) - Q_{biom}(t) + Q_{dbiom}(t_0) - Q_{dbiom}(t)]}] \quad (5.37)$$

The average rate of NH$_4$ production during the time step is:

$$k_{0,NH_4} = \frac{1}{t - t_0} \text{MAX}[0 \; ; \; M_N^{\Delta t}] \quad (5.38)$$

5.4.7.2 Nitrification

If the decomposition (oxidation) rate for NH_4 under aerobic condition equals $k_{1,NH4}$ then the decomposition rate for NH_4 under partial anaerobic conditions is given as:

$$k^*_{1,NH_4} = f_{ae} k_{1,NH_4} \tag{5.39}$$

in which:
f_{ae} = aerated fraction of the layer during the time step considered

$k^*_{1,NH4}$ is negative because all rate constants in the transport and conservation equation are expressed in terms of production. The calculation of f_{ae} is given in section 6.2 dealing with aeration and mineralization.

The production rate of NO_3 equals the average decomposition rate of NH_4 and is given by the expression:

$$k_{0,NO_3} = k^*_{1,NH_4} \bar{c}_{NH_4} \bar{\theta} = f_{ae} k_{1,NH_4} \bar{c}_{NH_4} \bar{\theta} \tag{5.40}$$

in which:
$\bar{c}_{NH4}$ = average concentration of NH_4-N in solution in kg.m^{-3}
$\bar{\theta}$ = average moisture fraction

The value of $\bar{c}_{NH4}$ follows from the transport and conservation equation when applied to NH_4.

5.4.7.3 Denitrification

The decomposition rate of NO_3 is determined by that part of the total decomposition of organic material that takes place under anaerobic conditions. Under these conditions, the oxygen demand is replaced by a nitrate demand.

Considering the anaerobic oxidation of glucose by biomass as the general form in which denitrification takes place then the following expression holds:

$$5C_6H_{12}O_6 + 24NO_3^- \rightarrow 30CO_2 + 18H_2O + 12N_2 + 24OH^- \tag{5.41}$$

From Eq. (5.41) it follows that for the oxidation of each mole of carbon 24/30 mole of NO_3 is needed. If the carbon content of an organic material is considered as a constant fraction fr_C, it follows that the maximum NO_3-N demand for denitrification, S_{NO3}, can be expressed as:

$$S_{NO_3} = 0.93 fr_C (1 - f_{ae}) J_{om} \quad (5.42)$$

The minimum production of nitrate is given as a zero order rate constant by combining Eq. (5.40) and Eq. (5.42), yielding:

$$k_{0,NO_3} = f_{ae} k_{nh} \overline{c}_{NH_4} \theta - 0.93 fr_C (1 - f_{ae}) J_{om} \quad (5.43)$$

$k_{0,NO3}$ can be positive as well as negative.

At low NO_3 concentrations the availability of nitrate can be the limiting factor for denitrification. So, denitrification has to be described as a function of both decomposition of organic material and nitrate availability. For the description of the nitrate reduction a first order rate constant for the nitrate consumption has been introduced in the transport and conservation equation.

Each time step the transport and conservation equation has to be applied for both organic materials and nitrate concentration, yielding two different values for the nitrate concentration and the anaerobic decomposition of organic materials. In the case that the nitrate concentration limits the decomposition of organic materials under anaerobic conditions the value of J_{om} as well as the quantities of organic materials present at the end of the time step have to be recalculated. For the anaerobic part of the decomposition, J_{om}^{an}, under nitrate limiting conditions the following expression holds:

$$J_{om}^{an} = \frac{k_{1,NO_3} \overline{c}_{NO_3} \theta}{0.93 fr_C} \quad (5.44)$$

in which:
$k_{1,NO3}$ = first order reduction rate of NO_3 in d^{-1}
$\overline{c}_{NO3}$ = the mean NO_3 concentration during the time interval in $kg.m^{-3}$
θ = the mean soil moisture content during the time interval in $m^3.m^{-3}$

The real value for anaerobic decomposition of organic materials, J_{om}^{an*}, is given by the condition:

$$J_{om}^{an*} = MIN[(1 - f_{ae}) J_{om} ; J_{om}^{an}] \quad (5.45)$$

The correction factor for the decomposition rate of organic materials, ξ_{om}, is under these conditions:

$$\xi_{om} = \frac{f_{ae} J_{om} + J_{om}^{an*}}{J_{om}} \quad (5.46)$$

5.4.7.4 Nitrogen fixation

In addition to animal slurry and inorganic nitrogen fertilizers another source of nitrogen for crops is available due to nitrogen fixation by free living soil bacteria and symbiosis with leguminous plants.

Rusch (1974) presents data of nitrogen fixation by different leguminosae in the order of up to 300 kg $N.ha^{-1}.a^{-1}$ and this author refers to Russian experience as presented by Fedorow (1960) that should indicate a nitrogen fixation by free living soil bacteria of 100 $kg.ha^{-1}.a^{-1}$. Anonymous (1972) indicates a maximum N-fixation by free living Azotobacter under optimum conditions of 35 kg $N.ha^{-1}.a^{-1}$. Bell and Nutmann (1971) present experimental data concerning the nitrogen fixation of lucerne at a level of 300 kg $N.ha^{-1}.a^{-1}$. Postgate (1974) presents similar values for lucerne and for clover and lupine, 250 and 150 $kg.ha^{-1}.a^{-1}$, respectively. So leguminous plants are potentially able to realize an N-fixation of 300 $kg.ha^{-1}$ within 3 to 4 month duration of their growing season.

Comm. Onderz. Biol. Landb. (1976) indicates that for the agricultural practice under the climatological conditions in the Netherlands the N-fixation will be from 100 to 200 kg $N.ha^{-1}.a^{-1}$ per leguminous crop. Rijtema (1978) assumed for non-fertilized grassland a maximum N-fixation rate of 4.5 $kg.ha^{-1}.a^{-1}$ per percent clover present in the botanical composition of grassland. Breazu et al. (1994) give for Rumanian conditions an average yearly N-fixation of 3.8 $kg.ha^{-1}$ per percent clover at N-fertilization levels from 0-150 $kg.ha^{-1}$. This results in a maximum N-fixation of 4.4 $kg.ha^{-1}$ per percent clover after correction for a N-fertilization level of 75 $kg.ha^{-1}$.

The nitrogen fixation is described as a function of the fraction of leguminosae present in the botanical composition, temperature, pH and N-fertilization. Effects of other environmental factors on nitrogen fixation are not considered.

- In the schematization of ANIMO, the N-fixation is, above a minimum temperature, linearly related to soil temperature. The description of soil temperature at a certain depth below the soil surface and at a certain day of the year can be given by a harmonic function (see section 6.3.2), so the following expression holds:

$$S_N(z,t) = \varsigma_N(z)(T(z,t) - T_{min}) = \varsigma_N(n)\left[(T - T_{min}) + T_{am}\exp[-\frac{z}{D_m}]\cos(\omega t + \phi - \frac{z}{D_m})\right] \quad (5.47)$$

where:
S_N = nitrogen fixation rate in $k.m^{-3}.d^{-1}$
ς_N = yearly average nitrogen fixation rate in $k.m^{-3}.°C^{-1}.d^{-1}$
T = yearly average temperature at the soil surface in °C
T_{am} = yearly temperature amplitude in °C
T_{min} = minimum temperature for N-fixation in °C
D_m = damping depth, set at 2.1 m

t	= day-number of the year in d
z	= depth below soil surface in m
ω	= frequency of temperature wave, equalling 0.01721 rad.d
ϕ	= phase shift, equalling -3.7721 rad

- The pH-dependent reduction factor for N-fixation is given by the same function, described in section 6.5, that is used for the decomposition of organic materials.

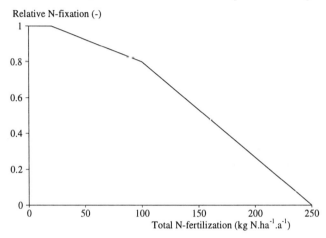

Fig. 5.6: Fertilization dependent activity coefficient for N-fixation after Janssen (1992).

- N-fixation decreases with increasing application of fertilizers. The reduction function for N-fixation in relation to fertilization is based on data presented by Janssen (1992). The relation is presented in Fig. 5.6. It is assumed that no fixation occurs at a N-fertilization level above 250 kg.ha^{-1}.
- Insufficient information is available concerning the distribution of the N-fixation in the rootzone of crops. Moreover, it is assumed that this nitrogen source becomes available for immediate uptake by the plants, so only data of total N-fixation over the whole rootzone will be considered. Taking the temperature in the top layer of the rootzone as the reference temperature for the N-fixation in the complete rootzone gives:

$$F_N(t) = \overline{\varsigma_N} \Delta z_r [(T - T_{min}) + 0.975 T_{am} \cos(0.01721 t - 3.7959)] \qquad (5.48)$$

in which:

F_N	= nitrogen fixation rate in kg.m^{-2}.d^{-1}
Δz_r	= layer thickness in root zone in m
$\overline{\varsigma_N}$	= average nitrogen fixation rate in kg.m^{-3}.$^{\circ}$C^{-1}.d^{-1}

Integration of Eq. (5.48) between t_0 and t gives:

$$F_N^{\Delta t} = \overline{\varsigma_N}\Delta z_r[(T - T_{\min})(t - t_0) + \frac{0.975 T_{am}}{0.01721}[\sin(0.01721 t - 3.7959) - \sin(0.01721 t_o - 3.7959)]] \quad (5.49)$$

with:
$F_N^{\Delta t}$ = total N-fixation in kg.m^{-2}

5.4.7.7 Volatilization

One important process in the nitrogen cycle has only been incorporated very roughly in the model. Volatilization of NH_3, formed from NH_4, is a process strongly dependent on short term weather conditions. When NH_4-containing material is added to the top of the soil and the weather is dry and warm, a major part of it gets lost through volatilization. If, however, the material is incorporated in the soil or precipitation falls immediately after application, the major part will be saved. Because these weather conditions cannot be foreseen, modelling of volatilization is user defined.

Roelsma (1997) gives, on the basis of a literature study, emission data for different land spreading techniques of animal slurry. These data are presented in Table 5.9.

Table 5.9: Emission fractions for ammonium volatilization of different land spreading techniques for animal slurry, after Roelsma (1997).

Application technique	Emission fraction	Application technique	Emission fraction
Grassland		Arable land	
Surface application	0.50	Surface application	0.50
Sprinkling irrigation	0.20	Ploughing within 36 hours	0.36
Shuffling feet	0.125	Ploughing within 12 hours	0.20
Sod slurry injection	0.10	Ploughing within 1 hour	0.075
Deep slurry injection	0.05	Ploughing in one work-load	0.05
		Deep slurry injection	0.05

5.4.8 Phosphorus processes

Less studies have been made of the phosphorus mineralization as a result of decomposition of organic materials and the formation of biomass than has been done for the nitrogen mineralization. Generally, it is assumed that the mineralization of phosphorus is based on the same principles as the nitrogen mineralization. There are indications, however, that the phosphorus mineralization and immobilization, compared with the nitrogen mineralization, is less dependent on the carbon. Phosphorus plays another role by the functioning of biomass then nitrogen. The distribution of phosphorus over the fractions of organic materials can differ considerably from the nitrogen distribution in these materials. Phosphorus mineralization is an exo-enzymatic process, outside the biomass, transforming organic phosphates through hydrolysis into anorganic phosphates.

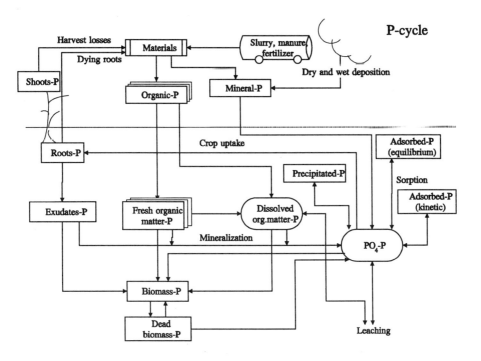

Fig. 5.7: Schematic presentation of the phosphate cycle in soils.

A presentation of the phosphate cycle is given in Fig. 5.7. Adsorption and precipitation of phosphorus play an important role in phosphorus availability.

According to Chauhan *et al.* (1979, 1981), as cited by Tate (1984), is the growth of biomass in soils less dependent on the availability and mineralization of organic phosphates, due to the availability of large quantities of anorganic phosphates. These authors also showed that the C/P-ratio of the biomass decreased when a large storage of anorganic phosphates was available. This indicates that when immobilization is larger than mineralization anorganic phosphates are used to bridge the gap. The net mineralization per time step follows from:

$$M_P^{\Delta t} = \sum_{f_n=1}^{n_f} \left[fr_P(f_n)[Q(f_n,t_0) + \Delta Q(f_n,t_0) - Q(f_n,t)]\right] +$$

$$fr_{P,om}^{dis}[V(t_0)c_{om}(t_0) - V(t)c_{om}(t) + (k_{0,om}^{dis} + q\,{}^{i}\overline{c}_{om}^{\,i} - q\,{}^{o}\overline{c}_{om}^{\,o})(t-t_0)]$$

$$+ fr_P^{ex}[Q_{ex}(t_0) - Q_{ex}(t) + Q_{ex}^{pr}(t-t_0)]$$

$$+ fr_P^{biom}[Q_{biom}(t_0) - Q_{biom}(t) + Q_{dbiom}(t_0) - Q_{dbiom}(t)] \quad (5.50)$$

in which:
$fr_P(f_n)$ = phosphorus fraction in organic material fraction f_n
fr_P^{ex} = phosphorus fraction in exudates
fr_P^{biom} = phosphorus fraction in biomass
$M_P^{\Delta t}$ = net mineralization of phosphorus in kg.m^{-2}

If $M_P^{\Delta t} < 0$, dissolved and adsorbed P will be immobilized. In that case:

$$M_P^{\Delta t*} = M_P^{\Delta t} + (\rho_d K_{d,P} + \theta(t_0))c_P(t_0) + [q\,{}^{i}\overline{c}_P^{\,i} - q\,{}^{o}\overline{c}_P^{\,o}](t-t_0) \quad (5.51)$$

If $M_P^{\Delta t*} \geq 0$ than the immobilization is controlled by the decomposition rate of the organic materials present, else if $M_P^{\Delta t*} < 0$, the decomposition rate of all organic materials will be reduced, due to lack of phosphorus. The reduction factor is given by:

$$\xi_P = \text{MIN}[1\,;\,1 - \frac{M_P^*}{fr_P^{biom}[Q_{biom}(t_0) - Q_{biom}(t) + Q_{dbiom}(t_0) - Q_{dbiom}(t)]}] \quad (5.52)$$

5.5 Additions to the soil and ploughing

In TRANSOL as well as in ANIMO, additions to the soil can be introduced at the start of any time step. The properties of the added materials and the way of addition in terms of surface application or depth of incorporation of the material have to be defined in the input. The materials are equally divided over the layers of incorporation. Production of above ground crop residues during the growing season have been left out of consideration. If relevant, these crop residues can be introduced as additions of fresh organic material to the top layer at any time step, to be defined by the model user. Only when plant roots are added, that is when the crop is harvested, is the distribution of the root material taken into account. Ploughing of the topsoil can be introduced in both models at the start of each time step. Depth of ploughing or the number of layers involved in ploughing have to be given as input. The simulation of ploughing in both models means the uniform redistribution of all species present in the participating layers over the layers for which ploughing is desired.

References

Alexander, M.A., 1977. *Introduction to soil microbiology*. 2nd ed. J. Wiley and Sons. New York, U.S.A.
Allison, F.E., 1973. Soil organic matter and its role in crop production. *Developments in Soil Science*, **3**. 637 pp. Elsevier Scientific Publishing Company: Amsterdam, The Netherlands.
Amato, M. and Ladd, J.N. 1980. Studies of nitrogen immobilization and mineralization in calcareous soils V. Formation and distribution of isotope-labelled biomass during decomposition of ^{14}C- and ^{15}N-labelled plant material. *Soil Biol. and Biochem.*, **12**, 405-411.
Anderson, D.W. and Domsch, K.H. 1989. Ratios of microbial biomass carbon to total organic carbon in arable soils. *Soil Biology and Biochemistry*, **21**, 471-479.
Anonymous, 1973. *Grundsätze für den Anbau von Obst aus naturgemassen, biologischen Anbau*. Anbaurichtlinien Fassung vom 15-1-1973. Arbeitsgemeinschaft für naturgemässen Qualitätsanbau von Obst und Gemüse.
Babiuk, L.A. and Paul, E.A. 1970 The use of fluorescein isothiocyanate in the determination of the bacterial biomass of grassland soil. *Can. J. Microbiology*, **16**, 57-62.
Barber, D.A. and Lynch, J.M. 1977. Microbial growth in the rhizosphere. *Soil Biol. and Biochem.*, **9**, 305-308.
Behra, B. and Wagner, G.H. 1974. Microbial growth rate in glucose amended soil. *Soil Sci. Am. Proc.*, **38**, 591-594.
Bell, F. and Nutmann, P.S. 1971. Experiments on nitrogen fixation by modulated lucerne. *Plant and Soil*, special volume: 231-264.
Bonde, T.A., Schnürer, J. and Rosswall, T. 1988. Microbial biomass as a fraction of potentially mineralizable nitrogen in soils from long-term field experiments. *Soil Biol. and Biochem.*, **20**, 447-452.
Breazu, I., Oprea, G., Chiper, C. and Panga, G.H. 1994. Contribution to the N supply in grassland of ten white clover varieties. In: *Proceedings of the 15th General Meeting of the European Grassland Federation, June 6-9, 1994*. (ed. L. 't Mannetje en J. Frame), Wageningen Pers, Wageningen, The Netherlands.
Brookes, P.C., Landman, A., Pruden, G. and Jenkinson, D.S. 1985. Chloroform fumigation and the release of soil nitrogen: A rapid direct extraction method to measure microbial biomass nitrogen in soil. *Soil Biology and Biochemistry*, **17**, 837-840.
Commissie Onderzoek Biologische Landbouw, 1976. *Alternatieve lanbouwmethoden: Inventarisatie, evaluatie en aanbevelingen voor onderzoek*. PUDOC: Wageningen, The Netherlands.
Deconinck, K., Vlassak, K. and Verstraeten, L.M.J. 1984. *The ATP content of Belgian soils*. DBG-Tagung, pp. 1-10, Trier, Germany
Dendooven, L., 1990. Nitrogen mineralization and nitrogen cycling. Thesis **191**. Faculty of Agricultural Sciences, University Leuven: Leuven, Belgium.
Fedorow, M.W. 1960. *Biologische Bindung des atmosphärischen Stickstoffs*. VEB Gustav Fischer Verlag, Jena.
Genuchten, M. Th van and Šimůnek, J. 1996 Evaluation of pollutant transport in the unsaturated zone. In *Regional approaches to water pollution in the environment*, (ed. P.E. Rijtema and V. Eliáš) NATO ASI Series 2: Environment, **20**, pp 139-172. Kluwer Academic Publishers: Dordrecht, The Netherlands.

Hämäläinen, M., 1991 *Principal variations in the chemical composition of peat.* Swedish Univ. Agric. Sci. Dept. Chem.: Uppsala, Sweden.
Hendriks, R.F.A., 1991. *Afbraak en mineralisatie van veen; Literatuur onderzoek.* Rapport **199**. SC–DLO: Wageningen, The Netherlands.
Janssen, B.H., 1984. A simple method for calculating decomposition and accumulation of 'young' soil organic matter. *Plant and Soil*, **76**, 297-304.
Janssen, B.H. 1986. Een één-parametermodel voor de berekening van de decompositie van organisch materiaal. *Vakblad voor Biologen*, **76**, 297-304.
Janssen, B.H., 1992. *Nutrients in Soil Plant Relationships.* College dictaat 06172205. Landbouwuniversiteit: Wageningen, The Netherlands.
Janssen, J. 1995. *Organische stof in de akker en tuinbouw; Een nieuwe benadering of oude wijn in een nieuwe fles.* Afd. Milieu Kwaliteit en Techniek, Informatie en Kennis Centrum Landbouw: Ede, The Netherlands.
Jenkinson, D.S. 1977. Studies on the decomposition of plant material in soil V. The effects of plant cover and soil type on the loss of carbon from ^{14}C-labelled rye grass decomposing under field conditions. *J. Soil Science*, **28**, 424-494.
Jenkinson, D.S. and Ladd, J.N. 1981. Microbial biomass in soil: Measurement and turnover. In Soil Biochemistry, (ed. E.A. Paul and J.N. Ladd), **5**, 415-471. Dekker, New York.
Jenkinson, D.S. and Rayner, J.H. 1977. The turnover of soil organic matter in some of the Rothamsted classical experiments. *Soil Science*, **123**, 298-305.
Kolenbrander, G.J., 1969. *De bepaling van de waarde van verschillende soorten organische stof ten aanzien van hun effect op het humusgehalte bij bouwland.* C 6988 Institute for Soil Fertility: Haren, The Netherlands.
Kolenbrander, G.J., 1974. Efficiency of organic manure in increasing soil organic matter content. *Trans. 10th Int. Congr. Soil Sci. vol.2.* Moscow, USSR.
McGill, W.B., Hunt, H.W., Woodmansee, R.G. and Reuss, J.O. 1981. Phoenix, a model of the dynamics of carbon and nitrogen in grassland soils. In *Terrestrial nitrogen cycles.* Ecol. Bull. (ed. F.E. Clark, and T. Rosswall), **33**: 49-115. Stockholm
Minderman, G., 1968 Addition, decomposition and accumulation of organic matter in forests. *J. Ecology*, **56**.
Paul, E.A., 1984. Dynamics of organic matter in soils. *Plant and Soil*, **76**. 275-285.
Postgate, J.R., 1974 New advances and future potential in biological nitrogen fixation. *J. of Applied Bacteriology*, **37**,185-202.
Rijtema, P.E., 1978. *Een benadering voor de stikstofemissie uit het graslandbedrijf.* Nota **982**. Institute Land and Water Management Research: Wageningen, The Netherlands.
Roelsma, J. 1997. *Vervluchtiging van ammoniak uit dierlijke mest; Literatuur onderzoek ten behoeve van de watersysteemverkenningen.* Rapport **442**. SC–DLO: Wageningen, The Netherlands.
Rusch, H.P., 1974. Bodenfruchtbarkeit. *Eine Studie biologischen Denkens.* Karl F. Haug Verlag. 2e Druck.
Russel, R.S. 1977. *Plant root systems.* McGraw-Hill: London.
Smith, J.L. 1989. Sensitivity analysis of critical parameters in microbial maintenance-energy models. Mineralization rate constants. *Biol. Fertil. Soils*, **8**, 7-12.
Smith, J.L., McNeal, B.L., Cheng, H.H., and Campbell, G.S. 1986. Calculation of microbial maintenance rates and net nitrogen mineralization in soil at steady state. *Soil Science Soc. Am. Proc.*, **50**, 332-338.

Tate, K.R. 1984. The biological transformation of P in soil. *Plant and Soil*, **76**, 245-256.
Veen, J.A. van, Ladd, J.N. and Amato, M. 1985. Turnover of carbon and nitrogen through the microbial biomass in a sandy loam and a clay soil incubated with [^{14}C(U)]-glucose and [^{15}N](NH$_4$)$_2$SO$_4$ under different moisture regimes. *Soil Biology and Biochemistry*, **17**, 747-756.
Verstraete, W. 1977. *Fundamentele studie van de opbouw en omzettingsprocessen in microbiële gemeenschappen*. Aggregaatswerk, 444 pp. Faculty Agric. Sciences: Gent, Belgium
Widmer, P., Brookes, P.C. and Parry, L.C. 1989. Microbial biomass nitrogen measurements in soils containing large amounts of inorganic nitrogen. *Soil Biology and Biochemistry*, **21**, 865-867.
Yang, H.S. 1996. *Modelling organic matter mineralization and exploring options for organic matter management in arable farming in northern China*. Ph.D Thesis, Agricultural University: Wageningen, The Netherlands.

CHAPTER 6

ENVIRONMENTAL INFLUENCES ON PROCESSES

The main environmental influences on the transformation processes are soil moisture, oxygen, temperature, pH and clay content of the soil.

- Soil moisture and oxygen are strongly related in soils and they are therefore generally treated together as a function of soil moisture content only. In literature the relation between the transformation rate and the soil moisture content is very often given as an optimum curve. Micro-organisms need moisture to perform their biological functions. Below wilting point these biological functions are already disturbed. At low moisture suctions reaction rates may slow down by dilution effects or, if oxygen is needed, by insufficient availability of oxygen.
- The transformation rate of chemical processes increases with temperature. In biochemical processes, which are often performed with enzymes, an optimum temperature range is present below and above which the process rate decreases. The reaction rates at different temperatures are expressed relative to the maximum rate found, or to rates found at a certain average temperature, so a correction factor for other temperatures on the reaction rate can be applied.
- The influence of pH is dependent on the type of reaction and on the preference of the micro-organisms involved. Measurements performed by several authors indicate a broad optimum pH range. The pH of most agricultural soils falls within this range. Special attention to pH, however, must be given, when dealing with nature reserves and with acidification in forest soils.
- Decomposition of organic compounds in clay containing soils can be reduced due to adsorption of exo-enzymes by the clay present in the soil system.

6.1 Soil moisture

The effect of soil moisture has to be divided into two different aspects:

- soil moisture suction to describe the reduction of processes related to stress conditions in dry soils.
- soil aeration, expressed in oxygen concentration, under influence of oxygen diffusion and oxygen consumption.

Generally, the effect of dry conditions in the soil are described in terms of relative moisture fractions related to field capacity. These relations are soil type dependent as the relative moisture content does not give a real indication of the energy status by which the water is bounded by the soil matrix. The aspect of soil moisture under different conditions has to be translated preferably into the physical soil properties really governing the biological processes, which are the soil moisture potential in dry soils and the oxygen supply under wet soil conditions. It can be expected that these reactions, related to soil moisture potential and aeration, are not or only weakly dependent on the type of soil.

6.1.1 Schematization in TRANSOL

The transformation rate is reduced under dry conditions. The influence of the moisture content is described with a relation given by Boesten and Van der Linden (1991) for the transformation of pesticides as:

$$f_\theta = \text{MIN}\left[1 \; ; \; \left(\frac{\theta}{\theta_{fc}}\right)^B\right] \qquad (6.1)$$

where:
f_θ = correction factor for dry conditions
θ = actual soil moisture volume fraction (-)
θ = soil moisture volume fraction at field capacity (-)
B = exponent depending on soil type (-)

Boesten (1986) presents, on the basis of a literature study, a large number of values for the exponent B. The values of B were highly variable depending on type of pesticide, soil type and experimental conditions. Boesten gives, on the basis of the frequency distribution of the exponent B found in literature, an average value of 0.65. Eq.(6.1) implies that the reduction occurs at volume fractions below field capacity. No reduction in transformation rate is present between saturation and field capacity.

6.1.2 Schematization in ANIMO

The influence of soil moisture conditions on transformation processes is given on the basis of a literature study, as reported by van Huet (1983). Such a comparison, however, is only useful when this is done on the basis of soil water potential instead of soil moisture contents, so the available data have been transformed, if necessary, from soil moisture contents into suctions using standardized soil moisture characteristics.

For ammonification, most authors found a relation with moisture suction showing an optimum near pF 3. However, if they used topsoil samples for their studies, which are often aerated in field situations, the decrease in mineralization rate at low suctions may be due to the necessity of adaptation of the micro-flora from aerobic to facultative anaerobic species, which in a poor growth medium takes time. In a subsoil sample, under wet conditions, the decomposition rate may be equal to the optimum found at pF 3. Also for nitrification, a relation with an optimum was found. The rate decrease was, however, determined by a shortage of oxygen.

For denitrification it is obvious that the rate of the process increases with lower moisture suction, because this implicates decrease of oxygen availability. Denitrification in unsaturated soil is a result of the presence of anaerobic soil parts, indicated as partial anaerobiosis. The soil texture plays an important role in the aeration of soils.

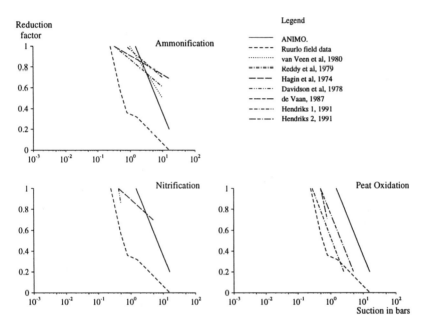

Fig. 6.1: The relation between the reduction factor and the soil moisture suction for different processes.

Fig. 6.1 gives the literature data collected by van Huet (1983) for dry soil conditions on the basis of soil moisture suction. The data presented by Hendriks (1991) for peat soils are also given in this figure.

Reduction in process rates due to high moisture contents is assumed to only take place if oxidation is hampered by lack of oxygen or a limited availability of nitrate.

The relation originally used in ANIMO was derived from the measured influence of moisture on the mineralization rate of soil organic nitrogen in field experiments on grassland at Ruurlo in the Netherlands over several years. The influence of sub-optimal moisture contents on the rate constants is described by multiplication of these rates with a reduction factor depending on the pF. For each layer the reduction factor must be determined from the average moisture fraction during the time step and the soil moisture characteristic. The reduction factor is determined by interpolation of the data presented in Table 6.1.

Table 6.1: Relation between the reduction factor for stress conditions and pF.

pF	Reduction factor
≤ 2.4	1.00
2.7	0.57
2.9	0.36
3.2	0.32
4.2	0.20

Summarizing, the reduction of mineralization at low moisture suctions is important, and both oxygen content and distribution, as well as oxygen diffusion, are important factors influencing the processes of mineralization, nitrification and denitrification.

6.2 Aeration and bio-chemical processes

6.2.1 General

The gas phase, commonly indicated as soil air, shows the same composition as the overlying atmosphere in the upper few millimetres of the top layer only. In soils many processes occur, of biological nature, which cause a continuous change in the composition of the soil air. Exchange with the atmosphere occurs by diffusion, which is a slow process as compared to mass flow. The fact that gaseous diffusion in the liquid phase proceeds roughly 10^4 times slower as does diffusion in the soil air explains the huge influence of soil moisture content on the gas phase composition and the exchange with the overlying atmosphere.

6.2.1.1 Schematization in TRANSOL

The quantity of persistent organic compounds, like oil products, polycyclic aromates and pesticides in the soil is generally too low to affect the total distribution of natural organic materials and the growth of the biodegrading population. For this reason steady state conditions for the distribution of natural organic materials and bio-mass are assumed.

The effect of the distribution of organic material with depth, the corresponding bio-mass activity, soil aeration and depth of the mean groundwater table are introduced by a lump sum reduction factor for the degradation rate of the persistent compounds.

The reduction factor equals one in the rootzone and decreases linearly with depth till zero at the mean groundwater table depth.

6.2.1.2 Schematization in ANIMO

Aeration in soils has a major influence in the unsaturated domain on transformation rates of all micro-biological processes in agricultural eco-systems. Since one of the model aims is to evaluate the environmental impacts of water management for a large number of soil types and a wide range of hydrological conditions, a detailed sub-model describing oxygen diffusion in the soil gas phase and in soil aggregates has been implemented.

In the ANIMO model, the partitioning between the aerobic soil fraction and the anaerobic soil fraction is determined by the equilibrium between the oxygen demand of organic conversion processes plus nitrification and the oxygen supply capacity of the soil air and soil water system. Both vertical diffusion in air filled pores and the lateral oxygen diffusion in aggregates and in the soil moisture phase are taken into consideration. The modelling of the oxygen distribution in soil air and water is described in this paragraph.

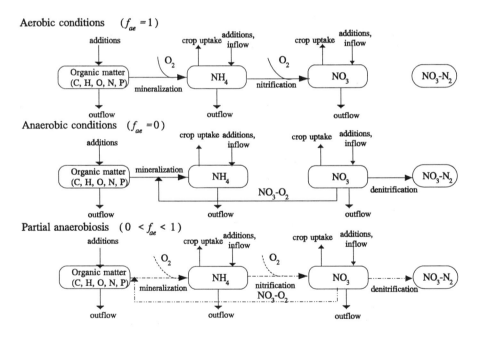

Fig. 6.2: Atmospheric oxygen and nitrate oxygen related processes in the ANIMO model.

The purpose of this modelling is to determine which part of the decomposition of organic compounds present in the soil, during a time step considered, occurs aerobically with oxygen present in the soil air and which part is decomposed with nitrate, resulting in denitrification. This partitioning is expressed by the aeration factor f_{ae} and $(1.0 - f_{ae})$, in which f_{ae} is the aerated fraction of the soil in any layer during the time step considered. Fig. 6.2 gives a description of the meaning of this aeration factor for both the atmospheric oxygen and nitrate oxygen related processes in the ANIMO model. At $f_{ae} = 1$, organic transformation and nitrification processes are optimal. For sub-optimal conditions, with $f_{ae} < 1$, the diffusive capacity of the unsaturated zone is insufficient to fulfil the oxygen requirement. In situations where partial anaerobiosis is present, the oxygen demand of the organic transformations is met by atmospheric oxygen as well as by nitrate-oxygen. The nitrification rate will be sub-optimal. Under these conditions, the available nitrate will be reduced partially or completely through denitrification. Under unfavourable wet conditions the upper layers consume all the oxygen, which can enter the soil profile by diffusion and the atmospheric oxygen will not penetrate into the lower part of the unsaturated zone.

6.2.2 Oxygen requirement in soils

Under complete aerobic conditions, the oxygen demand can be calculated, considering a carbon content fr_C in organic material. Assuming complete oxidation of the organic material to CO_2 results in the fact that for the oxidation of each mole of C, 1 mole of O_2 is needed, yielding:

$$S_{ox}^C = \frac{32}{12} fr_C J_{om} = 2.67 fr_C J_{om} \tag{6.2}$$

in which:
S_{ox}^C = oxygen demand for carbon oxidation per day in kg.m^{-2}.d^{-1}
J_{om} = net decomposition of organic material in kg.m^{-2}.d^{-1}

The quantity of NH_4 released during mineralization, as well as the NH_4 in solution and released from the soil complex, will be oxidized, according to:

$$2NH_4^+ + 4O_2 \rightarrow 2H_2O + 2NO_3^- + 4H^+ \tag{6.3}$$

This results in an oxygen demand for nitrification given by:

$$S_{ox}^N = \frac{128}{28} \Delta z k_{1,NH_4} \overline{c}_{NH_4} \theta = 4.57 \Delta z k_{1,NH_4} \overline{c}_{NH_4} \theta \tag{6.4}$$

in which:
S_{ox}^N = oxygen demand for nitrification in kg.m^{-2}.d^{-1}
$k_{1,NH4}$ = first order oxidation rate of NH_4 under aerobic conditions in d^{-1}

$\bar{c}_{NH4}$ = average ammonium concentration in kg.m^{-3}
θ = average soil moisture content (-)

The total oxygen demand under aerobic conditions is the sum of S_{ox}^C and S_{ox}^N. Expressed in terms of production of oxygen, it is:

$$S_{ox} = -\frac{1}{\Delta z}(S_{ox}^C + S_{ox}^N) \quad (6.5)$$

in which:
S_{ox} = oxygen production rate in kg.m^{-3}.d^{-1}

In order to express the oxygen demand in a volume of oxygen, the law of Boyle-Gay Lussac is applied, which is written as:

$$pV_g = mR_g(T + 273) \quad (6.6)$$

with: value
p = gas pressure in atm. 1.0
V_g = gas volume in l
m = number of moles 1.0
R_g = gas constant in l.atm.°C^{-1}.mole^{-1} 0.08205

The gas volume V_g is calculated as:

$$V_g = 0.08205(T + 273)\frac{1}{32}\text{l.g}^{-1} = 2.56\,10^{-3}(T + 273)\,\text{m}^3.\text{kg}^{-1} \quad (6.7)$$

The total oxygen demand per unit volume of soil is derived from Eq. (6.6) as:

$$Y_{ox} = -2.56\,10^{-3}(T + 273)S_{ox} \quad (6.8)$$

where:
Y_{ox} = oxygen demand in m^{3}gas.m^{-3} soil.d^{-1}

6.2.3 Vertical oxygen transport in soils

Oxygen supply from the air into the soil system mainly takes place through the process of diffusion through the air-filled pores. The diffusion coefficient can, according to Bakker (1965), be given as a function of the air filled pores, using the expression:

$$D_s^{ox} = D_a^{ox} \varsigma_1 \varepsilon_g^{\varsigma_2} \qquad (6.9)$$

in which:
D_s^{ox} = diffusion coefficient for O_2 in the soil gas filled pore space in $m^2.d^{-1}$
D_a^{ox} = diffusion coefficient for O_2 in the free atmosphere in $m^2.d^{-1}$
ε_g = volume fraction of air-filled pores in $m^3.m^{-3}$ soil
ς_1, ς_2 = empirical constants depending on soil type (-)

Bakker et al. (1987) give values of ς_1 and ς_2 for different soils. These data are summarized in Table 6.2. None of these relations, however, is valid for shrinking and cracking soils.

Table 6.2: Relative diffusion coefficient D_s/D_a for different soils after Bakker et al. (1987).

Soil type	Relative diffusion coefficient
Poorly loamy and humousless sands	$D_s/D_a = 1.5\ \varepsilon_g^{3.0}$
Structureless loamy sands	$D_s/D_a = 7.5\ \varepsilon_g^{4.0}$
Weakly aggregated topsoils of loamy sands, light clays and humous sands, subsoils of light clays	$D_s/D_a = 2.5\ \varepsilon_g^{3.0}$
Aggregated light clays	$D_s/D_a = 2.0\ \varepsilon_g^{2.5}$
Dense clay soils	$D_s/D_a = 0.3\ \varepsilon_g^{1.5}$

Because of the low air content of the soil matrix, oxygen diffusion in a cracking clay soil is largely determined by the presence of continuous air-filled macro-pores. In poorly structured heavy clay soils large vertical shrinkage cracks are the predominant type of macro-pores. When cracks in a soil column are continuous, the diffusion of oxygen through the column is mainly determined by the horizontal cross-section of vertical shrinkage cracks.

For these soils:

$$D_s^{ox} = \eta A_{cr}^{rel} D_a^{ox} \qquad (6.10)$$

with:
A_{cr}^{rel} = relative horizontal surface area of cracks (-)
η = labyrinth factor (-)

In many soils, normal isotropic shrinkage is the dominant shrinkage process (Bronswijk 1991). In that case A_{cr}^{rel} can be replaced by a function of the relative aggregate volume at different soil moisture contents (see section 2.1.2), so D_{sg}^{ox} can be given by the expression:

$$D_s^{ox} = \eta[1 - (\frac{V_{dry}}{V_{sat}})^{2/3}]D_a^{ox} \tag{6.11}$$

with:
V_{sat} = aggregate volume at complete saturation in m^3
V_{dry} = aggregate volume in unsaturated state in m^3

This function holds, with $\eta = 1.0$, for measured data in poorly structured heavy clay soils, but for well structured topsoils $\eta = 0.3$, to account for the longer diffusion pathway and the irregularity of the crack walls (Bronswijk 1991).

The diffusion coefficient D_a^{ox} of oxygen in air at a temperature of 0 °C is 1.5466 m^2.d^{-1}. Assuming no changes in the partial pressure of oxygen in air, the relation with temperature can be calculated following Bakker (1965) as:

$$D_a^{ox}(T) = D_a^{ox}(0)(\frac{273 + T}{273})^{1.75} \tag{6.12}$$

So for each soil layer and each time step a different value for D_s^{ox} is used, depending on the air filled pore space and on soil temperature. The vertical diffuse transport of oxygen in the air-filled pores in the soil system is given by the expression:

$$\frac{\partial \varepsilon_g c_g^{ox}}{\partial t} = \frac{\partial}{\partial z} D_s^{ox} \frac{\partial c_g^{ox}}{\partial z} - f_{ae} Y_{ox} \tag{6.13}$$

in which:
c_g^{ox} = the oxygen concentration in the soil air in m^{3}O$_2$.m^{-3}soil air
f_{ae} = reduction factor for partial anaerobiosis (-)

The factor f_{ae} accounts for partial anaerobiosis, which may cause a reduction of the respiratory activity. The factor is a function of the oxygen concentration in the gas phase, the potential oxygen consumption rate and soil moisture conditions related to soil type. In the ANIMO model, the partitioning between the aerobic soil fraction and the anaerobic fraction is determined by the equilibrium between oxygen demand from the organic conversion processes plus nitrification and the oxygen supply capacity of the soil air and soil water system. Both the vertical diffusion in the air-filled pores and the lateral oxygen diffusion in the soil moisture phase are taken into consideration. Fig. 6.3 depicts a schematic representation of the diffuse transport in both zones. The factor f_{ae} follows from a sub-model describing the oxygen diffusion in the soil-water phase, as is presented in section 6.2.4. The oxygen storage in soils is generally very small compared to the oxygen requirements, so the main part of the required oxygen must be supplied by diffusion. Sufficient aeration is present when the requirements can be supplied under steady state conditions.

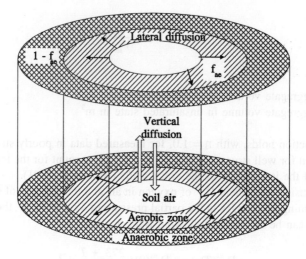

Fig. 6.3: Schematic representation of diffusive oxygen transport in the aeration module.

The diffusion equation reduces for steady state conditions to:

$$\frac{d}{dz} D_s^{ox} \frac{dc_g^{ox}}{dz} = f_{ae} Y_{ox} \qquad (6.14)$$

Because the diffusion of oxygen in water is about 10^4 times as slow as in the air, the consumption of oxygen below the groundwater level is neglected. In the model approach, the soil system is divided into N layers. Each layer k has its own thickness Δz, while D_s^{ox}, f_{ae} and Y_{ox} are considered to be constant per layer and per time step, expressing them as $D_s^{ox}(k)$, $f_{ae}(k)$ and $Y_{ox}(k)$, respectively. Since the factor f_{ae} depends on the oxygen concentration, an iterative computation scheme is necessary to determine its value. In the first iteration round, the factor is set to 1 and oxygen concentrations are calculated using the vertical diffusion model. Then new values of f_{ae} are calculated per layer using the radial diffusion model. This model uses the calculated oxygen concentration in the gas phase and the potential oxygen consumption rates as input. After determining new values of f_{ae} a new computation round can start. The iteration procedure is terminated when the summated difference between old and new values is less than a certain criterium.

If $Z(k-1)$ is the depth of the top of layer k from the soil surface in m, the following expression holds for layer k for $Z(k-1) \leq z \leq Z(k)$:

$$D_s^{ox}(k) \frac{d^2 c_g^{ox}}{dz^2} = f_{ae}(k) Y_{ox}(k) \qquad (6.15)$$

The boundary condition at the soil surface is:

$$k = 1 \; ; \; Z(k-1) = 0 \; ; \; c_g^{ox}(k-1) = c_a^{ox} \tag{6.16}$$

in which:
c_a^{ox} = concentration of oxygen in the free atmosphere in m^3.m^{-3} air

At each layer boundary the oxygen supply rate must be equal to the oxygen consumption rate below that boundary. This condition can be expressed as:

$$z = Z(k-1); \; c_g^{ox} = c_g^{ox}(k-1); \; \frac{dc_g^{ox}}{dz} = \frac{-1}{D_s^{ox}(k)}[Y_{ox}^{prof} - \sum_{i=1}^{k-1}\Delta z(i)f_{ae}(i)Y_{ox}(i)] \tag{6.17}$$

in which:
Y_{ox} = oxygen demand in m^{3}O$_2$.m^{-3}soil.d^{-1}
Y_{ox}^{prof} = the oxygen consumption rate of the complete soil profile in m^{3}O$_2$.m^{-2}.d^{-1}
i = number of the layers above layer k

Integration of Eq.(6.14) subject to the boundary condition in Eq. (6.16) yields:

$$\frac{dc_g^{ox}}{dz} = \frac{f_{ae}(k)Y_{ox}(k)}{D_s^{ox}(k)}(z - Z(k-1)) - \frac{1}{D_s^{ox}(k)}[Y_{ox}^{prof} - \sum_{i=1}^{k-1}\Delta z(i)f_{ae}(i)Y_{ox}(i)] \tag{6.18}$$

Integration of Eq. (6.18) subject to the boundary condition in Eq.(6.17) gives for the distribution of the oxygen concentration in layer k:

$$c_g^{ox} = c_g^{ox}(k-1) + \frac{f_{ae}(k)Y_{ox}(k)}{2D_s^{ox}(k)}(z - Z(k-1))^2 -$$

$$\frac{(z-Z(k-1))}{D_s^{ox}(k)}[Y_{ox}^{prof} - \sum_{i=1}^{k-1}\Delta z(i)f_{ae}(i)Y_{ox}(i)] \tag{6.19}$$

For $z = Z(k)$ then:

$$c_g^{ox}(k) = c_g^{ox}(k-1) + \frac{f_{ae}(k)Y_{ox}(k)}{2D_s^{ox}(k)}(\Delta z(k))^2 -$$

$$\frac{\Delta z(k)}{D_s^{ox}(k)}[Y_{ox}^{prof} - \sum_{i=1}^{k-1}\Delta z(i)f_{ae}(i)Y_{ox}(i)] \tag{6.20}$$

Eq. (6.20) can, subject to the surface boundary condition, be rewritten as:

$$c_g^{ox}(k) = c_a^{ox} + \sum_{k=1}^{k} \frac{f_{ae}(k)Y_{ox}(k)}{2D_s^{ox}(k)}(\Delta z(k))^2 -$$

$$\sum_{k=1}^{k} \frac{\Delta z(k)}{D_s^{ox}(k)}[Y_{ox}^{prof} - \sum_{i=1}^{k-1}\Delta z(i)f_{ae}(i)Y_{ox}(i)] \qquad (6.21)$$

If, however, the oxygen demands are high and/or the diffusion process is hampered by wet conditions, the oxygen profiles will show a very steep decline with depth, and in the deeper layers near the groundwater level anaerobic conditions may be present. At the depth of the boundary of the anaerobic zone (Z_{an}), both the diffusion flux and the concentration are equal to zero. The value of Z_{an} is given by the expression:

$$Z_{an} = Z(k-1) + \sqrt{\frac{2D_s^{ox}(k)c_g^{ox}(k-1)}{f_{ae}(k)Y_{ox}(k)}} \qquad (6.22)$$

The total oxygen consumption rate Y_{ox}^{prof} is calculated with the expression:

$$Y_{ox}^{prof} = [\sum_{k=1}^{k-1}\Delta z(k)f_{ae}(k)Y_{ox}(k)] + (Z_{an} - Z(k-1))f_{ae}(k)Y_{ox}(k) \qquad (6.23)$$

When the depth of the phreatic groundwater table, Z_{gw}, is situated within the layer considered, than the zero flux condition is valid. This gives the following expression:

$$Y_{ox}^{prof} = [\sum_{k=1}^{k-1}\Delta z(k)f_{ae}(k)Y_{ox}(k)] + (Z_{gw} - Z(k-1))f_{ae}(k)Y_{ox}(k) \qquad (6.24)$$

With the values of Z_{an}, Z_{gw} and Y_{ox}^{prof} the oxygen profile in the soil can be calculated. The average concentration in layer k is obtained by integration of Eq. (6.20) between the boundaries $Z(k-1)$ and $Z(k)$ and dividing by the layer thickness $\Delta z(k)$:

$$\overline{c}_g^{ox}(k) = c_g^{ox}(k-1) + \frac{f_{ae}(k)Y_{ox}(k)}{6D_s^{ox}(k)}(\Delta z(k))^2 -$$

$$\frac{\Delta z(k)}{2D_s^{ox}(k)}[Y_{ox}^{prof} - \sum_{i=1}^{k-1}\Delta z(i)f_{ae}(i)Y_{ox}(i)] \qquad (6.25)$$

For $\Delta z(k)$:

$$\Delta z(k) = \text{MIN}[Z(k) - Z(k-1); Z_{an} - Z(k-1); Z_{gw} - Z(k-1)] \qquad (6.26)$$

Eq. (6.20), Eq. (6.21), Eq. (6.23), Eq. (6.24) and Eq. (6.25) are used to calculate the oxygen concentration of a layer k from that of layer $k-1$, in a profile where the oxygen demands are fully met. If Z_{an} or Z_{gw} are greater than $Z(k)$ the boundary has not been reached and the procedure has to be repeated for the following layer.

6.2.4 Diffusion in the water phase

Another complication for some layers may be that, even if some oxygen is present in the air-filled pores, the diffusion into the water-filled soil aggregate pores may not be fast enough to make these aggregates completely aerobic. The result is that in the middle of the aggregates anaerobic conditions are present, while at the edges aerobic decomposition takes place. The soil layer is then called partially anaerobic. Because in the anaerobic parts decomposition of organic material takes place with nitrate consumption instead of oxygen, the oxygen consumption of such a layer will be smaller than $Y_{ox}(k)$, and a new oxygen profile has to be calculated with a reduced value for the oxygen consumption. The partial anaerobiosis fraction f_{ae} is calculated utilizing a sub-model which describes the diffusion in soil water. The diffusion process in the water phase depends on the type of soil. The transport in the water phase in each layer can be approximated with the general equation:

$$\frac{\partial c_w^{ox}}{\partial t} = r^{-m} D_{sw}^{ox} \frac{\partial}{\partial r} (r^m \frac{\partial c_w^{ox}}{\partial r}) + S_{ox} \qquad (6.27)$$

in which:
c_w^{ox} = concentration of oxygen in water in kg.m^{-3}
D_s^{ox} = diffusion coefficient for oxygen in saturated soil in m^2.d^{-1}
r = distance from the centre of air-filled pores or the centre of aggregates in m
m = aggregate shape parameter

Eq. (6.27) reduces under steady state conditions to:

$$r^{-m} D_{sw}^{ox} \frac{d}{dr} (r^m \frac{dc_w^{ox}}{dr}) = -S_{ox} \qquad (6.28)$$

This equation is subject to the following boundary conditions:

- the aqueous concentration at the air-water interface is in equilibrium with the concentration in the air-filled pores;
- at the distance R where the anaerobic conditions start both the flux and the concentration equal zero.

Integration of Eq. (6.28) yields different solutions depending on the value of m. If $m \neq 1$ the following expressions are obtained:

$$\frac{dc_w^{ox}}{dr} = \frac{-S_{ox}}{(m+1)D_{sw}^{ox}} r + K_1 r^{-m} \quad (6.29)$$

and

$$c_w^{ox} = \frac{-S_{ox}}{2(m+1)D_{sw}^{ox}} r^2 + \frac{K_1}{-(m-1)} r^{-(m-1)} + K_2 \quad (6.30)$$

If $m = 1$ the following expressions are obtained after integration:

$$\frac{dc_w^{ox}}{dr} = \frac{-S_{ox}}{2D_{sw}^{ox}} r + K_1 r^{-1} \quad (6.31)$$

and

$$c_w^{ox} = \frac{-S_{ox}}{4D_{sw}^{ox}} r^2 + K_1 \ln r + K_2 \quad (6.32)$$

6.2.4.1 Non-aggregated sandy soils

Fig. 6.4 gives a presentation of the boundary conditions in the model schematization for non-aggregated soils.

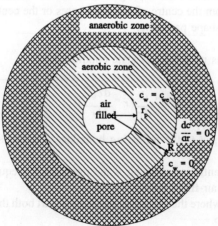

Fig. 6.4: Schematization of an aerobic and an anaerobic zone around an air filled pore.

Environmental Influences on Processes   181

In non-aggregated sandy soils the air-filled pore space can be considered as a number of vertical ventilation channels. The air-filled pores in a soil layer differ in their radius r_p. If ψ (mbar) is the suction in the layer, the corresponding smallest air-filled pore radius is calculated as $r_p = 0.0015\psi^{-1}$ m. If the air entry point of the soil is at soil moisture suction ψ_a (mbar), the corresponding radius of the biggest pore is $r_p^{\psi a}$. The average radius $\bar{r}_p$ of air-filled pores in the layer is calculated as the geometrical mean of $r_p^{\psi a}$ and r_p^{ψ}. The gas transport from these channels is subject to the conditions:

$$r = r_p \qquad c_w^{ox} = c_{we}^{ox}$$

$$r = R \qquad c_w^{ox} = 0 \qquad \frac{dc_w^{ox}}{dr} = 0 \qquad (6.33)$$

in which:
c_w^{ox} = oxygen concentration in water in kg.m^{-3}
c_{we}^{ox} = equilibrium oxygen concentration in water at the water/air boundary in kg.m^{-3}
r_p = radius of an air filled pore in m
R = distance from the centre of the pore where $c_w^{ox} = 0$ in m

The solubility of oxygen decreases when temperature increases. The equilibrium concentration at the air-water interface can be calculated from the oxygen concentration in gas, using the expression:

$$c_{we}^{ox} = \frac{K_B^{sol}}{2.564 \, 10^{-3}(T + 273.15)} c_g^{ox} \qquad (6.34)$$

where:
K_B^{sol} = Bunsen's coefficient of solubility in m^3.m^{-3}
T = temperature in °C

Table 6.3: Relation between temperature and the oxygen diffusion coefficient D_w^{ox} in water.

Temperature (°C)	Diffusion coefficient (m².d⁻¹)	Bunsen's coefficient (m³.m⁻³)
0	8.554 10⁻⁵	0.0489
5	1.097 10⁻⁴	0.0436
10	1.331 10⁻⁴	0.0394
15	1.572 10⁻⁴	0.0360
20	1.814 10⁻⁴	0.0333
25	2.056 10⁻⁴	0.0309
30	2.307 10⁻⁴	0.0290

Both the diffusion coefficient of oxygen in water and the Bunsen coefficient are dependent on temperature. The influence of soil temperature on the oxygen diffusion coefficient in water in each layer and each time step is obtained by interpolation between the data presented in Table 6.3, as given by Gliński and Stepniewski (1985).

The value of the diffusion coefficient in soil water D_{sw}^{ox} is approximated through the introduction of a labyrinth factor η_w^{lat} with the expression:

$$D_{sw}^{ox} = \eta_w^{lat} \theta_{sat} D_w^{ox} \qquad (6.35)$$

According to McCaskill and Blair (1990) 98 % of the variation in η_w^{lat} for soils widely differing in clay contents can be described by the expression:

$$\eta_w^{lat} = -0.00389 fr_{cl} + 1.28 \theta_{sat} \qquad (6.36)$$

in which:
fr_{cl} = weight fraction of clay (-)
θ_{sat} = volumetric soil moisture fraction at saturation (-)

The solution of Eq. (6.28) for the distance R from the soil pore centre to the place where $c_w = 0$ is then given by the expression:

$$\frac{4 D_{sw}^{ox} c_{we}^{ox}}{S_{ox} \bar{r}_p^2} = (\frac{R}{\bar{r}_p})^2 (1 - \ln(\frac{R}{\bar{r}_p})^2) - 1 \qquad (6.37)$$

R can be solved from Eq. (6.37) either by the method of the Newton-Raphson iteration or by interpolation from a calculated table, with $R/r_v \geq 1$. However, Eq. (6.37) can also be approximated by an empirical relation. The empirical relation which gives similar results as Eq. (6.37) is given by the expression:

$$\frac{R}{\bar{r}_p} = [1 - \frac{4 D_{sw}^{ox} c_{we}^{ox}}{S_{ox} \bar{r}_p^2}]^{0.391} \qquad (6.38)$$

The comparison of both equations is presented in Fig. 6.5.

The aerated area in horizontal direction A_{ae} (m^2) for each pore is:

$$A_{ae} = \pi R^2 \qquad (6.39)$$

If the volume of a single pore per unit depth is estimated as $\eta_p^{-1} \cdot \pi \cdot r_v^2$ and the soil moisture difference between the suction ψ_a and ψ is $\Delta\theta$, the number of air-filled pores can be approximated as:

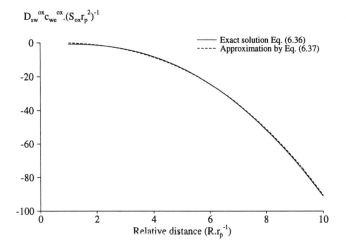

Fig. 6.5: The relation between the relative distance $R.\overline{r}_p^{-1}$ and the relative oxygen production $D_{sw}^{ox} c_{we}^{ox}.(S_{ox} \overline{r}_p^2)^{-1}$ for non-aggregated soils.

$$N = \frac{\eta_p \Delta\theta}{\pi \overline{r}_p^2} \qquad (6.40)$$

in which:
η_p = labyrinth factor in m soil depth per m pore length

If all air-filled pores are regularly distributed, without touching each other, the aerated soil fraction will be:

$$f_{ae} = NA_{ae} = \eta_p \Delta\theta \frac{R^2}{\overline{r}_p^2} \qquad (6.41)$$

In practice, the distribution of the pores is a random one, so that the aerated volume of soil will not increase linearly with the number of air-filled pores. Defining the chance that a new air-filled pore interferes with an already aerated soil part proportional to the aerated part, the total aerated soil volume per unit of soil volume with N air-filled pores is equal to:

$$f_{ae} = A_{ae} + (1-A_{ae})A_{ae} + (1-A_{ae})^2 A_{ae} + \cdots + (1-A_{ae})^N A_{ae} = 1 - (1-A_{ae})^N \qquad (6.42)$$

Since the number of air filled pores per m² will be very large, even at small suctions, the limit rule for large numbers can be applied:

$$f_{ae} \approx 1 - \lim_{N \to \infty}[1 - A_{ae}]^N \approx 1 - \exp[-A_{ae}N] \quad (6.43)$$

Substitution of Eq. (6.35), Eq. (6.38) and Eq. (6.41) into Eq. (6.43) gives:

$$f_{ae} = 1 - \exp\left[-\eta_p \Delta\theta(1 - \frac{4\eta_w^{lat}\theta_{sat}D_w^{ox}c_{we}^{ox}}{S_{ox}\bar{r}_p^2})^{0.782}\right] \quad (6.44)$$

6.2.4.2 Swelling and shrinking soils

The approximation of shrinkage is given in the schematization of ANIMO by complete saturation of aggregates until the value of the void ratio by wilting point has been reached, as was discussed in section 2.1.2 dividing the curve into two linear sections. The first one follows the saturation line until the void ratio at wilting point is reached. Considering the volume of solids V_s as being constant in a model layer gives automatically a correction for layer subsidence during shrinkage. The change in the volume of the aggregates in this schematization equals the change in moisture volume in the model layer, until the intersection point of both linear relations. A further decrease of the moisture volume until wilting point than takes place with a constant aggregate volume. The total number of soil aggregates is divided into n_{max} classes with radius $r_{sbl,n}$.

The volume of a soil aggregate $\upsilon_{ag,n}$ in aggregate class n is given as:

$$\upsilon_{ag,n} = 2^{m(n)} \frac{\pi^{\frac{1}{2}[\left(\frac{3}{m(n)}\right)-1]}}{m(n) + 1} r_{sbl,n}^{(m(n)+1)} \quad (6.45)$$

The volume of the anaerobic part of a soil aggregate $\upsilon_{an,n}$ in class n is given as:

$$\upsilon_{an,n} = \frac{2^{m(n)} \pi^{\frac{1}{2}[\left(\frac{3}{m(n)}\right)-1]}}{m(n) + 1} R_n^{(m(n)+1)} \quad (6.46)$$

Now it is assumed that the total number of aggregates per soil layer remains constant and independent of the gas-filled pore space ε_g. Swelling and shrinking of a soil is described in terms of a variable aggregate radius depending upon the gas-filled pore fraction ε_g. Shrinking stops if the gas-filled pore fraction equals the macro-pore fraction ε_e of the soil. If the condition $\varepsilon_g > \varepsilon_e$ is true, air will enter from the aggregate surface into the aggregates and this apparently reduces the effective aggregate radius.

The total volume per aggregate class n per m³ soil is given by the expression:

$$V_m(n) = fr(n)(1 - \varepsilon_g) \tag{6.47}$$

The total number of aggregates per class n per m³ of soil volume is calculated as:

$$N_a(n) = fr(n)(1.0 - \varepsilon_e)\frac{m(n) + 1}{2^{m(n)}\pi^{\frac{1}{2}[\left(\frac{3}{m(n)}\right)-1]} r_{ve}(n)^{m(n)+1}} \tag{6.48}$$

where:
$fr(n)$ = volume fraction of aggregate class n (-)
$r_{ve}(n)$ = aggregate radius if $\varepsilon_g = \varepsilon_e$ in m
ε_e = macro porosity of the soil in m³.m⁻³
ε_g = gas filled pore space in m³.m⁻³

The apparent aggregate radius for each aggregate class is calculated, by combining Eq. (6.47) and (6.48), using the expression:

$$r_v(n) = \sqrt[(m(n)+1)]{\left(\frac{1-\varepsilon_g}{1-\varepsilon_e}\right) r_{ve}(n)} \tag{6.49}$$

The value of *m* depends on the shape of the aggregates and soil blocks in shrinking soils:

- $m = 0$ holds for linear diffusion into soil plates,
- $m = 1$ holds for cylindrical diffusion into soil cylinders
- $m = 3$ holds for spherical diffusion.

Solutions of Eq. (6.29), Eq. (6.30), Eq. (6.31) and Eq. (6.32) are subject to the conditions:

$$r = r_{sbl} \qquad c_w^{ox} = c_{we}^{ox}$$
$$r = R \qquad c_w^{ox} = 0 \qquad \frac{dc_w^{ox}}{dr} = 0 \tag{6.50}$$

in which:
c_w^{ox} = oxygen concentration in water in kg.m⁻³
c_{we}^{ox} = equilibrium oxygen concentration in water at the water/air boundary in kg.m⁻³
r_{sbl} = radius or half thickness of the aggregates in m
R = distance from the centre of the aggregate where $c_w^{ox} = 0$ in m

- If $m \neq 1$ the following expression is obtained using Eq. (6.29) and Eq. (6.30):

$$\frac{2(m+1)c_{we}^{ox}D_w^{ox}}{S_{ox}r_{sbl}^2} = (\frac{R}{r_{sbl}})^2[1 + \frac{2}{m-1}(1 - (\frac{R}{r_{sbl}})^{(m-1)})] - 1 \qquad (6.51)$$

- If $m=1$ the following expression is obtained using Eq. (6.30) and Eq. (6.31):

$$\frac{4D_w^{ox}c_{we}^{ox}}{ar_{sbl}^2} = (\frac{R}{r_{sbl}})^2[1 - \ln(\frac{R}{r_{sbl}})^2] - 1 \qquad (6.52)$$

- In both cases $R/r_{sbl} \leq 1$

Eq. (6.51) and Eq. (6.52) can be approximated by equations expressing R/r_{sbl} as the dependent variable.

- If $m = 0$:

$$\frac{R}{r_{sbl}} = [1 - (\frac{-2c_{we}^{ox}D_w^{ox}}{S_{ox}r_{sbl}^2})^{0.5}] \qquad -0.5 \leq \frac{c_{we}^{ox}D_w^{ox}}{S_{ox}r_{sbl}^2} \leq 0 \qquad (6.53)$$

- If $m = 1$:

$$\frac{R}{r_{sbl}} = [1 - (\frac{-4c_{we}^{ox}D_w^{ox}}{S_{ox}r_{sbl}^2})^{0.4}]^{0.59} \qquad -0.25 \leq \frac{c_{we}^{ox}D_w^{ox}}{S_{ox}r_{sbl}^2} \leq 0 \qquad (6.54)$$

- If $m = 2$:

$$\frac{R}{r_{sbl}} = [1 - (\frac{-6c_{we}^{ox}D_w^{ox}}{S_{ox}r_{sbl}^2})^{0.333}]^{0.444} \qquad -0.1667 \leq \frac{c_{we}^{ox}D_w^{ox}}{S_{ox}r_{sbl}^2} \leq 0 \qquad (6.55)$$

Fig. 6.6 shows the effect of the shape factor of the aggregates on the distribution of the relative oxygen consumption. It appears that the expressions, presented in Eq. (6.53), Eq. (6.54) and Eq.(6.55), give an excellent approximation of the shape of the relations. If the relative oxygen production is smaller (more negative) than the lower boundary $R/r_{sbl} = 1$, the aggregates are completely saturated. The aerated soil fraction f_{ae} is calculated by combining Eq. (6.45) and Eq. (6.46):

$$f_{ae} = \sum_{n=1}^{n_{max}} (fr(n)(1 - \frac{v_{an,m}(n)}{v_{ag,m}(n)})) = \sum_{n=1}^{n_{max}} (fr(n)(1 - (\frac{R(n)}{r_{sbl}(n)})^{m(n)+1})) \qquad (6.56)$$

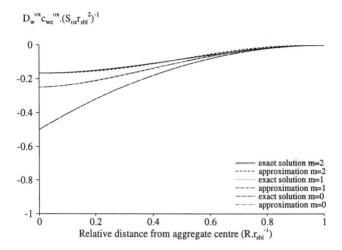

Fig. 6.6: The relation between the relative distance R/r_{sbl} and the relative oxygen production $D_w^{ox} c_{we}^{ox}/(S_{ox} r_{sbl}^2)$ for aggregated soils for different aggregate shapes.

6.2.3.3 Rainfall and aeration

During short-term periods with heavy rainfall, it is possible that the rootzone is temporarily saturated with water, and anaerobic conditions occur. In a model that uses steady state average weather conditions during time steps of one day or more, these periods will not be described. However, due to the relative abundance of organic materials in the topsoil, the oxygen demand remains high and denitrification during such short periods cannot be ignored. Therefore an estimate of the period and the depth over which the top-layers are saturated has to be made. This is done as follows: if the precipitation surplus q_r (m) of the time step is considered as a vertical column of water with area θ_{sat} then the height h (m) of this column is q_r/θ_{sat}. The relative duration τ of this temporary anaerobiosis caused by saturated transport of this water column is:

$$\tau = \frac{h}{k_{sat} t} \quad (6.57)$$

in which:
k_{sat} = saturated hydraulic conductivity in m.d^{-1}
τ = fraction of time step with anaerobiosis in top-layer(s)

The number of layers partaking in this temporary anaerobiosis is estimated as all those layers with $Z(k) < h$. This extra denitrification process has to be combined with the scheme given before. This can be done easily, because both τ and $(1 - A_{ae})$ have been defined as a fraction, although the first is expressed as a fraction of the time step and the second as a fraction of the layer volume. Both can be expressed as a fraction of the total amount of organic material decomposed, that is the fraction decomposed anaerobically. So, for the layers under consideration:

$$f_{ae}^* = f_{ae}(1 - \tau) \qquad (1 - f_{ae}^*) = 1 - f_{ae}(1 - \tau) \qquad (6.58)$$

The reduced oxygen production and demand rates are calculated, respectively, as:

$$S_{ox}^*(k) = -f_{ae}(1 - \tau)S_{ox}(k) \qquad (6.59)$$

and

$$Y_{ox}^*(k) = f_{ae}(1 - \tau)Y_{ox}(k) \qquad (6.60)$$

6.3 Influence of temperature

6.3.1 Temperature and biological processes

For chemical processes, the rate generally increases with temperature. For biological processes, often performed with enzymes, there is an optimum temperature range below and above which the rate decreases. Because the optimum temperature for biological processes is often around 30 °C or higher both chemical and biological processes in soils will increase with temperature. For some processes in the nitrogen cycle this influence has been especially studied. The reaction rates can be expressed relative to the maximum rate found, or to the rate found at a certain average temperature, so a correction factor for other temperatures on the reaction rate can be applied in calculations.

Rijtema presented, in Lammers (1983), data collected by Kolenbrander (personal communication, 1977) as shown in Fig 6.7. This figure shows a rise in the relative activity until about 26 °C. A maximum decomposition rate occurs between temperatures from 26 to 38 °C, followed by a sharp decrease at higher temperatures.

For temperatures from 0 to 26 °C this relation can be given by the Arrhenius equation as:

$$\xi_T = \exp[-\frac{a}{R_g}(\frac{1}{T + 273} - \frac{1}{T_{ref} + 273})] \qquad (6.61)$$

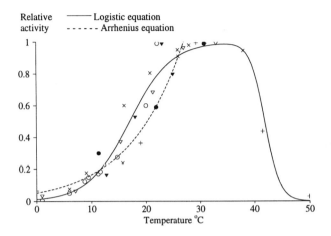

Fig. 6.7: The relation between the relative biological activity and temperature after data presented by Kolenbrander (pers. comm.) The curves are calculated using Eq. (6.61) and Eq. (6.62).

in which:
ξ_T = the relative activity coefficient
a = the molar activation energy in J.mole^{-1}
R_g = the gas constant in J.mole^{-1}.°K^{-1}
T_{ref} = reference temperature in °C

The curve in Fig. 6.7 is calculated with $a.R_g^{-1} = 9000$ °K.
The second curve is given by a double logistic function as:

$$\xi_T = \frac{1}{1 + \exp[-0.26(T - 17.0)]} - \frac{1}{1 + \exp[-0.77(T - 41.9)]} \quad (6.62)$$

For mineralization or mineralization combined with nitrification many studies of the relation between temperature and biomass activity exist, giving comparable results. It should be realized, however, that the mineralization rate is first of all determined by the decomposition rates of organic materials and by their nitrogen contents. So a straightforward relation for temperature influences should be obtained from decomposition studies of organic material. Van Huet (1983) collected both tabulated data and functions from literature for the effect of temperature on mineralization, nitrification and denitrification. The results for ammonification indicated as mineralization are presented in Fig. 6.8a. Nitrification also increases with temperature. However, the process is dominated by the presence of oxygen. Under aerobic conditions nitrification is so fast that ammonification becomes the rate limiting factor. The temperature dependency of nitrification is presented in Fig. 6.8b.

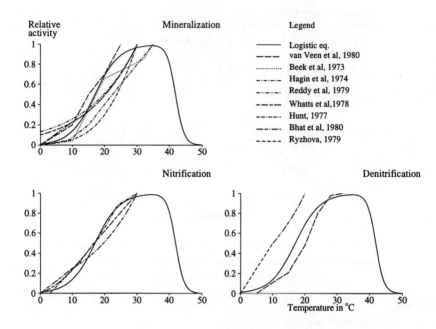

Fig. 6.8: The effect of temperature on different biological processes presented by different authors as collected by van Huet (1983).

Denitrification increases with temperature, but this is completely due to the increase of organic matter decomposition with temperature. The effect of temperature on denitrification is shown in Fig. 6.8c. Summarizing, it must be concluded that the most important temperature influences are those on organic matter decomposition. For this reason either the Arrhenius equation or the double logistic equation can be used.

6.3.2 Calculation of soil temperature

The differential equation describing the temperature course in soils with time and depth below the soil surface is given by:

$$C_h \frac{\partial T}{\partial t} = \lambda_T \frac{\partial^2 T}{\partial z^2} \tag{6.63}$$

where:
T = temperature as function of time t and depth z in °C
C_h = volumetric heat capacity in $J.m^{-3}.°C$
λ_T = thermal conductivity in $J.m^{-1}.d^{-1}.°C^{-1}$

A solution that is a harmonic function of time can be written as:

$$T(z,t) = \zeta(z)\exp[i\omega t] \tag{6.64}$$

in which:
ζ = a complex function of z only.

Substituting Eq. (6.64) in Eq (6.63) gives an differential equation with constant coefficients:

$$\frac{\lambda_T}{C_h}\frac{d^2\zeta}{dz^2} - i\omega\zeta = 0 \tag{6.65}$$

The solution of this differential equation gives as the general solution of Eq. (6.63):

$$T(z,t) = \left[A_1\exp[(1+i)z\sqrt{\frac{\omega C_h}{2\lambda_T}}] + A_2\exp[-(1+i)z\sqrt{\frac{\omega C_h}{2\lambda_T}}]\right]\exp[i\omega t] \tag{6.66}$$

When the model is used for prediction purposes, the course of the temperature in the future years is not known. In this case a simple sinus-wave model for the air temperature, with a damping effect for depths below the soil surface, is used. Depending on the available soil data in regional studies the soil will be considered either as a homogeneous one layer soil or as a multi-layered system. The description of soil temperature at a certain depth z (m) below the soil surface and at a certain day of the year for a homogeneous soil is given by the expression:

$$T(z,t) = \overline{T} + A_0\exp[\frac{z}{D_m}]\cos(\omega t + \phi - \frac{z}{D_m}) \tag{6.67}$$

in which:
$T(z,t)$ = temperature at depth z and time t in °C
t = time of the year in d
$\overline{T}$ = average yearly temperature at the soil surface in °C
A_0 = amplitude of the temperature wave at the soil surface in °C
D_m = damping depth, equalling A_0/e, in m
ω = frequency of the temperature wave in rad.d^{-1}
ϕ = phase shift in rad

Van Duin (1956) showed that the yearly temperature wave lags $\pi/4$ behind the heat flux wave in the Netherlands. If the minimum of the heat flux wave is on the 22nd of December, then the minimum of the temperature wave at the soil surface is on the 7th of February. As a consequence the phase shift ϕ equals -3.7721 rad.

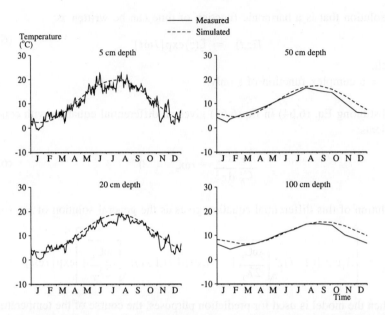

Fig. 6.9: Comparison between measured temperature and temperature simulated with Eq. (6.66).

For the yearly temperature wave the frequency $\omega = 2.\pi/365 = 0.01721$ rad.d^{-1}. $D_{m,i}$ is calculated by:

$$D_{m,i} = \sqrt{\frac{2\lambda_{T,i}}{\omega C_{h,i}}} \qquad (6.68)$$

in which:
λ_T = the thermal conductivity in J.m^{-1}.d^{-1}.°C^{-1}
C_h = the volumetric specific heat in J.m^{-3}.°C^{-1}

An example of the comparison between the temperature course simulated with Eq. (6.67) as function of time and depth and measured data is presented in Fig. 6.9. For non-homogeneous soils, the soil profile is schematized in a multi-layered profile with n homogeneous layers of thickness Δz_i. According to Van Wijk and Derksen (1966), Eq. (6.66) can be rewritten for each layer i ($Z_{i-1} \leq z \leq Z_i$) as:

$$T_i(z,t) - \overline{T} = A_{i,1}\exp[-\frac{z}{D_{m,i}}]\sin(\omega t - \frac{z}{D_{m,i}} + \phi_{i,1}) +$$
$$A_{i,2}\exp[\frac{z}{D_{m,i}}]\sin(\omega t + \frac{z}{D_{m,i}} + \phi_{i,1}) \qquad (6.69)$$

For the last layer $z > Z_{n-1}$:

$$T_n(z,t) - T = A_{n,1} \exp\left[-\frac{z}{D_{m,n}}\right] \sin\left(\omega t - \frac{z}{D_{m,n}} + \phi_{n,1}\right) \quad (6.70)$$

The solutions for the different layers have to satisfy the following boundary conditions:

$$z = 0 \rightarrow T_1 - T = A_0 \sin(\omega t) \; ; \; z = Z_{i-1} \rightarrow T_{i-1} - T = T_i - T$$

$$z = Z_{i-1} \rightarrow -\lambda_{i-1}\frac{\partial T_{i-1}}{\partial z} = -\lambda_i \frac{\partial T_i}{\partial z} \; ; \; z \rightarrow \infty \rightarrow T_n - T = 0 \quad (6.71)$$

Complex variables $\xi_{i,1}$ and $\xi_{i,2}$ are introduced to facilitate the derivation of expressions for the unknown amplitudes $A_{i,1}$ and $A_{i,2}$ and for the phase shifts $\phi_{i,1}$ and $\phi_{i,2}$:

$$\xi_{i,1} = A_{i,1}\exp[i\phi_{i,1}] \quad \xi_{i,2} = A_{i,2}\exp[i\phi_{i,2}] \quad (6.72)$$

Substitution of these variables in Eq. (6.66) and Eq. (6.68) gives the following set of equations subject to the boundary conditions (6.71):

$$\begin{pmatrix} 1 & 1 & 0 & 0 & 0 & \cdot \\ a_{2,1} & a_{2,2} & a_{2,3} & 0 & 0 & \cdot \\ a_{3,1} & a_{3,2} & a_{3,3} & 0 & 0 & \cdot \\ 0 & 0 & a_{4,1} & a_{4,2} & a_{4,3} & \cdot \\ 0 & 0 & a_{5,1} & a_{5,2} & a_{5,3} & \cdot \\ \cdot & \cdot & \cdot & \cdot & \cdot & \cdot \\ \cdot & \cdot & \cdot & \cdot & \cdot & \cdot \\ \cdot & \cdot & \cdot & \cdot & \cdot & \cdot \end{pmatrix} \begin{pmatrix} \xi_{1,1} \\ \xi_{1,2} \\ \xi_{2,1} \\ \xi_{2,2} \\ \xi_{3,1} \\ \cdot \\ \xi_{n,1} \\ \xi_{n,2} \end{pmatrix} = \begin{pmatrix} A_0 \\ 0 \\ 0 \\ 0 \\ 0 \\ \cdot \\ 0 \\ 0 \end{pmatrix} \quad (6.73)$$

with:

$$a_{2,1} = \exp\left[-(1+i)\frac{\Delta z}{D_{m,1}}\right] \; ; \; a_{2,2} = \exp\left[(1+i)\frac{\Delta z}{D_{m,1}}\right] \; ; \; a_{2,3} = -1$$

$$a_{3,1} = -\frac{\lambda_1}{D_{m,1}}\exp\left[-(1+i)\frac{\Delta z}{D_{m,1}}\right] \; ; \; a_{3,2} = \frac{\lambda_1}{D_{m,1}}\exp\left[(1+i)\frac{\Delta z}{D_{m,1}}\right] \; ; \; a_{3,3} = \frac{\lambda_2}{D_{m,2}}$$

$$a_{n,1} = -\frac{\lambda_{n-2}}{D_{m,n-2}}\exp\left[-(1+i)\frac{\Delta z}{D_{m,n-2}}\right]; a_{n,2} = \frac{\lambda_{n-2}}{D_{m,n-2}}\exp\left[(1+i)\frac{\Delta z}{D_{m,n-2}}\right] \; ;$$

$$a_{n,3} = \frac{\lambda_{n-1}}{D_{m,n-1}} \quad (6.74)$$

This set of $2n-1$ linear equations with $2n-1$ unknown variables ξ can easily be solved.

The amplitudes and the phase shifts are derived from the complex variables as:

$$A_{i,1} = \sqrt{\text{Re}_{\xi_{i,1}}^2 + \text{Im}_{\xi_{i,1}}^2} \qquad \phi_{i,1} = \arctan(\frac{\text{Im}_{\xi_{i,1}}}{\text{Re}_{\xi_{i,1}}})$$

$$A_{i,2} = \sqrt{\text{Re}_{\xi_{i,2}}^2 + \text{Im}_{\xi_{i,2}}^2} \qquad \phi_{i,2} = \arctan(\frac{\text{Im}_{\xi_{i,2}}}{\text{Re}_{\xi_{i,2}}}) \qquad (6.75)$$

in which:
Re = the real part of ξ
Im = imaginary part of ξ

When the model is used for evaluating processes in the past, measured temperature data can be used in the calculations. To derive a temperature course from a limited number of data the method of a Fourier analysis is used. If 52 weekly measurements of air temperatures and corresponding day numbers are known, which must be equidistant in time, the model describes the soil temperature with the expression:

$$T(z,t) = \overline{T} + \sum_{n=1}^{6} \left[A(n) \exp[-z \sqrt{\frac{n}{D_m}}] \sin(n\omega t + \phi(n) - z \sqrt{\frac{n}{D_m}}) \right] \qquad (6.76)$$

in which the Fourier coefficients $A(n)$ and $\phi(n)$ (n from 0 to 5) are given as:

$$\phi(n) = \arctan(a(n)/b(n)) \; ; \qquad A(n) = b(n)/\cos\phi(n)$$

$$a(n) = \frac{2}{52} \sum_{i=1}^{52} [T(0,i)\cos(n\omega t(i))]; \quad b(n) = \frac{2}{52} \sum_{i=1}^{52} [T(0,i)\sin(n\omega t(i))] \qquad (6.77)$$

The average temperature $\overline{T}$ is calculated as $a(0)/2$.

6.3.3 Thermal properties of soils

The thermal conductivity (λ_T) and the volumetric specific heat (C_h) are not constant for a certain type of soil but they change with the composition of the soil and with moisture content. Before temperature profiles can be calculated, the volumetric specific heat and the thermal conductivity of the soil need to be known. Values of C_h and λ_T can be approximated from the basic physical properties of the soil. The volumetric specific heat of a soil is the sum of the volumetric specific heats of all soil constituents, so:

$$C_h = C_q \upsilon_q + C_{cl} \upsilon_{cl} + C_{om} \upsilon_{om} + C_w \theta + C_a \varepsilon_g \qquad (6.78)$$

where C is the volumetric specific heat and υ is the volume fraction of the component indicated by the subscript. The subscripts q, cl, om, w and a indicate quartz, clay minerals, organic matter, water and air, respectively. The thermal conductivity λ_T of a soil depends on its bulk density, quartz content, clay content, organic matter content and water content. Table 6.4 gives typical values for C_h and λ_T for each of these soil components. The contribution of the soil air to the heat capacity is generally ignored. At low water contents in the soil the air space controls the thermal conductivity. At high water content, the thermal properties of the different components become more important.

Table 6.4. Thermal properties of soil materials after de Vries (1963).

Material	Density	Specific heat	Thermal conductivity	Volumetric specific heat
	Mg.m^{-3}	J.g^{-1}.°K^{-1}	W.m^{-1}.°K^{-1}	MJ.m^{-3}.°K^{-1}
Quartz	2.66	0.80	8.80	2.13
Clay minerals	2.65	0.90	2.92	2.39
Organic matter	1.30	1.92	0.25	2.50
Water	1.00	4.18	0.57	4.18
Air (20 °C)	0.0012	1.01	0.025	0.0012
Ice	0.92	1.88	2.18	1.73

McInnes (1981), as cited by Campbell (1985), derived by curve fitting an expression for the thermal conductivity which is given as:

$$\lambda_T = \varsigma_1 + \varsigma_2 \theta - (\varsigma_1 - \varsigma_4) \exp[-(\varsigma_3 \theta)^{\varsigma_5}] \qquad (6.79)$$

where θ is volumetric water content and ς_1, ς_2, ς_3, ς_4 and ς_5 are empirically derived constants. The parameters ς_1, ς_2, ς_3, ς_4 and ς_5 are calculated using the expressions:

$$\varsigma_1 = \frac{0.57 + 1.73\upsilon_q + 0.93\upsilon_{cl}}{1 - 0.74\upsilon_q - 0.4\upsilon_{cl}} - 2.8\upsilon_{cl}(1 - \upsilon_{cl})$$
$$\varsigma_2 = 1.06(2.66\upsilon_q + 2.65\upsilon_{cl} + 1.3\upsilon_{om})$$
$$\varsigma_3 = 1 + 2.6[\frac{2.65\upsilon_{cl}}{2.66\upsilon_q + 2.65\upsilon_{cl} + 1.30\upsilon_{om}}]^{-1/2}$$
$$\varsigma_4 = 0.03 + 0.7(\upsilon_q + \upsilon_{cl} + \upsilon_{om})^2 \quad \varsigma_5 = 4 \qquad (6.80)$$

Rijtema et al. (1997) give calculated thermal properties for a range of soil types with different combinations of organic matter and clay contents. For practical applications soil moisture contents between saturation and wilting point have to be considered. It appeared from these calculations that both λ_T and C_h decrease with decreasing soil moisture contents. This means that the damping depth becomes less dependent on soil moisture content than λ_T and C_h, with the exception of sandy soils that do not contain organic matter and with extremely low moisture contents at wilting point. So, for forecasting purposes a constant value of the damping depth D_m will be used for each soil type, depending on yearly average soil moisture conditions. Characterizing the soil type by the sum of 2.5 times organic matter fraction plus clay fraction gives a reasonable relation between soil type and damping depth, as is presented in Fig 6.10, with the exception of a number of data with extremely low soil moisture contents at wilting point. These values near wilting point do not play an important role in regional studies, so the relation between damping depth and soil type is given by:

$$D_m = 1.45 + 1.35\exp[-1.6(2.5fr_{om} + fr_{cl})] \qquad (6.81)$$

in which:
fr_{om} = weight fraction of organic matter (-)
fr_{cl} = weight fraction of clay in the mineral parts (-).

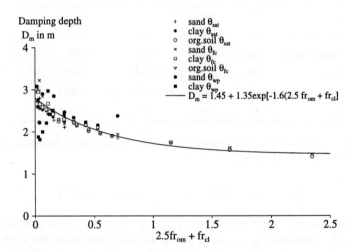

Fig. 6.10: The relation between damping depth and the sum of 2.5 times organic matter fraction plus clay fraction of the soil; the curve has been calculated ignoring the data at wilting point.

In multi-layered soil systems the ratio between heat conductivity and damping depth in Eq. (6.74) also plays a role. This ratio can be given as:

$$\frac{\lambda_T}{D_m} = \sqrt{\frac{\omega C_h \lambda_T}{2}} \qquad (6.82)$$

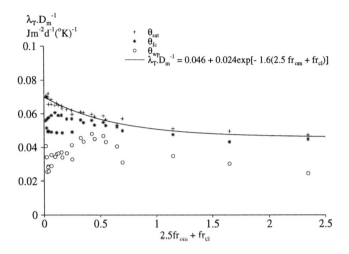

Fig. 6.11: Relation between the function ($\lambda_T.D_m^{-1}$) and the sum of 2.5 times organic matter content and the clay content; the curve is calculated for saturated soils.

The relation between λ_T/D_m and $2.5fr_{om} + fr_{cl}$ is presented in Fig. 6.11 for soils at saturation, field capacity and wilting point, respectively. For saturated soils the relation is given as:

$$\frac{\lambda_T}{D_m} = 0.046 + 0.024\exp[-1.6(2.5fr_{om} + fr_{cl})] \qquad (6.83)$$

The scatter in the data depends on the degree of saturation of the soil, as appears from Fig. 6.12, showing the reduction in the function in relation to the relative saturation. This relation is given by the expression:

$$f_{red} = 0.35 + 0.65(\frac{\theta}{\theta_{sat}})^{0.75} \qquad (6.84)$$

Combination of Eq. (6.83) and Eq (6.84) gives:

$$\frac{\lambda}{D_m} = [0.046 + 0.024\exp[-1.6(2.5fr_{om} + fr_{cl})]][0.35 + 0.65(\frac{\theta}{\theta_{sat}})^{0.75}] \qquad (6.85)$$

Fig. 6.13 shows the relation between the value of λ/D_m with Eq. (6.85) and the data given by Rijtema et al. (1997). It appears that in regional studies the thermal properties of soils can be approximated with the relations presented in Fig. 6.10, Fig. 6.11 and Fig. 6.12

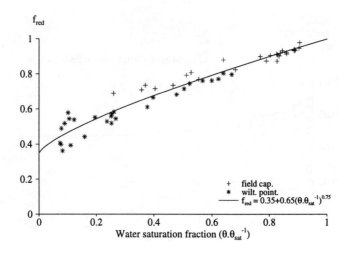

Fig. 6.12: Relation between the reduction function and relative water saturation for unsaturated soils.

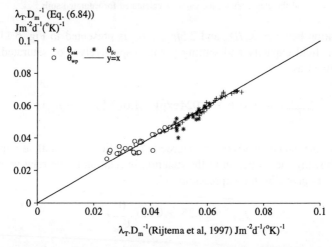

Fig. 6.13: Relation between the function $\lambda_T . D_m^{-1}$, calculated from the original soil data given by Rijtema *et al.* (1997), and the value calculated using Eq. (6.84).

6.4 Influence of clay content

Hendriks (1991), discussing the oxidation of peat, concluded on the basis of both literature data and measurements that the presence of clay reduces the decomposition of peat. Hendriks assumed that the lower decomposition data presented by Otten (1985), based on respiration measurements, are more realistic for the decomposition of peat than those given by Kolenbrander (1969, 1974). The data given by Kolenbrander were based on decomposition of peat with a destroyed structure both during composting and as addition to mineral soils. This change in decomposition can be explained either by a possible alteration in configuration during the aggregate formation through which organic compounds are structurally encapsulated in the aggregates, or by adsorption to clay minerals of the exo-enzymes, formed by the biomass, reducing their activity. Hansen et al. (1990) used the relation presented in Fig. 6.14 for the description of the reduction in decomposition of organic material as a function of the clay content of the soil. This function is also introduced in ANIMO.

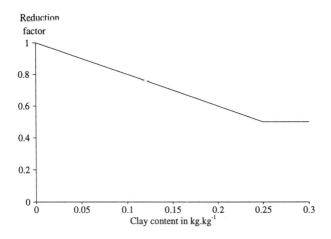

Fig. 6.14: The relation between the reduction factor for decomposition of organic material and the clay content of the soil after Hansen et al. (1990).

6.5 Influence of pH

The influence of pH depends on the type of reaction and on the preference of the micro-organisms involved. The acidity affects the enzymatic activity, the composition of the biomass population and the shape and magnitude of organic molecules. Under combined acid and anaerobic conditions decomposition of organic material will be reduced almost to zero.

Measurements by several authors indicate a broad optimum pH range for mineralization, nitrification and denitrification. The pH of most agricultural soils fall within this range. Data collected by van Huet (1983) from literature are presented in Fig. 6.15. The effect of pH on reaction rates is given in ANIMO by one general function for the organic transformation processes and the processes in the nitrogen cycle.

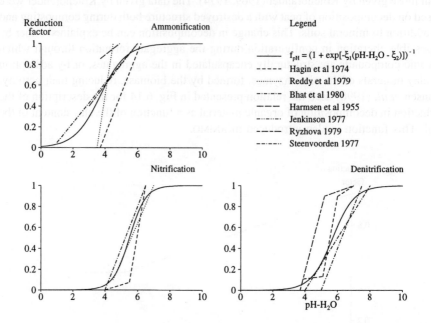

Fig. 6.15: The effect of pH on the reduction of the rate coefficients for different processes.
Ammonification: $\xi_1 = 1.24$; $\xi_2 = 3.7$;   Nitrification: $\xi_1 = 2.00$; $\xi_2 = 5.5$;
Denitrification: $\xi_1 = 1.50$; $\xi_2 = 5.7$.

The multiplier factor f_{pH} for the influence of pH on processes is given as:

$$f_{pH} = \frac{1}{1 + \exp[-\xi_1(pH - \xi_2)]} \tag{6.86}$$

Values of ξ_1 and ξ_2 for some processes are given in Table 6.5.

Table 6.5: Values of the constants ξ_1 and ξ_2 of Eq. (6.86) for different processes.

Process	ξ_1	ξ_2
Mineralization	1.24	3.7
Nitrification	2.00	5.5
Denitrification	1.50	5.7

It appears from Fig. 6.15 and Table 6.5 that the given processes have different pH relations. This particularly plays a role under partial or complete anaerobic conditions in the soil. Mineralization of organic material and denitrification are, under these conditions, coupled processes. Describing these processes, due to denitrification, in terms of mineralization of organic materials requires the introduction of an additional pH dependent reduction factor for the anaerobic process, which can be given as:

$$f_{red}^{den} = \frac{f_{pH}^{den}}{f_{pH}^{min}} = \frac{1 + \exp[-1.24(pH - 3.7)]}{1 + \exp[-1.50(pH - 5.7)]} \qquad (6.87)$$

The relation between f_{red}^{den} and pH is graphically presented in Fig. 6.16.

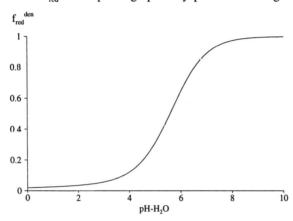

Fig. 6.16: The fraction of mineralization of organic material by denitrification under anaerobic conditions in relation to pH.

The relation has been based on soil water quality data, so pH-values have to be considered as pH-H_2O values. Very often, only pH-KCl values are available from soil information systems. A linear relation has been fitted between pH-KCl and pH-H_2O values, yielding for different soil types the regression equations presented in Table 6.6.

Table 6.6: Transfer functions for assessment of pH-values from soil chemical properties.

Soil type	Relation
Sandy soils	pH-H2O = 0.726pH-KCl + 2.116
Peat soils	pH-H2O = 0.851pH-KCl + 1.384
Sandy loam soils	pH-H2O = 0.782pH-KCl + 1.977
Clay soils	pH-H2O = 0.762pH-KCl + 2.252

It has been assumed that, under optimal agricultural practices, the pH-value will not change and the seasonal fluctuations have been ignored. In that situation time independent pH values have to be defined by the model user for each soil horizon.

References

Bakker, J.W., 1965. *Luchthuishouding van bodem en plantewortels; een literatuurstudie.* Nota 302. Institute Land and Water Management Research: Wageningen, The Netherlands.

Bakker, J.W., Boone, F.R. en Boekel, P. 1987. *Diffusie van gassen in grond en zuurstofdiffusiecoëfficienten in Nederlandse akkerbouwgronden.* Rapport **20**. Institute Land and Water Management Research: Wageningen, The Netherlands.

Beek, J. and Frissel, M.J. 1973. *Simulation of nitrogen behaviour in soils.* PUDOC: Wageningen, The Netherlands.

Bhat, K.K., Flowers, T.H. and O'Callaghan, J.R. 1980. A model for the simulation of the fate of nitrogen in farm wastes on land application. *J. Agricultural Science*, **94**, 183-193.

Boesten, J.J.T.I. 1986. *Behaviour of herbicides in soil: simulation and experimental assessment* Ph.D. Thesis, Agricultural University: Wageningen, The Netherlands.

Boesten, J.J.T.I. and Linden, A.M.A. van der, 1991. Modelling the influence of sorption and transformation on pesticide leaching and persistence. *J. Environmental Quality*, **20**, 425-435.

Bronswijk, J.J.B. 1991. *Magnitude, modelling and significance of swelling and shrinkage processes in clay soils.* Ph. D. Thesis. Agricultural University, Wageningen, The Netherlands.

Campbell, G.S. 1985. Soil physics with basic transport models for soil-plant systems. Elsevier Science Publishers: Amsterdam, The Netherlands.

Davidson, J.M., Greatz, D.A., Rao P.S.C. and Selim, H.M. 1978. Simulation of nitrogen movement, transformation and uptake in plant root zone. Ecological Research Series 600/3-78-029. Environmental Protection Agency: USA.

Duin, W.R. van, 1956. *Over de invloed van grondbewerking op het transport van warmte, lucht en water in de grond.* Ph. D. Thesis. Agricultural University: Wageningen, The Netherlands.

Gliński, J. and Stepniewski, W. 1985. *Soil aeration and its role for plants.* CRC Press: Boca Raton.

Hagin, J. and Amsberger, A. 1974. *Contribution of fertilizers and manures to the N- and P-load of waters.* Final Report to the Deutsche Forschungsgemeinschaft.

Hansen, S., Jensen, H.E., Nielsen, N.E. and Svendsen, H. 1990. *Daisy; a soil plant system model.* The Royal Veterinary and Agricultural University, Dept. agric. Sci., Section Soil and Water and Plant Nutrition: Copenhagen, Denmark.

Hendriks, R.F.A., 1991. *Afbraak en mineralisatie van veen; Literatuur onderzoek.* Rapport **199**. SC–DLO: Wageningen, The Netherlands.

Huet, H. van, 1983. *Kwantificering en modellering van de stikstofhuishouding in bodem en grondwater na bemesting.* Nota **1426**. Institute Land en Water Management Research: Wageningen, The Netherlands.

Hunt, H.W., 1977. A simulation model for decomposition in grasslands. *Ecology*, **58**, 469-484.

Kolenbrander, G.J., 1969. *De bepaling van de waarde van verschillende soorten organische stof ten aanzien van hun effect op het humusgehalte bij bouwland.* C 6988 Institute for Soil Fertility: Haren, The Netherlands.

Kolenbrander, G.J., 1974. Efficiency of organic manure in increasing soil organic matter content. *Trans. 10th Int. Congr. Soil Sci.* Vol.2. Moscow, USSR.

Lammers, W.H., 1983. *Gevolgen van het gebruik van organische mest op bouwland.* Consulentschap voor Bodemaangelegenheden in de landbouw: Wageningen, The Netherlands.

McCaskill, M.R. and Blair, G.J. 1990a. A model of S, P, and N uptake by a perennial pasture. 1. Model construction. *Fertilizer Research*, **22**, 161-172.

McCaskill, M.R. and Blair, G.J. 1990b. A model of S, P, and N uptake by a perennial pasture. 2. Calibration and predictions. *Fertilizer Research*, **22**, 173-179.

McInnes, K.J. 1981. Thermal conductivities of soils from dry-land wheat regions of Eastern Washington. M.S. Thesis. Washington State University, Pullman: Washington.

Otten, W., 1985. *Nader onderzoek naar oxydatie van veengronden, literatuuroverzicht en metingen aan veenmonsters.* Nota **1620**. Institute Land and Water Management Research: Wageningen, The Netherlands.

Reddy, K.R., Khaleel, R., Overcash, M.R. and Westerman, P.W. 1979. A non-point source model for land areas receiving animal wastes: 1: Mineralisation of organic nitrogen. *Transactions of the ASAE*, **22**: 863-874.

Rijtema, P.E., Groenendijk, P, Kroes, J.G. and Roest, C.W.J. 1997. *Modelling the nitrogen and phosphorus leaching to groundwater and surface water; Theoretical backgrounds and future developments of the ANIMO model.* Report **30**. SC–DLO: Wageningen, The Netherlands.

Ryzhova, I.M., 1979. Effect of nitrate concentration on the rate of soil denitrification. *Soviet Soil Science*, **11**, 168-171.

Vaan, G.J. de, 1987. *Maaiveldszakking van veenweidegebieden, oxydatiemetingen aan veenmonsters.* Stageverslag, Hogere Agrarische School: 's Hertogenbosch, The Netherlands.

Veen, J.A. van and Frissel, M.J. 1981. Simulation model of N- behaviour of N in soil. In *Simulation of nitrogen behaviour of soil-plant systems; papers of a workshop: Models for the behaviour of nitrogen in soil and uptake by plant; comparison between different approaches.* (ed. M.A. Frissel and J.A. van Veen). PUDOC: Wageningen, The Netherlands.

Vries, D.A. de, 1963. Thermal properties of soils. In *Physics of Plant Environment* (ed. W.R. van Wijk), pp. 210-235. North-Holland Publishing Co.: Amsterdam, The Netherlands.

Watts, D.G. and Hanks, R.J. 1978. A soil-water model for irrigated corn on sandy soils. *Soil Science Soc. Amer. Proc.*, **42**, 492-499.

CHAPTER 7

WATER, NUTRIENT UPTAKE AND CROP PRODUCTION

Plants play an important role in the nutrient and carbon cycles of soils. The description of plant development is important in nutrient and organic carbon balance studies as:

- nutrient uptake by plants is strongly related to development and production;
- mineral nutrients can only be taken up from those layers in which roots are present;
- root exudation is related to root growth;
- part of the roots die off during growth;
- remaining crop losses as root mass and stubble on arable land are available for decomposition after harvesting the crop.

In particular, when policy strategies affect both the level of fertilization as well as the application of animal slurries, sewage sludge and (treated) waste water in agriculture, related changes in nutrient uptake and dry matter production of crops might become important factors in the carbon and nutrient balances of the soil. Models used for the evaluation of the environmental effects of management strategies and different policies require at least a brief formulation of the effects of water and nutrient availability on crop development, nutrient uptake and production.

7.1 Transport to roots

The soil-plant continuum can be considered as a series of nutrient fluxes from the soil to the plant pool, where inorganic nutrients are transferred into organic compounds. The uptake of nutrients can be limited by:

- the transport to the roots in the soil system;
- the root uptake capacity;
- the crop demand.

The nutrient uptake from the soil is highly dependent on the root development of the plant. In two stages of crop development, depending on crop species, nutrient uptake can lag behind the demand for a maximum growth rate. This can occur during initial growth, when the demand per unit length of root is usually high, as well as during the second half of the growing period when root growth stops and nutrients in the root environment have been partially depleted by the preceding demand. Root systems of finite density can only extract a fraction of the nutrients present in the soil at the rate required for a non-restricted crop growth. The rate of availability of the remaining part is determined by transport processes in the soil.

7.1.1 Nutrient transport to plant roots

De Willigen and van Noordwijk (1987, 1994a,b) developed a model for nutrient uptake, based on mass flow and diffusion of nutrients to a root. The plant root system is schematized by a number of vertical hollow cylinders, with a radius of r_r, each of them having their own sphere of influence. The nutrient transport to a single root per m root length can be described by a mass conservation equation, assuming the absence of gradients in the vertical direction. The general expression is given by the equation:

$$\frac{\partial(\theta c)}{\partial t} + \rho_d K_d \frac{\partial c}{\partial t} = r^{-1}\frac{\partial}{\partial r}[-r((\eta \theta D_w + \Lambda q_r)\frac{\partial c}{\partial r} + q_r c)] \quad (7.1)$$

subject to the boundary conditions:

$$r = r_{max} \quad q_r = 0 \quad \frac{\partial c}{\partial r} = 0$$

$$r = r_r \quad q_r = \frac{q_t}{2\pi r_r \Pi_r Z_r} \quad c = c_r \quad (7.2)$$

where:
- r = radial distance to the root in m
- r_r = radius of the root in m
- r_{max} = radius of the maximum sphere of influence = $1/(\pi \Pi)^{0.5}$ in m
- Π_r = root density in m^{-2} = m root length per m^{-3} soil
- D_w = diffusion coefficient in water in m^2.d^{-1}
- η = labyrinth factor (-)
- θ = moisture content in m^3.m^{-3}
- q_t = transpiration rate in m.d^{-1}
- q_r = radial flux to a plant root in m.d^{-1}
- Λ = dispersion length in m
- Z_r = depth of rooting in m

Eq. (7.1) can be solved numerically, considering the flow rate of a chemical compound to the plant roots as the flow through a series of cylindrical compartments and proportional to the concentration gradient. The uptake models developed by de Willigen and van Noordwijk (1987, 1994a,b) are based on steady-rate approximations considering a regular root distribution. Heinen (1997) concluded that under a non-uniform distribution of the root system and root clustering the actual uptake will be less than calculated with a uniform root distribution. For an approximation of the effect of nutrient diffusion on crop uptake in regional models, it is assumed that the solution in the compartment is mixed by vertical incoming and outgoing fluxes, so it is presumed that the radial dispersion can be ignored.

In the analysis of water uptake by a single plant root, as presented by Gardner (1960; 1964) and Rijtema (1965), it is assumed that the water participating in the flow comes from the outer boundary of the sphere of influence of each root. Moreover it has been assumed that capillary conductivity is a constant related to the mean suction in the root zone. Following these conditions also for the transport of nutrients with inflow of material only from the outer boundary and considering steady state conditions gives:

$$U_r = -r(\eta \theta D_w \frac{dc}{dr} + q_r c) \tag{7.3}$$

in which:
U_r = the rate of compound uptake at the root surface in kg.m^{-1}.d^{-1}.

Eq. (7.3) is subject to the boundary conditions:

$$r = r \quad q_r = \frac{q_t}{2\pi \Pi_r Z_r} \frac{1}{r} \quad c = c$$

$$r = r_r \quad q_r = \frac{q_t}{2\pi \Pi_r Z_r} \frac{1}{r_r} \quad c = c_r$$

$$r = r_{max} \quad q_r = \frac{q_t}{2\pi \Pi_r Z_r} \frac{1}{r_{max}} \quad c = c_{rmax} \tag{7.4}$$

Eq. (7.3) can be rewritten as:

$$\frac{dc}{dr} + \frac{q_t}{2\pi \Pi_r Z_r} \frac{1}{\eta \theta D_w} \frac{c}{r} = -\frac{U_r}{\eta \theta D_w} \frac{1}{r} \tag{7.5}$$

Introduction of the parameters:

$$\varsigma_1 = \frac{q_t}{2\pi \Pi_r Z_r} \frac{1}{\eta \theta D_w} \quad \varsigma_2 = \frac{U_r}{\eta \theta D_w} \quad \frac{\varsigma_2}{\varsigma_1} = \frac{J_{cr}}{q_t} \quad y = \ln r \tag{7.6}$$

transforms the steady state differential equation into:

$$\frac{dc}{dy} + \varsigma_1 c = -\varsigma_2 \tag{7.7}$$

Integration of Eq. (7.7) yields:

$$c = -\frac{\varsigma_2}{\varsigma_1} + \varsigma_3 r^{-\varsigma_1} \tag{7.8}$$

Eq. (7.8) is subject to the boundary conditions:

$$y = \ln r \qquad c = c$$
$$y_o = \ln r_r \qquad c = c_r$$
$$y_a = \ln r_{max} \qquad c = c_{rmax} \tag{7.9}$$

Substitution of the boundary conditions in Eq. (7.8) gives:

$$c = c_{rmax}(\frac{r}{r_{max}})^{-\varsigma_1} + \frac{J_{cr}}{q_t}[(\frac{r}{r_{max}})^{-\varsigma_1} - 1] \tag{7.10}$$

in which:

$J_{cr} = 2\pi\square\,Z_r U_r$ = the sink term for crop uptake rate in kg.m^{-2}.d^{-1}.

The average concentration is written as:

$$\bar{c} = \frac{1}{\pi(r_{max}^2 - r_r^2)} \int_{r_r}^{r_{max}} 2\pi c r dr =$$

$$\frac{1}{\pi(r_{max}^2 - r_r^2)} \int_{r_r}^{r_{max}} 2\pi[(c_{rmax} + \frac{J_{cr}}{q_t})(\frac{r}{r_{max}})^{-\varsigma_1} - \frac{J_{cr}}{q_t}]rdr =$$

$$[\frac{2}{2-\varsigma_1}[(c_{rmax} + \frac{J_{cr}}{q_t})\frac{r_{max}^{2-\varsigma_1} - r_r^{2-\varsigma_1}}{r_{max}^{-\varsigma_1}(r_{max}^2 - r_r^2)}] - \frac{J_{cr}}{q_t}] \tag{7.11}$$

or rewriting gives:

$$c_{rmax} = \frac{2-\varsigma_1}{2}(\bar{c} + \frac{J_{cr}}{q_t})\frac{r_{max}^{-\varsigma_1}(r_{max}^2 - r_r^2)}{r_{max}^{2-\varsigma_1} - r_r^{2-\varsigma_1}} - \frac{J_{cr}}{q_t} \tag{7.12}$$

Substitution of Eq. (7.12) into Eq. (7.10) yields for $r = r_r$:

$$U_{cr} = -J_{cr} = [\frac{1 - \frac{c_r}{\bar{c}}\xi_{so}}{1 - \xi_{so}}]\bar{c}q_t = \sigma_{so}\bar{c}q_t \tag{7.13}$$

in which:

σ_{so} = transpiration stream concentration factor for compound uptake (-)

The soil-plant coefficient depends on the root radius, root density and rooting depth, as well as on moisture content and diffusion coefficient and the ratio of the nutrient concentration at the root surface over the average nutrient concentration in the soil.

The factor ξ_{so} reads as:

$$\xi_{so} = \frac{2}{2-\varsigma_1}\left(\frac{1-(\frac{r_r}{r_{max}})^{2-\varsigma_1}}{1-(\frac{r_r}{r_{max}})^2}\right)\left(\frac{r_r}{r_{max}}\right)^{\varsigma_1} \tag{7.14}$$

The differential equation (7.5) yields, when q_t equals 0:

$$U_r = -r\eta\theta D_w \frac{dc}{dr} \tag{7.15}$$

subject to the boundary conditions:

$$\begin{array}{ll} r = r & c = c \\ r = r_r & c = c_r \\ r = r_{max} & c = c_{rmax} \end{array} \tag{7.16}$$

Integration yields:

$$c = c_{rmax} + \frac{J_{cr}}{2\pi \varPi_r Z_r \eta \theta D_w}(\ln r_{max} - \ln r) \tag{7.17}$$

For the average concentration the following expression holds:

$$\bar{c} = \frac{1}{\pi(r_{max}^2 - r_r^2)} \int_{r_r}^{r_{max}} 2\pi c r \, dr =$$

$$c_{rmax} + \frac{J_{cr}}{2\pi \varPi_r Z_r \eta \theta D_w}[\ln r_{max} - \frac{r_{max}^2 \ln r_{max} - r_r^2 \ln r_r}{(r_{max}^2 - r_r^2)} + \frac{1}{2}] \tag{7.18}$$

Combination of Eq. (7.17) and Eq. (7.18) gives, if r is set equal to r_r:

$$U_{cr} = -J_{cr} = -\frac{(2\pi \varPi_r Z_r \eta \theta D_w)(1-(\frac{r_r}{r_{max}})^2)}{\ln(\frac{r_r}{r_{max}}) + \frac{1}{2}[1-(\frac{r_r}{r_{max}})^2]}(\bar{c} - c_r) \tag{7.19}$$

De Willigen and Noordwijk (1987) give a review on root characteristics of different crops. The range for root diameters in the root zone varies from 0.1 to 0.2 cm and the root length density $\varPi_r$ varies from 0.8 to 20 cm.cm^{-3}. This gives the following parameter values: $r_r = 0.0005 - 0.001$ m with a geometric mean radius $\bar{r}_r = 0.00071$ m; $r_{max} = 0.00125 - 0.0063$ m and $\pi_r = 8\,000 - 200\,000$ m^{-2}. The approximation used for the evaluation of π_r and r_{max} for different crops as a function of time and depth is discussed in section 7.3.2, resulting in a linear relation between root density and depth in the root zone.

Some results of the relation between the specific uptake rate $U_{cr}.\bar{c}^{-1}$ in m.d^{-1} and the concentration ratio $c_r.\bar{c}^{-1}$ for some root densities and soil moisture conditions are presented in Fig.7.1.

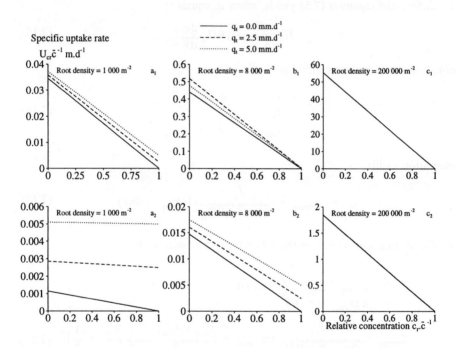

Fig. 7.1: The relation between relative uptake rate $U_{cr}.\bar{c}^{-1}$ and the concentration ratio $c_r.\bar{c}^{-1}$ for different conditions of root density and transpiration. a_1, b_1 and c_1: wet soils; a_2, b_2 and c_2: dry soils.

The calculations have been performed for a rootzone of 0.5 m and transpiration rates of 0, 2.5 and 5.0 mm.d^{-1}. The data show that the crop uptake is completely controlled by the diffusion process at root densities exceeding 8 000 m^{-2} and wet soil conditions. The transpiration rate has a great influence on the specific uptake rate in relatively dry soils, combined with a medium to low root density.

The relation between the soil-plant dependent transpiration stream concentration factor σ_{so} and the concentration ratio $c_r.\bar{c}^{-1}$ for different root densities, transpiration rates and soil moisture conditions is given in Fig. 7.2 for crops with a rooting depth of 0.5 m. It must be concluded from Fig. 7.2 that with high values for the root density the value of the coefficient σ_{so} can increase to very high values. It must also be concluded that it is not necessary to consider the concentration at the root surface under all conditions equal to zero, to realize sufficient nutrient uptake by the plant root system. The question remains as to which values of the concentration ratio can be expected under field conditions, since this factor determines largely the value of the transpiration stream concentration factor σ_{so}.

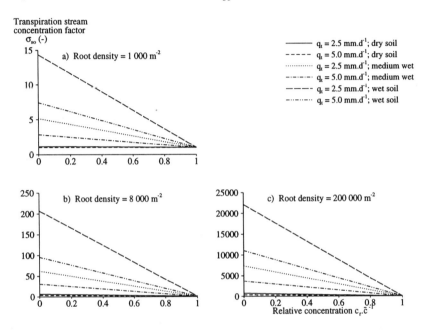

Fig. 7.2: The relation between the soil-plant coefficient and the concentration ratio $c_r . \bar{c}^{-1}$ with transpiration rates of 2.5 and 5.0 mm.d^{-1} and different soil moisture conditions; the root densities are given for Π_r equalling 1 000, 8 000 and 200 000 m^{-2}.

For a further analysis of realistic values of σ_{so}, Eq. (7.13) can be rewritten as a diffusion type expression:

$$U_{cr} = \frac{\xi_{so}^{-1}\bar{c} - c_r}{\Upsilon_{so}} \qquad (7.20)$$

with:

$$\Upsilon_{so} = q_t^{-1}[\xi_{so}^{-1} - 1] \qquad (7.21)$$

If $q_t = 0$ then $\xi_{so} = 1.0$ and the soil resistance becomes:

$$\Upsilon_{so} = -\frac{\ln(\frac{r_r}{r_{max}}) + \frac{1}{2}[1 - (\frac{r_r}{r_{max}})^2]}{2\pi \mathit{\Pi}_r Z_r \eta \theta D_W [1 - (\frac{r_r}{r_{max}})^2]} \quad (7.22)$$

7.1.2 Nutrient uptake by plant roots

The selection of nutrient uptake by the plant occurs in the root system, where a forced transport through the plant cells is present due to the Casparian strips. The nutrient transport into the plant is considered as a diffusion type transport process into the roots.

$$-U_r = \frac{c_r - c_{pl}}{\Upsilon_{cpl}} \quad (7.23)$$

where:
c_{pl} = concentration in the plant in kg.m^{-3}
Υ_{cpl} = compound dependent resistance of the crop per m root length in d.m^{-2}

Combination of Eq. (7.20) and Eq. (7.23) gives:

$$U_{cr} = \frac{\xi_{so}^{-1}\bar{c} - c_r}{\Upsilon_{so}} = 2\pi \mathit{\Pi}_r Z_r \frac{c_r - c_{pl}}{\Upsilon_{cpl}} = \frac{\xi_{so}^{-1}\bar{c} - c_{pl}}{\Upsilon_{so} + \Upsilon_{cpl}(2\pi \mathit{\Pi}_r Z_r)^{-1}} \quad (7.24)$$

Substitution of Eq (7.21) in Eq. (7.24) and rearranging of terms yields:

$$U_{cr} = \frac{1 - \xi_{so}\frac{c_{pl}}{\bar{c}}}{1 - \xi_{so}[1 - q_t \Upsilon_{cpl}(2\pi \mathit{\Pi}_r Z_r)^{-1}]}\bar{c}q_t = \sigma_{pl}\bar{c}q_t \quad (7.25)$$

The maximum uptake is present when $c_{pl} = 0$. Eq(7.25) can then be rewritten as:

$$U_{cr}^{max} = \frac{1}{1 - \xi_{so}(1 - q_t \Upsilon_{cpl}(2\pi \mathit{\Pi}_r Z_r)^{-1})}\bar{c}q_t = \sigma_{pl}^{max}\bar{c}q_t \quad (7.26)$$

Now, it is essential to estimate a realistic value of the plant resistance Υ_{cpl}. This is done by rewriting Eq. (7.24) under the condition that $c_{pl} = 0$, which gives:

$$U_{cr}^{max} = \frac{c_r^{min}}{\Upsilon_{cpl}(2\pi\varPi_r Z_r)^{-1}} = \frac{\xi_{so}^{-1}\bar{c}}{\Upsilon_{so} + \Upsilon_{cpl}(2\pi\varPi_r Z_r)^{-1}} \quad (7.27)$$

or:

$$\Upsilon_{cpl} = \frac{2\pi\varPi_r Z_r[1-\xi_{so}]c_r^{min}\bar{c}^{-1}}{q_t[1-\xi_{so}c_r^{min}\bar{c}^{-1}]} \quad (7.28)$$

The relation between Υ_{cpl} and the ratio $c_r^{min}.\bar{c}^{-1}$ is presented in Fig. 7.3 for different values of root density, transpiration rate and soil moisture conditions.

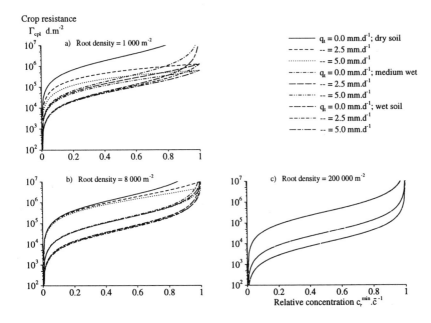

Fig. 7.3: The relation between Υ_{cpl} and the ratio $c_r^{min}.\bar{c}^{-1}$ for different conditions of root density, transpiration rate and soil moisture, and a rooting depth of 0.5 m.

It appears from Fig.7.3 that only at low root densities and dry soil conditions does the transpiration rate affect the relation between Υ_{cpl} and $c_r^{min}.\bar{c}^{-1}$ to a certain extent. At the highest root density the relations are completely controlled by the moisture conditions in the soil and the transpiration rate has no influence. The relation between the maximum value of the transpiration stream concentration uptake rate σ_{cpl}^{max} and the plant resistance Υ_{cpl} is obtained from Eq. (7.26) when $q_t > 0$ and by combination of Eq. (7.24) and Eq. (7.22) with $c_{pl} = 0$, yielding, respectively, the expressions:

$$\sigma_{cpl}^{max} = \frac{q_t}{1 - \xi_{so}(1 - q_t \Upsilon_{cpl}(2\pi \Pi_r Z_r)^{-1})} \qquad (7.29)$$

and:

$$\sigma_{cpl}^{max} = \left[-\frac{\ln(\frac{r_r}{r_{max}}) + \frac{1}{2}[1 - (\frac{r_r}{r_{max}})^2]}{2\pi \Pi_r Z_r \eta \theta D_W [1 - (\frac{r_r}{r_{max}})^2]} + \Upsilon_{cpl}(2\pi \Pi_r Z_r)^{-1} \right]^{-1} \qquad (7.30)$$

These relations are presented for different conditions in Fig. 7.4.

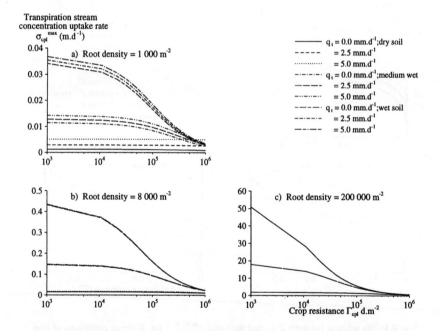

Fig. 7.4: The relation between the transpiration stream concentration uptake rate σ_{cpl}^{max} and the plant diffusion resistance Υ_{cpl} with different soil moisture, root density and transpiration conditions.

It appears from Fig. 7.4 that the value of σ_{cpl} slightly depends on the transpiration rate only at small values of the root density. This is in agreement with the results of the root uptake model given by de Willigen and van Noordwijk (1987). It is reasonable to assume that the crop resistance Υ_{cpl} depends on the molecular size or ionic size of the chemical compound considered, resulting in a different uptake rate σ_{cpl} for different compounds.

Combination of Eq (7.25) and Eq. (7.26) gives:

$$U_{cr} = U_{cr}^{max}(1 - \xi_{so}\frac{c_{pl}}{c}) \qquad (7.31)$$

The nutrient uptake will be less than the required one if $U_{cr}^{req} > U_{cr}^{max}$. The crop tends to reach a situation that $U_{cr} = U_{cr}^{req}$ by building up a compound concentration in the plant as long as $U_{cr}^{max} > U_{cr}^{req}$.

Nutrient requirements (U_{cr}^{req} kg.m^{-2}.d^{-1}) in the plant during different growth stages are discussed in section 7.5.

7.2 Gross photosynthesis

The synthesis of compounds other than carbohydrates may quantitatively be ignored during growth of a field crop. This means that dry matter production is mainly the result of net photosynthesis. Photosynthesis is a photochemical process using light energy for the reduction of CO_2. It must be taken into account that high photosynthesis rates can only be obtained when only a small fraction of dry matter is lost by respiration and when the process in the leaves is not adversely affected by:

- shortage of water or minerals;
- low or high temperature;
- the age of the leaves.

The process is affected by a diffusion process transporting CO_2 from the external air to the chloroplasts. The rate of the diffusion process depends mainly on the CO_2 concentration in the external air and in the leaf mesophyll, as well as on the resistances in the pathways.

Plant water relations are determined by the balance of water uptake through the roots and water loss from the leaves, in a situation where the flow of water through the plant is orders of magnitude larger than growth requires. Moisture stress affects photosynthesis mainly indirectly through an influence on the stomatal resistance, in a similar way as it affects the transpiration rate. In early stages of growth moisture stress also affects photosynthesis by reduced cell elongation, resulting in a reduced development of the leaf area index. Similar to water relations, plant nutrient relations determine in many ways the primary production and carbon partitioning processes. But there is a significant difference between the nutrient related and water related processes. Storage of water and its incorporation in growth processes is quantitatively insignificant compared to the total water flux in the plant. In the case of nutrients, the uptake of nutrients may be at a different time than the actual requirement, so the storage of inorganic nutrients in the plant can play an important role. Additionally, the stored nutrients may or may not be accessible. The nutrient relations control photosynthesis via protein synthesis. In most species the capacity for fixing CO_2 is related to the nitrogen content of the leaf. Lack of nitrogen in plants results in a translocation of nutrients from older leaves, which are photosynthetically non-active, to younger active leaves and to seeds. This translocation of nutrients requires additional energy resulting in an increased respiration rate. Lack of nitrogen can be described as a reduced efficiency for photosynthesis.

7.2.1 Gross photosynthesis of a standard crop

Since light energy is the factor which can be determined with the greatest ease and accuracy, it is useful to take this variable as the main factor in a production function while other limiting production factors are used as reduction functions. Photosynthesis of a canopy does not only depend on light intensity and the photosynthesis function of the single leaves, but also on factors that affect the distribution of light over the leaves of the canopy. The most important of these latter factors are the number and size of the leaves and their position with respect to the soil and to each other, the transmission and the reflection of the leaves and the ratio between diffuse and direct light and the height of the sun. The daily total gross photosynthesis of a canopy under standard conditions, as defined by De Wit (1965), can be approximated for any arbitrary place on earth and any time of the year by the following expressions:

- For perfectly clear days:

$$P_{cst} = 435 \cos(\frac{\pi \Phi}{180})^{-0.72} (\cos \frac{\pi (23.45 \sin(2\pi \frac{t+284}{365}) - \Phi)}{180}) (1 + 0.005|\Phi|) \quad (7.32)$$

- For completely overcast days:

$$P_{ost} = 229 \cos(\frac{\pi \Phi}{180})^{-0.75} (\cos \frac{\pi (23.45 \sin(2\pi \frac{t+284}{365}) - \Phi)}{180}) (1 + 0.008|\Phi|) \quad (7.33)$$

- Moreover:

$$\cos(23.45 \sin(\frac{t+284}{365} 2\pi)) \leq 0 \quad P_{cst} = 0 \quad P_{ost} = 0 \quad (7.34)$$

in which:
P_{cst} = photosynthesis on perfectly clear days for a standard crop (kg CH_2O ha^{-1}.d^{-1})
P_{ost} = photosynthesis on overcast days for a standard crop (kg CH_2O ha^{-1}.d^{-1})
Φ = latitude, positive for northern and negative for southern latitudes, in degrees
t = day number of the year.

The comparison between the data for photosynthesis, given by De Wit (1965), for both perfectly clear and completely overcast skies and the calculated photosynthesis with Eq. (7.32) and Eq. (7.33) is presented in Fig. 7.5.

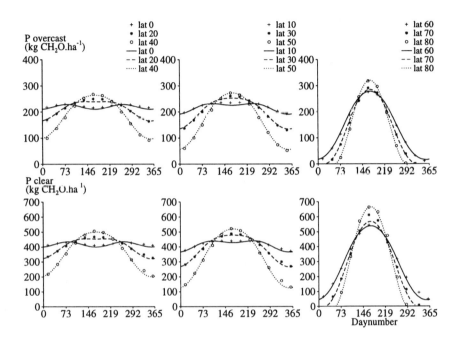

Fig. 7.5: Photosynthesis of a standard crop for both perfectly clear and completely overcast skies as a function of latitude and day number of the year. The markers are the data given by De Wit (1965).

The expressions give reliable results between 80° N and -80° S. Using the relative duration of bright sunshine (χ) as a distribution factor for partly clouded days gives for the standard crop production:

$$P_{st} = \chi P_{cst} + (1 - \chi) P_{ost} \qquad (7.35)$$

P_{st} = photosynthesis of standard crop in kg $CH_2O.ha^{-1}.d^{-1}$

7.2.2 Gross photosynthesis under non-standard conditions

The calculation of gross photosynthesis for a standard crop is well defined in relation to maximum assimilation rate (A_{max}), leaf area index (L_{ai}) under conditions of optimum water and nutrient supply. Actual crop conditions generally deviate strongly from these standard crop conditions. Nevertheless, photosynthesis of a standard crop is of general use in calculating crop production, provided some correction functions are derived.

7.2.2.1 Maximum assimilation rate

The maximum photosynthesis under standard crop conditions and normal CO_2 concentrations at the leaf surface is 20 kg CH_2O $ha^{-1}.h^{-1}$. The maximum assimilation rate for leaves of several plant species is about the same, but it may differ considerably for other agriculturally important species in one or other direction. The gross photosynthesis for deviating values of A_{max} can be calculated using two expressions derived from the De Wit (1965) data.

- For perfectly clear days:

$$P_c^m = 0.211 A_{max}^{0.52} P_{cst} \qquad (7.36)$$

- For completely overcast days:

$$P_o^m = 0.426 A_{max}^{0.285} P_{ost} \qquad (7.37)$$

in which:
A_{max} = maximum rate of photosynthesis in kg CH_2O $ha^{-1}.h^{-1}$
P_c^m = maximum photosynthesis on perfectly clear days in $kg.ha^{-1}.d^{-1}$
P_o^m = maximum photosynthesis on completely overcast days in $kg.ha^{-1}.d^{-1}$

7.2.2.2 Leaf area index

Standard gross photosynthesis is calculated for a crop completely covering the soil with a leaf area index (L_{ai}) equalling 5. Partial soil cover and low values of L_{ai} result in a waste of light available for photosynthesis during early stages of growth. For the effects of partial soil cover and small values of L_{ai} an efficiency factor has been introduced depending on light adsorption. This efficiency function is given by the expression:

$$P_a = P_a^m (1 - \exp[-\kappa L_{ai}]) \qquad (7.38)$$

with:
P_a = actual photosynthesis in kg $CH_2O.ha^{-1}.d^{-1}$
P_a^m = maximum photosynthesis in kg $CH_2O.ha^{-1}.d^{-1}$
κ = light absorption coefficient (-)

The value of $L_{ai}(t)$ is calculated from the quantity of dry matter present in the shoots of the standing crop, using the equation:

$$L_{ai}(t) = \frac{Q_{sh}(t)}{\overline{z}_{sh}(M_s \rho_{dm} + (1 - M_s)\rho_w)} \qquad (7.39)$$

in which:
L_{ai} = leaf area index
Q_{sh} = dry matter present in the shoots in kg.m^{-2}
$\overline{z}_{sh}$ = mean thickness of the shoots in m
M_s = fraction of dry matter present in the shoots in kg.kg^{-1}
ρ_{dm} = specific weight of dry matter in kg.m^{-3}
ρ_w = specific weight of water in kg.m^{-3}

7.2.2.3 Soil moisture stress

Soil moisture stress affects crop development and photosynthesis by affecting stomatal opening and cell elongation. Rijtema (1965) showed a direct relation between the soil moisture potential in the rootzone, the transpiration rate and the canopy resistance of grass. This relation has been confirmed for other crops by Rijtema and Ryhiner (1968), Endrödi and Rijtema (1969) and Feddes (1971). Roest et al. (1993) extended this relation, taking into account the effects of the osmotic potential. Though the canopy resistance in these studies is described as a partial closing of the stomata it seems more likely that under stress conditions the stomata alternately close and completely open.

This phenomenon is difficult to describe quantitatively. For this reason the relative transpiration $q_t \cdot q_{tp}^{-1}$ is used as the relative time that the stomata are fully open (Rijtema and Aboukhaled, 1975). Gross photosynthesis is under these conditions calculated as:

$$P_a = \frac{q_t}{q_{tp}} P_a^m (1 - \exp[-\kappa L_{ai}]) \qquad (7.40)$$

with:
q_t = actual transpiration in m.d^{-1}
q_{tp} = potential transpiration in m.d^{-1}

Soil moisture stress also affects cell elongation. As a result of this effect an increase in dry matter content will be obtained. This is very important during the early stages of growth when partial soil cover is present, since it reduces the development of L_{ai} with time.

The relation between dry matter fraction and soil moisture suction can be approximated using the expression:

$$fr_{dm} = fr_{dm}^{min} + (1 - fr_{dm}^{min})(\psi 10^{-4})^\varsigma \qquad (7.41)$$

in which:
fr_{dm}^{min} = minimum dry matter fraction at zero suction (0.07 for grass shoots)
ς = empirical constant (0.21 for grass shoots)
ψ = soil moisture suction in bar

An example for grass shoots is given in Fig. 7.6.

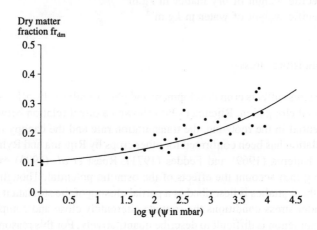

Fig. 7.6: Relation between soil moisture suction and dry matter content of grass shoots.

7.3 Dry matter production of arable crops

7.3.1 Distribution of dry matter

The relations that govern the translocation of carbohydrates are little known in the quantitative sense, so simulation is generally short-circuited by introducing a partitioning coefficient for growth over roots and shoots (De Wit and Penning de Vries, 1983). In case water is the controlling factor, the coefficient is assumed to be dependent on the relative water content of the crop. This relative water content is in its turn a state variable, which treats the water balance between water uptake by the roots and transpiration losses by the shoots in well known physical terms. The form of the partitioning curves reflects the observations that with decreasing relative water content the relative growth rate of the shoots decreases and that of the roots increases. The shoot/root ratio decreases with increasing evaporative demand. Defoliation leads to a reduced transpiration rate and to an increase in the shoot/root ratio.

Instead of using the relative water content of the crop as the steering variable, it is more convenient to use the maximum crop water potential as the steering variable. This factor has been introduced by Rijtema (1965) in an analysis of actual evapotranspiration. Roest et al. (1993) extended this relation for the effects of osmotic potential in connection with soil salinity. This maximum leaf water suction can be given by the expression:

$$\psi_l = q_{tp}(\Upsilon_{pl} + \frac{m_r}{k}) + \psi_m + \psi_{os}(\frac{\theta_{fc}}{\theta}) \qquad (7.42)$$

in which:
ψ_l = maximum leaf water suction in bar
ψ_m = mean soil water suction in the root zone in bar
ψ_{os} = mean osmotic pressure in the root zone at field capacity in bar
Υ_{pl} = crop resistance for liquid flow from root surface to the leaves in bar.d.m^{-1}
m_r = geometry factor for the root system in bar
k = capillary conductivity in m.d^{-1}
θ_{fc} = soil moisture content in the root zone at field capacity
θ = actual soil moisture content.

Based on data for different crops, Feddes and Rijtema (1972) showed that for practical purposes the geometry factor m_r of the root zone can be approximated by:

$$m_r = 1.275 Z_r^{-1} \qquad (7.43)$$

in which:
Z_r = effective depth of the root zone in m.

Feddes and Rijtema (1972) also showed that the value of Υ_{pl} increased with limiting soil moisture conditions in the root zone, caused by a changing uptake pattern when water becomes more and more limiting. Although the available information on the relation between Υ_{pl} and moisture conditions is limited Rijtema and Aboukhaled (1975) derived as a crude approximation:

$$\Upsilon_{pl} = MAX[0.5; 0.763 \ln \psi_m + 1.493] \qquad (7.44)$$

Rijtema and Aboukhaled collected data of the critical leaf water potential at which transpiration starts to reduce due to stomatal closure from literature. Rijtema and El Guindi (1986) found a good correlation between critical leaf water potential and the maximum salt concentration a crop could withstand. Based on their results the critical leaf water potential ψ_l^c for major field crops have been summarized in Table 7.1.

Table 7.1: Values of ψ_l^c in bars for a number of field crops derived from the analysis by Rijtema and El Guindi (1986).

Crop	ψ_l^c (bars)	Crop	ψ_l^c (bars)
Barley	13.8	Forage maize	7.7
Cotton	13.4	Cabbage	5.9
Sugar beet	11.9	Potatoes	4.9
Grass	10.0	Flax	4.9
Wheat	9.9	Carrot	4.0
Clover	9.4	Onions	3.7
Alfalfa	7.7	Beans	3.2

It is assumed that effects of crop water stress start at values above ψ_l^c. Taking $\Delta\psi_l = \psi_l - \psi_l^c$ as the driving force in the partitioning of dry matter between roots and shoots, gives as a general approximation for the relative shoot production the expression:

$$pc_{sh} = pc_{sh}^{max}(1 - (1 + \zeta_1 \exp[-\zeta_2 \Delta\psi_l])^{-1}) \; ; \quad pc_{ro} = 1 - pc_{sh} \qquad (7.45)$$

in which:

pc_{sh} = partition coefficient for shoot production (-)
pc_{sh}^{max} = maximum shoot partition coefficient under non-stress conditions (-)
pc_{ro} = partition coefficient for root production (-)
ζ_1, ζ_2 = empirically derived constants (-)

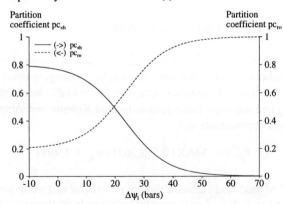

Fig. 7.7: The relation between the distribution of dry matter production and crop stress.

The dry matter distribution between shoots and roots, following Eq. (7.45), is presented in Fig. 7.7. After flowering and fruit setting, shoot and root growth stops and the net dry matter production is completely used for seed production.

7.3.2 Production of roots and exudates

It appears from literature data as presented in Table 7.2 that production respiration losses are 25 to 32 % of the gross photosynthesis. Root production is in the order of 30 % of net photosynthesis, which results in a dry matter production in the root system of about 6000 kg during growth. According to Sauerbeck et al. (1975), 75 % of the root dry matter of wheat was lost during growth by exudate production, dying of root hairs and root respiration. Similar data for maize were given by Louwerse et al. (1990).

Table 7.2: Relative production respiration and relative root production of some crops.

Author	Crop	Relative production respiration	Relative root production
Rijtema and Endrödi (1970)	Potatoes	0.32	-
Sauerbeck et al. (1975)	Wheat	0.30	0.30
Deinum (1985)	Grass	0.25	0.275
Louwerse et al. (1990)	Maize	0.30	0.30

Part of the shoots already dies off during the growing season and can be added as fresh organic material at the top of the soil system. Considering the loss of root mass during the growing season as being proportional to the root mass present, gives for the change of dry matter present in living roots the expression:

$$\frac{dQ_r}{dt} = P_r^{gr} - k_r Q_r \tag{7.46}$$

Integration of Eq. (7.46) yields as dry matter present in living roots at the end of each time interval $t - t_0$:

$$Q_r(t) = \frac{1}{k_r} P_r^{gr} + (Q_r(t_0) - \frac{1}{k_r} P_r^{gr}) \exp[-k_r(t-t_0)] \tag{7.47}$$

in which:
$Q_r(t)$ = dry weight of living roots at time t in kg.m^{-2}
$Q_r(t_0)$ = dry weight of living roots at time t_0 in kg.m^{-2}
P_r^{gr} = mean gross root production rate during the time interval in kg.m^{-2}.d^{-1}
k_r = rate coefficient for root mass consumption in d^{-1}

The root mass consumption during the growth of a crop is the sum of root respiration, dying roots and root hairs and exudate production. Root respiration and mineralization of dead roots and exudates result in oxygen consumption. Considering the total root mass consumption as exudation release, gives as the average exudate production rate $\bar{J}_e$, in kg.m^{-2}.d^{-1}, during the time interval:

$$\bar{J}_e = \frac{1}{t-t_0}\int_{t_0}^{t} k_r Q_r \mathrm{d}t = P_r^{gr} + (Q_r(t_0) - \frac{1}{k_r}P_r^{gr})\frac{1 - \exp[-k_r(t-t_0)]}{t-t_0} \quad (7.48)$$

The description of root development is important because:

- mineral nutrients can only be taken up from those layers in which roots are present, while for arable land root development as a function of time is important;
- root exudation is related to root growth;
- remaining root mass on arable land is available for decomposition after the harvest.

Development of root mass, root distribution and rooting depth are functions of plant species and environmental conditions, which are very difficult to describe. Therefore a standard development per plant species is introduced, using measured data from literature, collected by Berghuijs-van Dijk et al. (1985).

It appears from the data presented in Fig. 7.8, that a reasonable relationship between depth of rooting and day number after planting for different crops is present. Cereals, sown in autumn, show a deviating picture.

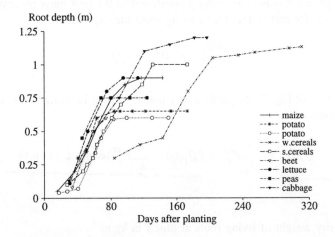

Fig. 7.8: The relation between depth of rooting and the number of days after planting for different crops.

Relating rooting depth with the dry matter present in the root system also gives a good approximation, as is shown in Fig. 7.9. The advantage of this approach is that the depth of rooting is directly related to root production under different stress conditions.

If the soil profile is not limiting the depth of rooting this relation can be approximated by considering three types of crops, respectively winter cereals, summer crops and potatoes.

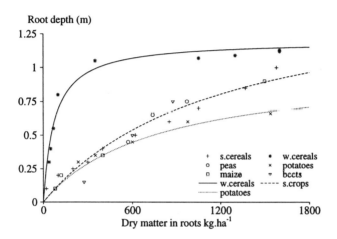

Fig. 7.9: The relation between the dry matter present in roots and depth of rooting.

The relations are given by the following expressions:

- winter cereals:

$$Z_r(t) = 1.2(1 - \frac{1}{135Q_r(t) + 1}) \qquad (7.49)$$

- summer crops:

$$Z_r(t) = 1.6(1 - \frac{16}{135Q_r(t) + 16}) \qquad (7.50)$$

- potatoes:

$$Z_r(t) = 1.0(1 - \frac{10}{135Q_r(t) + 10}) \qquad (7.51)$$

in which:
Q_r = aeric root mass present in living roots in kg.m^{-2}

It is assumed that a linear relation is present between the depth z in the root zone and the root mass distribution, given by the expression:

$$Q_r^V(z,t) = \frac{2Q_r(t)}{Z_r(t)}(1 - \frac{z}{Z_r(t)}) \qquad (7.52)$$

in which :
Q_r^V = volumic root mass in kg.m^{-3}
Z_r = depth of rooting in m
z = depth in root zone, subject to the condition $0 \le z \le Z_r$, in m

Denoting the specific weight of root dry matter by ρ_{dm} and the dry matter fraction of roots by M_r gives the following expression for the root volume intensity:

$$\upsilon_r(z,t) = \frac{Q_r^V(z,t)}{[M_r \rho_{dm}^{-1} + (1 - M_r)\rho_w^{-1}]} \qquad (7.53)$$

in which:
$\upsilon_r(z,t)$ = root volume in m^3.m^{-3}
M_r = dry weight fraction of roots in kg.kg^{-1}
ρ_{dm} = specific weight of dry matter in kg.m^{-3}
ρ_w = specific weight of water in kg.m^{-3}

The root length density, introduced in section 7.1.1, is calculated by:

$$Д_r(z,t) = \frac{\upsilon_r(z,t)}{\pi \bar{r}_r^2} \qquad (7.54)$$

and the radius of the sphere of influence by:

$$r_{max}(z,t) = \frac{\bar{r}_r}{\sqrt{\upsilon_r(z,t)}} \qquad (7.55)$$

The volumic exudate mass present $Q_{ex}^V(z,t)$, in kg.m^{-3}, is given by:

$$Q_{ex}^V(z,t) = \frac{2Q_{ex}(t)}{Z_r(t)}(1 - \frac{z}{Z_r(t)}) \qquad (7.56)$$

In water quantity models, very often the so-called effective root zone is used. This is the layer containing approximately 90 % of the total root mass. This effective root zone can vary for different soil profiles, depending on the root penetration resistance in the distinct profile layers. This soil constraint can be taken into account by taking under these conditions the maximum depth of rooting as:

$$Z_r = \text{MIN}[Z_r(t) ; Z_r^{max}] \quad (7.57)$$

in which:
Z_r^{max} = maximum depth of effective root zone due to soil constraints in m

7.3.4 Crop residues at harvest

Although crop yields can be calculated on the basis of net photosynthesis, this is not the case with crop residues remaining at the field after harvest. These residues are important, as they come available for decomposition in the organic matter cycle. Quantitative information concerning crop residues remaining at fields after harvesting the crop is only available from data obtained from field measurements. Table 7.3 shows a summary of crop residues present at harvest (de Jonge, 1981).

Table 7.3: Crop residues in kg dry matter.ha^{-1} (de Jonge, 1981).

Crop	Residues at harvest (kg DM.ha^{-1})		
	Below ground	Above ground	Total
Winter wheat	1 600	3 600	5 200
Spring wheat	1 400	3 800	5 200
Winter barley	1 400	3 600	5 000
Spring barley	1 000	3 200	4 200
Oats	1 400	3 600	5 000
Rye	1 200	3 600	4 800
Sweet maize	2 000	5 000	7 000
Silage maize	1 500	500	2 000
Potatoes			4 000
Beets			1 500
Beets including heads+leafs			6 000
Cabbage	1 000	4 000	5 000
Sprouts	1 000	500	1 500
Peas (+foliage)	400	1 600	2 000

7.4 Dry matter production of grassland

The utilization of grassland can be divided into:

- hay-winning;
- rotational grazing;
- continuously grazing.

For hay-winning, the grass sward is cut when the dry matter production of shoots equals 3 500 kg.ha^{-1}, above a 4 cm stubble. The remaining stubble, after cutting, has a dry matter weight of 750 kg.ha^{-1} (Lantinga, 1985).

The herbage intake by grazing cattle in a rotational grazing system is generally estimated by the sward cutting technique, i.e. estimating the amount of non-interrupted dry matter production of the sward before grazing starts. The starting date for grazing is determined by a threshold sward weight of 2 500 kg.ha^{-1}. When this criterium is not reached before a certain critical date, grazing will start at that date. If the grazing period is longer than one day, herbage production during the grazing period must be taken into account. After the grazing period, the stubble height has the same value as after hay harvesting. In situations with continuous grazing, the average minimum sward height equals 7-8 cm, which corresponds with a dry matter weight of the grass sward equalling 1 650 kg. ha^{-1}.

The continuously grazing system is also used in regional studies. However, part of the grassland area is used for hay-winning. Mixing both types of land use to a single system results in a grass production model in which the dry matter present in the sward increases, depending on the grazing intensity. In the regionalized model approach, the

Table 7.4: Default values for dry matter present in grass swards for different types of grassland utilization.

Land use	Default values of dry matter in kg.m^{-2}
Farm application	
1. Hay-winning	0.350
2. Rotational grazing	0.250
3. Continuously grazing	0.165
4. Stubble present at end of land use 1 and 2	0.075
5. Stubble present at end grazing season	0.075
Regional application	
1. Hay-winning	0.400
2. Stubble present after hay-winning	0.165
3. Stubble present at end of grazing season	0.075

grass sward always increases between two cuttings, due to the low regional and seasonal mean value of live-stock units per ha. When the total shoot production reaches a value of 4 000 kg.ha^{-1}, a grass cutting for hay-winning is introduced reducing the grass sward again to the minimum value of 1 650 kg.ha^{-1}, required for continuous grazing. At the end of the grazing season the remaining grass sward is cut and added to the first soil layer, leaving a stubble of 4 cm with a dry matter weight of 750 kg.ha^{-1}. Table 7.4 gives default values of the dry matter present in the grass sward at the start or at the end of the grassland use activity. These data can be set user defined in ANIMO.

7.4.1 Grazing and dry matter production

The dry matter shoot production for all three grass utilization systems can be described by the same production function:

$$\frac{dQ_{sh}}{dt} = \sigma_{dm} P_{st} \frac{q_t}{q_{tp}} (1 - \exp[-\kappa_{max} Q_{sh}/Q_{shmax}]) - U_i^{gr} \quad (7.58)$$

in which:
σ_{dm} = efficiency coefficient for dry matter production
κ_{max} = maximum light absorption factor if $Q_{sh}(t) = Q_{shmax}$
U_i^{gr} = total gross dry matter intake of grass by grazing cattle in kg.ha^{-1}.d^{-1}

The light absorption coefficient was determined by means of a number of test runs. It appeared that realistic production rates were obtained, assuming a light absorption of 90 % when Q_{shmax} = 3 500 kg.ha^{-1}, giving κ = 2.30. The efficiency coefficient σ_{dm} depends on the dry matter used for the formation of new roots, the production respiration and the maintenance respiration. According to Deinum (1985), 25 % of the gross photosynthesis is used for the production respiration and the maintenance respiration can be set at 12.5 %. The dry matter used for the formation of new roots is 27.5 %. So the efficiency coefficient for shoot production σ_{sh}^{dm} = 0.45 and for root production σ_r^{dm} = 0.17.

The dry matter intake by grazing cattle is about 14 kg.cow^{-1}.d^{-1} (Lantinga, 1985). The total daily dry matter uptake, including grazing losses U_i^{gr}, equals:

$$U_i^{gr} = \frac{1}{1 - fr_l} u n_c \quad (7.59)$$

in which:
fr_l = fraction of grazing losses with a default value of 0.25
u = dry matter herbage intake during grazing in kg.cow^{-1}.d^{-1}
n_c = number of animals in cow.ha^{-1}

Integration of Eq. (7.58) over a short time step gives for the shoot production:

$$Q_{sh}(t) = \frac{Q_{shmax}}{\kappa} \ln[\frac{\sigma_{sh}^{dm} P_{st}}{\sigma_{sh}^{dm} P_{st} - U_i^{gr}} +$$

$$(\exp[\frac{\kappa_{max} Q_{sh}(t_0)}{Q_{shmax}}] - \frac{\sigma_{sh}^{dm} P_{st}}{\sigma_{sh}^{dm} P_{st} - U_i^{gr}}) \exp[\frac{\kappa_{max}(\sigma_{sh}^{dm} P_{st} - U_i^{gr})(t - t_0)}{Q_{shmax}}]] \qquad (7.60)$$

in which:
$\bar{P}_{st}$ = average standard dry matter production during the period in kg.ha^{-1}.d^{-1}
Q_{shmax} = maximum dry matter present in shoots, default 3500 kg.ha^{-1}
$Q_{sh}(t_0)$ = dry matter in shoots present at t_0 in kg.ha^{-1}
$Q_{st}(t)$ = dry matter in shoots present at time t in kg.ha^{-1}
κ_{max} = relative light absorption if Q_{shmax} = 3500 kg.ha^{-1}, default value = 2.30 (-)

7.4.2 Root dry matter production

The amount of roots in grassland can be highly variable. On comparable natural grasslands the root mass is generally higher than on culture grasslands. The quantity is influenced by soil properties, botanical composition, age, cutting regime, time of the year, fertilization and water management (Dirven and Wind, 1982). When new grass is sown, the quantity of roots will generally increase during the first years, after which a certain stabilization takes place (Schuurman, 1973). When fertilization takes place, the equilibrium can be reached sooner (Schuurman and Knot, 1970). During one year, however, there is a general trend in root mass fluctuation, as is illustrated in Fig. 7.10.

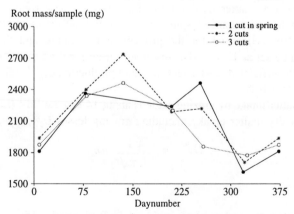

Fig. 7.10: Relation between dry matter present in grass roots and day number of the year for different treatments (Schuurman, 1973).

Throughton (1982) measured the average life-span of roots of some grasses under defoliation and suggested that the root mass is turned over between once and twice a year, depending on the intensity of cutting. For intensively used culture grasslands this means that the root mass is renewed twice a year, whereas it is once a year in extensively used natural grasslands. The fluctuation of the root mass during spring and summer is generally explained by the assumption that the growth rate is initially higher than the death rate, and later on the situation gets reversed (Goedewagen and Schuurman, 1950; Throughton, 1957). Therefore it is necessary to relate the death rate of roots to temperature, in the same way as the rate constants for organic matter transformation.

The gross production of new roots is given by the expression:

$$Q_r^{pr} = \left(\frac{\Delta Q_r}{\Delta t}\right)_{pr} = 0.17 P_{st}(1 - \exp[-\kappa_{max} Q_{sh}/Q_{shmax}]) \quad (7.61)$$

The assumption that the death rate of roots during the time step is proportional to the quantity of roots present results in the expression:

$$\frac{dQ_r}{dt} = Q_r^{pr} - k_r Q_r \quad (7.62)$$

Integration of Eq. (7.62) yields:

$$Q_r(t) = \frac{1}{k_r} Q_r^{pr} + (Q_r(t_0) - \frac{1}{k_r} Q_r^{pr}) \exp[-k_r(t - t_0)] \quad (7.63)$$

in which:
$Q_r(t)$ = quantity of dry matter present in living roots at time t in kg.ha^{-1}
Q_r^{pr} = mean gross production rate of new roots in kg.ha^{-1}.d^{-1}
k_r = death rate factor in d^{-1}, default 0.0055 for culture grassland and 0.00275 for natural grassland.

7.5 Nutrient uptake by crops

7.5.1 General aspects

Nutrients are absorbed by a plant root mainly as ions. If, therefore, total uptake of cations and anions is stoichiometrically unequal, then the plant should become electrically charged. Although small differences in electric potential do exist between the plant and its root environment and also between different compartments within a plant, such a nutrient uptake pattern would severely disturb the charge balance within the plant and in the rooting medium. Plants can grow at a rather wide range of pH values without suffering harmful effects of high H^+/OH^- concentrations. Physiological effects of pH on roots usually do not occur in the range between pH 4 and pH 8, but more extreme pH values can cause damage. The H^+ ions in the soil solution of very acid soils may replace Ca on the plasmalemma of root cell membranes. The plasmalemma then becomes permeable and no longer forms a diffusion barrier against passive influx or efflux of ions, so leakage of nutrients from the root cells may occur or harmful ions may enter from the soil solution into the root cells. Furthermore, cation uptake can be reduced at low pH through competition with hydrogen ions for adsorption sites on the root cell membranes. The same holds for anions in the extreme alkaline range through competition with OH^- ions.

If nitrogen is disregarded in determining the ionic uptake balance, cation uptake usually exceeds anion uptake (van Beusichem, 1984). Nitrogen, the nutrient taken up in largest amounts, can be absorbed both as anion (NO_3^-) and as cation (NH_4^+), besides other uncharged forms such as urea or fixed N_2. The form in which nitrogen is taken up will therefore have a major effect on the ionic balance of the plant. When all nitrogen is taken up in the ammonium form, the surplus of positive charge due to uptake of other ions will be further increased, leading to increased acidification of the rooting medium. If N is taken up as nitrate the overall charge balance will vary from slightly positive at low levels of nitrate uptake, to negative at higher levels, resulting in acidification or alkalinization of the rooting medium respectively. For model calculations it is essential to define the optimum N and P concentration for growth in respectively foliage, seeds and tubers. The crop growth period of arable crops is subdivided into two periods, generally from planting till flowering and from flowering till harvest. During each period an optimum concentration for both N and P in the soil solution, resulting in optimum growth, is defined as:

$$\begin{aligned} c_{optP} &= c_{1,P} & c_{optN} &= c_{1N} & t_p &< t < t_c \\ c_{optP} &= c_{2,P} & c_{optN} &= c_{2N} & t_c &< t < t_h \end{aligned} \quad (7.64)$$

where:

c_{optP} = optimum concentration for P in the soil solution
c_{optN} = optimum concentration for N in the soil solution
t_p = day-number of planting time
t_c = day-number of transition between both growing periods
t_h = day-number of harvest

Based on these optimum concentrations, the optimum N and P uptake is calculated as:

$$\sum_{t=t_p}^{t} U_{cr}^{opt} = \sum_{t=t_p}^{t} \sigma_{cpl}^{max}(t) c_{opt}(t) \Delta t \quad (7.65)$$

The uptake by the crop during each time step is calculated as:

$$U_{cr}(t) = \sum_{n=1}^{n_r} [\sigma_{cpl}(n)\overline{c}(n)] \qquad (7.66)$$

in which σ_{cpl} (n) follows from Eq (7.25) and (7.26) as:

$$\sigma_{cpl}(n) = \sigma_{pl}(n)q_t(n) = \sigma_{cpl}^{max}(n)(1 - \xi_{so}(n)\frac{c_{pl}}{\overline{c}(n)}) \qquad (7.67)$$

Luxurious consumption can take place under conditions of excessive nutrient supply in the soil, so under conditions that $\overline{c}(n) > c_{opt}$ and $U_{cr}^{max} > U_{cr}^{opt}$. By building up a nutrient concentration in the plant under these conditions the value of U_{cr} tends after some time to reach the situation that $U_{cr} = U_{cr}^{req}$. The uptake requirement is calculated as:

$$U_{up}^{req}(t) = \sigma_{cpl}^{max}(t)c_{opt}(t) - MAX\left[\sum_{t=t_p}^{t-1} U_{cr} - \sum_{t=t_p}^{t-1} U_{cr}^{opt} ; 0.0\right] \qquad (7.68)$$

The nutrient uptake will be less than the required one if $U_{cr}^{max} < U_{cr}^{req}$.
Dry matter production is reduced in the model simulation by:

- moisture stress conditions resulting in stomatal closure;
- deficit in nutrient uptake.

Reduction in growth due to water stress automatically reduces the nutrient demand by the crop, since it is assumed that the nutrient requirement of the crop is related to the actual growth rate. This means that under growth limiting conditions first the effects of water stress on dry matter production are taken into account and after that the availability of nutrients. It is assumed in the model that only mineral nutrients are taken up by the crops.

Reduction in dry matter production due to nutrient shortage is assumed when the nutrient supply in the crop is not sufficient for potential growth. The growth reduction is taken proportional to this shortage according to the condition:

$$\xi_Q(t) = MIN\left[\frac{U_{cr}(t)}{U_{cr}^{req}(t)} ; 1\right] \qquad (7.69)$$

Under conditions of nutrient shortage in the early stage of growth, photosynthesis will also be reduced due to poor leaf development.

7.5.2 Nitrogen requirements of arable crops

When both ammonium and nitrate are available in the soil preferential absorption of ammonium might be present. However, in relatively dry soils nitrate will be taken up at a much higher rate than ammonium. This can be explained by the fact that due to nitrification hardly any ammonium is present in the soil solution.

The fact that ammonium is preferentially absorbed does not mean that the plants grow best when the larger part of their nitrogen is taken up as ammonium. Generally dry matter production strongly decreases with increasing ammonium supply, while nitrate is virtually absent, and the carboxylate production by the plant was reduced when more than about 80 % of the N was taken up as ammonium. Carboxylates are needed for the regulation of the cytoplasmic pH of the plant and severe reduction in carboxylate concentration may lead to inability to control the cytoplasmic pH. With ammonium nutrition under acid soil conditions there is a heavy demand for carboxylates to eliminate protons arising from ammonium assimilation and from passive proton influx from the soil solution

Considering the effect of differences in ammonium contribution to total N uptake on the carboxylate concentration in the plant and on the rhizosphere pH, three situations can be distinguished (Gijsman, 1990). When more than 65 % of the N is absorbed as nitrate, the carboxylate concentration is at the proper level for regulating cytoplasmic pH and the rhizosphere is alkalized. This is a favourable situation under acid conditions. If nitrate contributes between about 20 and 65 % of the total N uptake, the carboxylate concentration in the plant is still at a proper level, but with decreasing nitrate contribution, the roots excrete increasing amounts of protons, leading to acidification of the rhizosphere. When less than about 20% of the amount of N absorbed is in the nitrate form, the carboxylate concentration in the plant decreases and physiological disorders may occur. The roots excrete protons and thus acidify the rhizosphere. With a low availability of nitrate in the total N supply reduction in dry matter production must be expected. The required nutrient availability is considered, under non-limiting nutrient conditions, as being proportional to the maximum rate of net photosynthesis.

The nitrogen uptake by a crop, dividing the rootzone into layers, is approximated as:

$$U_{cr}^{N}(t) = \sum_{n-1}^{n_r} \left[\sigma_{cpl}^{N}(n) \left(\overline{c}_{N-NO_3}(n) + \overline{c}_{N-NH_4}(n) \right) \right] \quad (7.70)$$

The nitrogen requirement of crops depends on the growing stage of the crop itself. In the early stage of growth, shoot and leaf production is predominant and this requires a relatively high nitrogen availability. Next to the N present in the foliage, the N present in the roots should be taken into account. Practical experience shows that the nitrogen content of the roots is about 1/2 to 2/3 of the value in the shoots. In the generative phase of growth when the photosynthetic products are mainly transported to the seeds, tubers or beets the nitrogen requirement reduces.

Water, Nutrient Uptake and Crop Production

In the generative phase reallocation of nitrogen from foliage to seeds can be present under conditions of N shortage in this phase of growth.

Because the N requirement in the first part of the growing season is often higher than in the second part, it is necessary to take into account the distribution of dry matter over the various plant parts. Nitrogen contents in foliage of different crops show, depending on the level of fertilization, a variation between 1.6 and 2.2 percent and for roots between 0.6 and 0.8 percent (Verveda, 1984).

Some data of N-contents for different crops at harvest are presented in Table 7.5.

Table 7.5: N-content of crop residues at harvest in %.

Crop	Foliage	Straw	Stubble	Roots	Tubers
Barley		0.94-1.07	0.74-0.88	0.60	
Winter wheat		1.25-1.42	1.02-1.17	0.80	
Potatoes	1.59-2.07			0.79-0.83	1.40
Maize	1.90-2.20				

7.5.3 Nitrogen requirement and gross production of grass

The amount of mineral nitrogen in the soil required for the production of 1 kg dry matter (dN_s/dt) can be considered as a function of the yield deficit. Rijtema (1980), using experimental data published by Frankena (1939), Mulder (1949), Oostendorp (1964) and Boxem (1973), formulated this relation as:

$$\frac{dN_s}{dQ_{sh}} = \frac{62.7}{Q_{shmax} - Q_{sh}} + 0.04 \tag{7.71}$$

where:
Q_{shmax} = maximum annual dry matter production in kg.ha^{-1}.a^{-1}
Q_{sh} = the actual annual dry matter production of grass in kg.ha^{-1}.a^{-1}

Integration of Eq. (7.56) yields a relation between the quantity of mineral nitrogen present in the soil (N_s) and gross production of grass. This relation is given by:

$$N_s = 0.04 Q_{sh} - 62.7 \ln \frac{Q_{shmax} - Q_{sh}}{Q_{shmax}} \tag{7.72}$$

Van Steenbergen (1977) showed that the N-uptake by the crop remains proportional to nitrogen availability over a much larger range than dry matter production, resulting in an excess uptake and nitrate accumulation in grass. It appeared from the available data, that 53 % of the mineral nitrogen was taken up in the grass sward. Taking 0.53 as the recovery coefficient in these experiments gives in combination with Eq. (7.72) the following relation between shoot dry matter production and N- uptake:

$$N_s = 0.0212 Q_{sh} - 33.2 \ln \frac{Q_{shmax} - Q_{sh}}{Q_{shmax}} \tag{7.73}$$

Next to the N present in the grass sward, the N present in the roots should be taken into account. Practical experience shows that the nitrogen content of the roots is about 1/2 to 2/3 of the value in the grass sward. From Eq. (7.73) it also follows that the minimum required quantity of nitrogen present in the shoots equals 2.12 %.

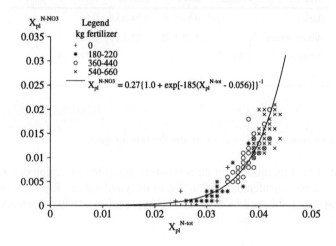

Fig. 7.11: The relation between total nitrogen uptake by grass and the accumulation of NO_3-N in the shoots.

The relation between total nitrogen uptake in grass shoots and the nitrate accumulation in the shoots is presented in Fig. 7.11, obtained from grassland fertilization experiments performed at the Experimental Farm at Ruurlo in the Netherlands from 1980 until 1984.

The relation can be given by the expression:

$$X_{pl}^{N-NO_3} = \frac{0.27}{1 + \exp[-185(X_{pl}^{N-tot} - 0.056)]} \tag{7.74}$$

in which:
$X_{pl}^{N-NO_3}$ = nitrate fraction in kg $N-NO_3$ per kg dry matter
X_{pl}^{N-tot} = total nitrogen fraction in kg N per kg dry matter.

7.5.4 Phosphorus uptake by crops

The solubility of phosphorus in the soil is strongly pH dependent. In the slightly alkaline to slightly acid range, P concentrations in the soil solution usually increase with increasing acidity. At higher or lower pH, precipitation of phosphorus, with calcium and aluminium respectively, can reduce the P concentration again. Another aspect of P availability at different pH values is the change in relative distribution of ionic species of phosphorus. The absorption rate of $H_2PO_4^-$ ions seems to be significantly greater than that of HPO_4^{--} ions, presumably due to the more rapid uptake of monovalent ions (Gijsman,1990). The phosphorus uptake by the plant is given by:

$$U_{cr}^P(t) = \sum_{n-1}^{n_r}\left[\sigma_{cpl}^P(n)\left(\bar{c}_{P-H_2PO_4}(n) + \bar{c}_{P-HPO_4}(n)\right)\right] \quad (7.75)$$

For phosphorus no accumulation of inorganic phosphate in the plant is assumed. The uptake of phosphorus seems to be correlated with the nitrogen uptake, so a relation between phosphorus uptake and nitrogen uptake is used. This relation between the N-fraction and the P-fraction for different crops is derived from data presented by Bosch and de Jonge (1989), Westerdijk et al. (1991) and Asijee (1993). The relation between the P-fraction present in dry matter and the N-fraction is given by the expression:

$$X_{pl}^P = 0.0238(X_{pl}^{N-tot})^{0.506} \quad (7.76)$$

The results are presented in Fig. 7.12.

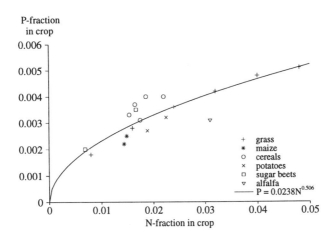

Fig. 7.12: The relation between N- fraction and P-fraction in dry matter of some crops derived from data given by Bosch and de Jonge (1989), Westerdijk et al. (1991) and Asijee (1993).

7.5.5 Gross production and nitrogen requirement of forest plantations

For the Netherlands detailed information exists on the growth rate of stemwood of different tree species as a function of soil type dependent suitability classes for tree growth. This information was summarized by de Vries (1991), who also gave the average growth rate per tree species based on the occurrence on various soil types. All other data needed to calculate maintenance growth and net growth were also given by this author, based on a compilation of various literature reviews.

The yearly increase in stemwood and branches (net growth) and the yearly production of dry matter present in foliage and fine roots (maintenance growth) as well as the nitrogen present for the main tree species in the Netherlands were for steady state conditions presented by Rijtema and de Vries (1994). These data are summarized for spruce trees and deciduous trees in Table 7.6.

Table 7.6: Dry matter production and nitrogen uptake of stemwood, foliage and fine roots by deciduous trees and spruce trees in the Netherlands after Rijtema and De Vries (1994).

Material	Deciduous trees	Spruce trees
Stemwood and branches		
Dry matter production (kg.ha^{-1}.a^{-1})	7300	3950
N- increase (kg.ha^{-1}.a^{-1})	16	6
Foliage		
Dry matter production (kg.ha^{-1}.a^{-1})	2950	2850
N- increase (kg.ha^{-1}.a^{-1})	90	80
Fine roots		
Dry matter production (kg.ha^{-1}.a^{-1})	7900	7750
N- increase (kg.ha^{-1}.a^{-1})	70	60

Table 7.6 shows that the main part of the nutrient uptake in forests returns to the soil system due to litter fall and root turnover. The nitrogen needed for stemwood growth should, however, enter into the soil system to avoid N depletion of the soil. Furthermore, N inputs should also compensate N losses by denitrification and leaching, since only part of the mineral nitrogen available in the soil is taken up by the trees. The main sources of additional N supply in natural systems are atmospheric deposition and biological N fixation. In ANIMO the effect of N shortage will induce a proportional reduction in dry matter production.

References

Asijee, K.(ed), 1993. Handboek voor de rundveehouderij. Informatie en Kennis Centrum Veehouderij, Lelystad, The Netherlands. Publicatie nr 35: 629 pp.
Berghuijs-van Dijk, J.T., Rijtema P.E. and Roest, C.W.J. 1985. *ANIMO: agricultural nitrogen model.* Nota 1671. Institute for Land and Water Management Research: Wageningen, The Netherlands.
Beusichem, M.L. van, 1984. *Non-ionic nitrogen nutrition of plants* Ph.D. Thesis. Agricultural University: Wageningen, The Netherlands.
Bosch, H. and Jonge, P. de 1989. Handboek voor de Akkerbouw en de Groenteteelt in de volle grond. Publikatie, 47,252 pp. Proefstation en Consulentschap Akkerbouw en Groenteteelt Volle Grond: Lelystad, The Netherlands.
Boxem, T. 1973. Stikstofbemesting en bruto opbrengst van grasland. *Stikstof*, **7**, 536-545.
Deinum, B., 1985. Root mass of grass swards in different grazing systems. *Neth. J. Agricultural Science*, **33**, 377-384.
Dirven, J.G.P. and Wind, K. 1982. *De invloed van bemesting op de beworteling van verschillende grassoorten en -rassen.* Mededeling 61. Vakgroep Landbouwplantenteelt en Graslandkunde, Agricultural University: Wageningen, The Netherlands.
Endrödi, G. and Rijtema, P.E. 1969. Calculation of evapotranspiration from potatoes. *Neth. J. Agricultural Science*, **18**, 26-36.
Feddes, R.A. 1971. *Water, heat and crop growth.* Ph.D. Thesis Agricultural University. Mededelingen Landbouwhogeschool 71.12. Veenman: Wageningen, The Netherlands.
Feddes, R.A. and Rijtema, P.E. 1972. Water withdrawal by plant roots. *J. Hydrology*, **17**, 33-59.
Frankena, H.J. 1939. *Over stikstofbemesting op grasland; Verslag van vier behandelingsproefvelden.* Verslagen Lanbouwkundig Onderzoek 45.11.
Gardner, W.R. 1960. Dynamic aspects of water availability to plants. *Soil Science*, **89**, 63-73.
Gardner, W.R. 1964, Relation of root distribution to water uptake and availability. *Agronomy J.*, **56**, 41-45.
Gijsman, A., 1990. *Nitrogen nutrition and rhizosphere pH of Douglas-fir.* Ph D Thesis, University Groningen, Groningen, The Netherlands.
Goedewaagen, M.A.J. and Schuurman, J.J. 1950. Wortelproductie van bouw- en grasland als bron van organische stof in de grond. *Landbouwkundig Tijdschrift*, **62**, 469-482.
Heinen, M. 1997. *Dynamics of water and nutrients in closed, recirculating cropping systems in glasshouse horticulture; With special attention to lettuce grown in irrigated sand beds.* Ph.D Thesis, Agricultural University, Wageningen, The netherlands
Jonge, P. de, 1981. *PAGV-handboek.* Publicatie nr 16. Proefstation voor de Akkerbouw en de groenteteelt in de vollegrond: Lelystad, Netherlands.
Lantinga, A.E., 1985. Simulation of herbage production and herbage intake during a rotational grazing period: An evaluation of Linehan's formula. *Neth. J. Agric. Science*, **33**, 385-403.
Louwerse, W., Sibma L. and Kleef, J. van, 1990. Crop photosynthesis, respiration and dry matter production of maize. *Neth. J. Agricultural Science*, **38**, 95-108.
Mulder, E.G. 1949. *Onderzoekingen over de stikstofvoeding van landbouwgewassen; I Proeven met kalkammonsalpeter op grasland.* Verslagen Lanbouwkundig onderzoek 55.7.
Oostendorp, D. 1964. Stikstofbemesting en bruto opbrengst van grasland. *Stikstof*, **4**, 192-202.
Rijtema, P.E., 1965. *An analysis of actual evapotranspiration* PH.D Thesis Agricultural University, Agricultural Research Reports 659. PUDOC, Wageningen, The Netherland.

Rijtema, P.E. 1980. Nitrogen emission from grassland farms - a model approach. In *On the role of nitrogen in intensive grassland production; Proceedings Int. Symp. Eur. Grassland Fed.*, pp. 137-147. PUDOC: Wageningen, The Netherlands.
Rijtema, P.E. and Aboukhaled, A. 1975. Crop Water use. In *Research on crop water use, salt affected soils and drainage in the Arabic Republic of Egypt*. pp. 5-61. FAO Near East Regional Office, Cairo.
Rijtema, P.E. and El Guindi, S.M. 1986. Some aspects of crop salt tolerance and water management. Nota 1724. Institute for Land and water Management Research: Wageningen, The Netherlands.
Rijtema, P.E. and Endrödi, G. 1970. Calculation of production of potatoes. *Neth. J. Agricultural Science*, **18**, 26-36.
Rijtema, P.E. and Ryhiner, A.H. 1968. Lysimeters in the Netherlands (III): Physical aspects of evapotranspiration and results of investigations. *Verslagen en mededelingen Hydrological Committee*, **14**, 86-149. TNO: The Hague, The Netherlands.
Rijtema P.E. and Vries, W. de, 1994. Differences in precipitation excess and nitrogen leaching from agricultural lands and forest plantations. *Biomass and Bioenergy*, **6**, 103-113.
Roest, C.W.J., Rijtema, P.E., Abdel Khalek, M.A., Boels, D., Abdel Gawad, S.T. and El Quosy, D.E. 1993. *Formulation of the on-farm water management model FAIDS*. Report 24, Reuse of Drainage Water Project. DRI: Cairo, Egypt. SC–DLO: Wageningen, The Netherlands.
Sauerbeck, D., Johnen, B. and Six, R. 1975. Atmung, Abbau und Ausscheidungen von Weizenwurzeln im Laufe ihrer Entwicklung. *Landwirtschaftliche Forschung, Sonderheft* **32/1**: 49-58.
Schuurman, J.J. 1973. *Overzicht van de resultaten van het bewortelingsonderzoek by grassen en op grasland aan het Instituut voor Bodemvruchtbaarheid*. IB-rapport 10-1973, Instituut voor Bodemvruchtbaarheid: Haren, The Netherlands.
Schuurman, J.J., and Knot, L. 1970. *Vergelijking van de wortelontwikkeling van drie grassoorten en zomertarwe*. Versl. Landbouwk. Onderz. 745. PUDOC: Wageningen, The Netherlands.
Steenbergen, T. van, 1977. Invloed van de grondsoort en jaar op het effect van stikstofbemesting op de graslandopbrengst. *Stikstof*, **8**, 9-16.
Throughton, A. 1957. *The underground organs of herbage grasses*. Bulletin 44 Commonwealth Bureau of Pastures and Field Crops: Huley, Berkshire.
Troughton, A., 1981. Length of life of grass roots. *Grass and Forage Science*, **36**, 117-120.
Verveda, H.W., 1984. *Opbouw en afbraak van jonge organische stof in de grond en de stikstofhuishouding onder een vierjarige vruchtwisseling met groenbemester*. Interne Mededeling 58. Vakgroep Bodemkunde en Plantevoeding, Agricultural University: Wageningen, The Netherlands.
Vries, W. de, 1991. *Methodologies for the assessment and mapping of critical loads and the impact of abatement strategies on forest soils*. Report 46 SC–DLO: Wageningen, The Netherlands.
Westerdijk, C.E., Alblas, J., Titulaer, H.H.H. and Dongen, G.J.M. van 1991. *Stand van onderzoek stikstofvoorziening en -emissie*. Internal Publication nr 826. Proefstation Akkerbouw en Groenteteelt Volle Grond: Lelystad, The Netherlands.
Willigen, P. de, and Noordwijk, M. van, 1987. *Roots, plant production and nutrient use efficiency*. PH. D Thesis. Agricultural University: Wageningen, The Netherlands.
Willigen, P. de and Noordwijk, M. van, 1994a. Mass flow and diffusion of nutrients to a root with constant or zero-sink uptake 1. Constant uptake. *Soil Science*, **157**, 162-170.

Willigen, P. de and Noordwijk, M. van, 1994b. Mass flow and diffusion of nutrients to a root with constant or zero-sink uptake II. Zero-sink uptake. *Soil Science*, **157**, 171-175

Wit, C.T. de, 1965. *Photosynthesis of leaf canopies*. Agricultural Research Reports 663. PUDOC: Wageningen, the Netherlands.

Wit, C.T. de and Penning de Vries, F.W.T. 1983. Crop growth without hormones. *Neth. J. Agricultural Science*, **31**, 313-323.

CHAPTER 8

MODEL VALIDATION AT FIELD SCALE

8.1 Validation of TRANSOL

The model approach and its implementation in the model TRANSOL were verified against an analytical solution. These results have been presented in section 3.1.4. and showed a good agreement. Moreover, a comparison has been made between the simulation results obtained with the model TRANSOL and those obtained with the pesticide leaching model PESTLA.

Boesten (1986) and Boesten and van der Linden (1991) developed the model PESTLA for the analysis of pesticide leaching under standard conditions for pesticide registration. The model PESTLA is based on detailed studies by Boesten (1986), Boesten (1994) and van den Bosch and Boesten (1994). PESTLA is internationally accepted and recognized as a model for a standardized approach to characterize pesticide behaviour in soils under standard conditions in pesticide registration procedures. The first versions of PESTLA were unable to simulate both the effect of fluctuating groundwater levels on pesticide behaviour as well as the effect of lateral outflow to drain systems to calculate the surface water load through field drainage.

A workshop at the end of 1994 (Crum and Deneer,1995) gave rise to an integration of the models PESTLA and TRANSOL. It was decided that an update of TRANSOL should be able to serve as a new basis for future PESTLA versions. A new version of TRANSOL (version 2.9) was implemented and documented (Kroes and Rijtema, 1996). At present both models have been integrated and the TRANSOL version 2.9 and the PESTLA version 2.9 have the same functionality.

Kroes and Rijtema (1996) used part of the data set given by Van den Bosch and Boesten (1994) for an analysis of the comparison of the simulation results with the model TRANSOL with both measured field data and the results obtained with the model PESTLA. The results obtained for a tracer with TRANSOL did show a good agreement with both the measurements and the simulations with PESTLA.

The simulation, using the data of the pesticide Ethoprophos, was carried out to test the sorption and transformation formulations. Ethoprophos is a nematicide which has a large sorption coefficient (K_{om} = 0.079 m^3.kg^{-1}) and a half-life time of 78 d in the upper 30 cm of the soil. The simulated mass balance of the upper 30 cm of the soil showed that during the first year after the application 62% of the applied dose disappeared by conversion, 3.3% was taken up by the crop, 0.004% leached at a depth of 30 cm, 0.8% was stored in the liquid phase and 33.9% in the solid phase.

244 *Environmental Impact of Land Use in Rural Regions*

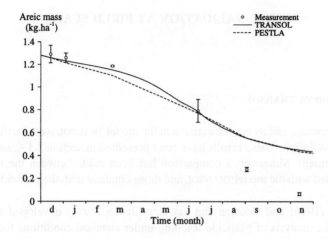

Fig. 8.1: Aeric mass of Ethoprophos in the soil layer 0-30 cm, from 1 December 1989 until 30 November 1990.

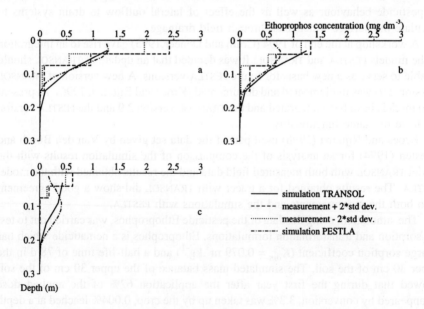

Fig. 8.2: Concentration profiles of Ethoprophos, 103 days (a), 214 days (b) and 278 days (c) after application.

To test the conversion of the pesticide, the aeric mass in the upper 30 cm of the soil was compared with measured data and with previous PESTLA simulations. The results are shown in Fig 8.1. The agreement between the simulation results of both models is reasonable; small deviations are probably due to small differences in model input. The results for measured and simulated data of the adsorbing pesticide did show that both models overestimate the aeric mass in the upper 30 cm. This might be due to uncertainties in the conversion rate of the pesticide.

The measured and simulated concentration profiles, presented in Fig 8.2, show that the penetration depth of the pesticide front is overestimated by both models after about 250 days. The uncertainty analysis presented by Kroes and Rijtema (1996) showed that the influence of the spatial variability of input parameters on the simulated aeric mass rate could not explain the deviation between measured and simulated values.

This confirms the suggestion by van den Bosch and Boesten (1994) to simulate the sorption process with a non-equilibrium approach. Such an approach retains the pesticide in the upper part of the soil for a longer period and reduces the penetration front. The present version of TRANSOL (version 2.9) includes such an approach for non-equilibrium sorption.

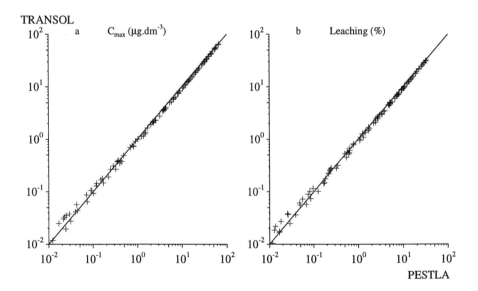

Fig. 8.3: Maximum concentration in the soil layer of 1.0 to 2.0 m below surface (a) and leaching percentage passing through the depth of 1 m below surface (b), as simulated by TRANSOL and PESTLA for 168 different combinations of degradation and adsorption.

Kroes and Boesten (1993) showed that for 168 different combinations of pesticide properties, concerning adsorption and degradation, the models TRANSOL and PESTLA produce similar results for the leaching of pesticides towards groundwater in spite of the different model approaches. Results of this comparison are presented in Fig. 8.3.

The deviations between both models are due to the following differences:

- Temperature correction: In the TRANSOL simulation a sinus-model was used to calculate the yearly temperature course with depth, whereas in PESTLA measured values were used.
- Dispersion length: The dispersion length in the top 10 cm was 2.5 cm in TRANSOL, layer thickness 5 cm, whereas in PESTLA a dispersion length of 5 cm was used.

8.2 Validation of ANIMO

8.2.1 Introduction

The model ANIMO has the ability to make reliable predictions for a wide range of different environmental conditions. This was illustrated by a number of case-studies which included comparisons between measured and simulated values of the soil-water-plant system at field scale. In some case-studies a calibration phase as well as a validation phase were distinguished, whereas other studies combined both phases. Validations on the scale of sub-models or process formulations are not available. In this chapter, the more practical perspective of model validation has been adopted.

The ANIMO-model was initially developed as a leaching model for nitrogen and therefore the first validations mainly focused on the leaching of nitrate (Kroes 1988, Reiniger *et al.* 1991). Recent validations included other processes of the nitrogen cycle and a validation of the leaching of mineral P and total P. In this validation procedure, results of field experiments at different locations in The Netherlands have been used. A prerequisite of an appropriate validation is a good description of water and nutrient movement in the soil. It must be realized, due to the huge amount of data required for this type of dynamic simulation model, that complete sets of field data are scarce and the opportunities to conduct a thorough model validation are often limited. Examples will be presented of recent validations using data-sets from different Dutch field experiments. The different locations are presented in Fig. 8.4. Jansen (1991a) applied the model using data-sets from different European countries. Depending on the availability and the suitability of experimental data, the behaviour of one or more compounds in the modelled soil-water-plant system was validated.

Model Validation at Field Scale 247

Fig. 8.4: Location of field experiments used to validate the ANIMO -model.

The description in this chapter is limited to the first 6 locations covering a fairly wide range of land use on different soil types, as is shown in Table 8.1.

Table 8.1: Validated variables of ANIMO model at different locations in The Netherlands.

Experiment		Soil-water-plant system			
nr	Location	Soil	Ground water	Surface water	Plant
1	Heino	mineral-N	nitrate-N		N-uptake
2	Ruurlo	mineral-N	nitrate-N		N-uptake
3	Nagele	mineral-N			N-uptake
4	Lelystad			nitrate-N	
5	Putten	sorbed-P	ortho-P	ortho-P	
6	St.Maartensbrug		total-P	total-P	

At some locations other entities were measured (total-N, ammonium-N, P-uptake, etc.) and used to validate the model. Due to their incidental measurements they have been left out of this chapter, but have been described in the pertinent project reports:

- Vredepeel and Borgerswold (location 7a and 7b): Different management options within integrated arable farming affecting nitrate leaching (Dijkstra and Hack-ten Broeke 1995) (ANIMO version 3.4)
- Cranendonck (location 8): High doses of cattle slurry applied in forage maize on sandy soil (Kroes *et al.* 1996) (ANIMO version 3.5)
- Hengelo (location 9): Effects of different management options for grazing cattle within dairy farming (Hack-ten Broeke and Dijkstra 1995) (ANIMO version 3.4)

In all validations, data from long-term field experiments where water fluxes and leached quantities were measured at different levels of fertilization management were used. The field experiments differ in soil type, land use and fertilizer management (Table 8.2).

Table 8.2: Main characteristics of field experiments used to validate the ANIMO model.

Experiment nr location	Soil texture	Land use	Groundwater level (m-surface)	Artificial fertilizer level (kg ha^{-1} a^{-1})	Animal fertilizer level (kg ha^{-1} a^{-1})	Nr. of variants
1 Heino	sand	forage maize [1]	0.5 - 1.6	20 - 140 N	180 N	9
2 Ruurlo	loamy sand	grassland	0.2 - 1.7	0 - 160 N	0 - 400 N	8
3 Nagele	silty loam	winter wheat	0.6 - 1.5	110 - 150 N	-	5
4 Lelystad	clay	grassland (grazed)	0.5 - 2.0	0 N	220 N	1
5 Putten	sand	grassland (grazed)	0.3 - 1.3	0 P	45 P	2
6 St.Maartensbrug	calc. weakly humous sand	flower bulbs	0.6 - 1.0	0 P	0 P	2

[1] some plots combined forage maize with a nitrogen catch crop in winter

The experiments in Heino and Ruurlo were used to validate grassland and forage maize on leaching of nitrate into groundwater, crop uptake of N and soil storage of mineral N. The experiment at Nagele was used to validate arable crop uptake of N and soil storage of mineral N. The experiment in Lelystad was used to validate leaching of nitrate to surface waters under grassland. The experiment in Putten validated leaching of mineral phosphorus to field ditches. The experiment in St.Maartensbrug validated the leaching of mineral phosphate and dissolved organic phosphorus in calcareous weakly humous sandy soils to tile drains and open field drains.

Each experiment was modelled by separate simulations for the hydrological and nutrient cycle of the soil-water-plant system. All hydrological simulations were carried out using the SWAP-model (Belmans *et al.* 1983, Feddes *et al.* 1978, van Dam *et al.* 1997). Results of the hydrological simulations were used as input to the ANIMO-model. Unless explicitly given, all simulations were carried out with ANIMO, version 3.5.

8.2.2 Forage maize and catch crops on a sandy soil

Between 1988 and 1994 research at the experimental farm in Heino was performed to establish the effect of nitrogen catch crops after forage maize on nitrogen leaching. Field data were reported by Schröder et al. (1992) and Van Dijk et al. (1995). Nitrate concentrations in soil moisture were analyzed and the nitrate leaching mass flux was computed by model simulations of combined water and nitrogen movement. A detailed report of the validation was presented by Kroes et al. (1996). Section 8.2.2.2 presents results of the comparison between simulated and measured values.

8.2.2.1 Method

Hydrology was simulated by the SWAP-model using measured groundwater levels as a lower boundary condition. Soil physical properties were partly measured and partly derived from standard soil moisture retention curves for the three soil horizons. The hydrology of the soil-water-plant system was simulated for three different cropping patterns:

- forage maize without a catch crop;
- forage maize with grassland as catch crop ;
- forage maize with rye as catch crop .

Hydrological simulations were validated by comparing the simulated and measured pressure heads at a depth of 15 cm.

The simulation of the nutrient processes was performed in two phases: a calibration and a validation phase. During the calibration phase only one field plot (forage maize without catch crop, fertilization level of 180 kg ha^{-1} N) was simulated and input parameters were adapted to obtain a reasonable fit of the simulation results. Only those parameters influencing denitrification and mineralization needed to be adjusted, on the basis of the parameters for oxygen diffusion and for the decomposition rate of organic matter in the subsoil.

During the validation phase the other variants were simulated, keeping the process parameters identical to the calibration phase, except, of course, for the meteorological data and the input data of the fertilizer management. Simulation results were evaluated by comparing measured and simulated values of:

- nitrate concentrations at a depth of 1 m below the soil surface;
- crop uptake by forage maize (nitrogen in harvested parts of crop);
- mineral nitrogen present in the upper 60 cm of the soil.

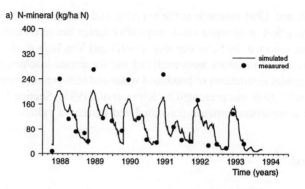
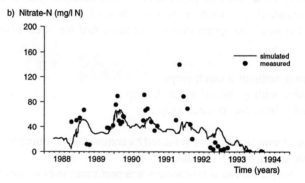
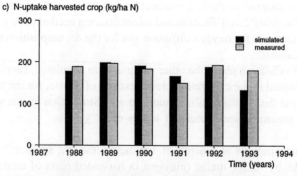

Fig. 8.5: Calibration Heino: measured and simulated mineral N in 0-60 cm-surface (a), concentration of nitrate at 1 m-surface (b) and uptake of N by forage maize without catch crop and with a fertilizer treatment of 200 kg ha^{-1} N (c).

8.2.2.2 Results

Simulated mineral N in the layer 0-60 cm-surface, presented in Fig. 8.5a, showed a fairly good comparison between simulated and measured values; largest deviations occur in the summers of the first four years. However, simulated nitrate concentrations, shown in Fig. 8.5b, at 1 m-surface are on the average only 9% below the measured values. Simulated crop uptake of nitrogen is in good agreement with the measured uptake; only in the wet year of 1993 is the simulated value too low. From a statistical evaluation, using all data of the different treatments (Kroes *et al.* 1996) it appeared that the average deviation for mineral-N, nitrate-N and crop uptake is 33, 65 and 12 %, respectively.

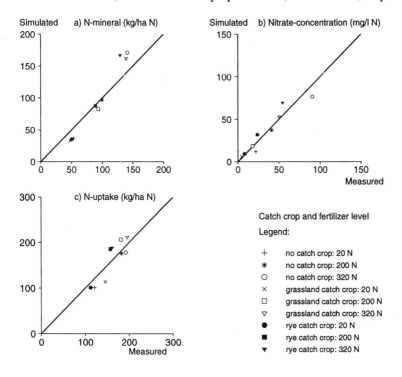

Fig. 8.6: a) measured and simulated mineral N in 0-60 cm-surface; b) concentration of nitrate at 1 m-surface; c) uptake of N by crop; forage maize with different catch crops and fertilizer treatments.

Results of the validation are presented in Fig. 8.6 as time averaged results of simulated and measured values for the whole period 1988-1994. Mineral N and nitrate N are based on comparisons for those time steps where measurements took place; crop uptake is based on a comparison of annual harvested crop yields. The computed average nitrate concentrations (Fig 8.6b) show a good agreement with the measured values. The largest deviations are found for variants with high inorganic fertilizer levels. Simulated mineral N (Fig.8.6a) and crop uptake (Fig.8.6c) tend to underestimate and overestimate measurements at respectively low and high fertilization levels. Other validations for the crop forage maize (Kroes *et al.* 1996) did not confirm this pattern. It has been concluded that the model performance is good, but the nutrient uptake by arable crops requires further study and other sources of validation data.

8.2.3 Non-grazed grassland on a sandy soil

Between 1980 and 1984 field experiments at an experimental farm in Ruurlo aimed to quantify the influence of different application techniques on soil fertility, crop yields and nitrate leaching. The data sets have been utilized by Jansen (1991a) to evaluate the performance of the ANIMO model within the framework of a comparison of six nitrate leaching models. The main characteristics of the Ruurlo field plots are:

- soils are classified as loamy sand soils;
- different fertilization levels and application techniques (both injection and surface applications of slurry);
- land use is grassland without grazing;
- high water holding capacity.

Field data were collected during the period 1980-1984, extensively published by Fonck (1982a, 1982b, 1986a, 1986b, 1986c), Wadman and Sluysmans (1992) and Jansen (1991b). Nitrate concentrations in soil water were measured and the nitrate leaching mass flux was computed by means of model simulations of combined water and nitrogen movement. A more detailed report of the validation was given by Jansen *et al.* (1991) and Kroes *et al.* (1996).

8.2.3.1 Method

Hydrological simulations were carried out by Jansen (1991a) with the SWAP model. ANIMO simulation results were verified with measured values of:

- nitrate concentrations at 1 m depth;
- N-uptake as found in all grassland cuttings;
- mineral nitrogen present in the upper 50 cm of the soil.

Nitrogen simulations were calibrated using the data of one fertilization experiment (800 kg ha^{-1} N; 50% as artificial fertilizer and 50% as animal manure). Animal manure (cattle slurry) was injected into the upper 10 cm of the soil. The data of the other seven experiments were used for model validation. The parameter adaptations during the calibration phase were related to the decomposition rate of organic matter in the subsoil.

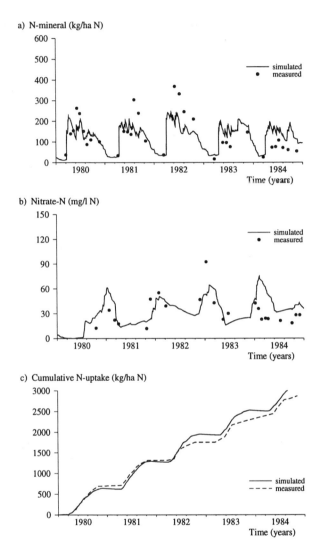

Fig. 8.7: Grassland with sub-surface injection of cattle slurry 80 ton/ha and artificial fertilizer level of 400 kg/ha (total fertilizer level of 800 kg ha^{-1} N). a) Measured and simulated mineral N in the layer 0-50 cm -surface, b) concentration of nitrate at 1 m-surface and c) uptake of N by crop during the period 1980-1984.

8.2.3.3 Results

Results of the calibration phase (Fig. 8.7) generally exhibit a good agreement between simulated and measured values. The time series of mineral nitrogen (Fig.8.7a) shows some large deviations during peak periods but the average values fit relatively well. The difference between measured and simulated nitrate concentrations at 1 m depth (Fig.8.7b) does not exceed 11%, which is acceptable. The calculated cumulative nitrogen uptake by the grass shoots is in good agreement with the measured data (Fig.8.7c). A more detailed comparison of 33 grassland-cuttings (Kroes *et al.* 1996) has proved that relatively large deviations per cutting can occur, but the average measured and computed values differed only by 9.5 % and the results were therefore quite acceptable.

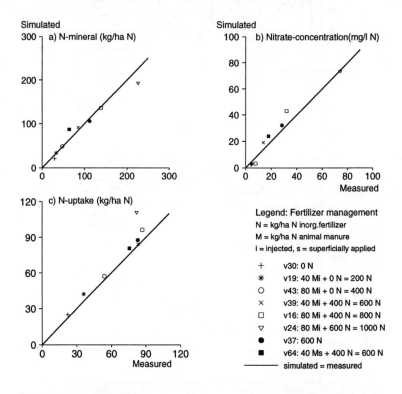

Fig. 8.8: Validation Ruurlo: a) measured and simulated mineral N in 0-50 cm-surface,
b) concentration of nitrate at 1 m-surface and c) uptake of N by crop;
average values for period 1980-1984; grassland with different fertilizer management.

Results of the model validation are presented in Fig. 8.8 showing the comparison of time averaged simulated and measured results with respect to N-mineral, NO_3-N concentration and N-uptake. The data relate to the whole experimentation period: 1980-1984. The storage of mineral N in the top-layer (Fig. 8.8a) is simulated fairly well; only two variants exhibit a deviation between measured and simulated values greater than 20%. The overall performance of the model is quite acceptable. Measured and simulated nitrate concentrations also agree well. Only measured and simulated values at plots with a very low fertilizer level are below the EC drinking water standard (11.3 mg.l^{-1} NO$_3$-N). N-yield of the grassland-cuttings is simulated fairly well. The model shows a slight tendency to overestimate crop uptake. A more detailed validation study (Kroes *et al.* 1996) comprising all individual grassland-cuttings showed that relatively large deviations can occur, but the average values fit fairly well.

8.2.4 Winter wheat on silty loam soil

Data of extensive field experiments were collected by Groot *et al.* (1990). The objective of these experiments was to obtain reliable data sets on soil nitrogen dynamics, nitrogen uptake, crop growth and crop development in different nitrogen treatments for winter wheat during the growing season. Data from two experiments were used to validate the ANIMO model on crop uptake by winter wheat and storage of mineral nitrogen (Rijtema and Kroes 1991). A summary of the validation, using data from the field experiments in Nagele, is presented.

8.2.4.1 Method

The experimental fields at Nagele can be characterized as:

- soil unit: silty loam;
- groundwater level: 0.6 - 1.5 m below soil surface;
- fertilization level: 110 and 150 kg ha^{-1} N, applied as artificial fertilizer.

Crop and soil samples were taken at three week intervals before anthesis and at two week intervals after anthesis. Groundwater levels were measured during the growing season of the year 1984. Daily weather data were available from the nearest meteorological station. Soil moisture retention curves and hydraulic conductivity data were reported by Groot *et al.* (1990). At each sampling date 0.5 m^2 crop was harvested in eight replicates for detailed plant analyses.

The SWAP model was applied to provide the hydrological data required by the ANIMO model. For this purpose measured soil physical data have been used as input. The SWAP model was applied using measured groundwater levels as a lower boundary condition. The nitrogen dynamics were simulated with version 3.0 of the ANIMO model.

8.2.4.2 Results

The measured and simulated uptake of nitrogen during growth are presented in Fig. 8.9a for the field with a fertilizer level of 110 kg ha^{-1} (N1). It appears that measured and simulated nitrogen uptake during growth agree reasonably well. Results of measured and simulated nitrogen uptake of two experimental field plots are presented in Fig. 8.9b. The two fields (N2 and N3) received equal doses of nitrogen fertilizer (150 kg ha^{-1}). The measured data of both fields have been plotted in Fig. 8.9b to give an impression of the spatial variability of the measured data. The simulated data fit reasonably well with the measured results.

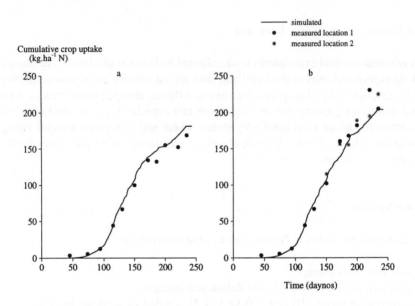

Fig. 8.9: Measured and simulated N-uptake of winter wheat at Nagele;
a) fertilizer levels of 110 kg ha^{-1} N, b) fertilizer level of 150 kg ha^{-1} N.

The total mineral nitrogen present in the soil of the N1 field during growth is given for the layer 0-40 cm in Fig. 8.10a and for the layer 0-100 cm in Fig. 8.10b. Fig. 8.10c shows the measured and simulated data of NO_3-N for the layer 0-100 cm. Fig 8.10d gives the comparison of the measured and simulated data of mineral N_4-N for the same layer. Attention is focused on the fertilizer addition of 60 kg.ha^{-1} N applied at daynumber 132 which is not followed by an increase in mineral nitrogen in the measured data. The simulated data, however, show an increase in total mineral nitrogen immediately after the application, followed by a sharp reduction in the days after application. Next to an increased crop uptake after this N-dose, also an increased denitrification caused by partial anaerobiosis due to heavy summer rains reduced the quantity of mineral nitrogen in the soil. The calculated total reduction in mineral nitrogen in the soil follows the measured data reasonably.

Fig. 8.10e and Fig. 8.10f present the mineral nitrogen storage in the soil profile for the layers 0-40 and 0-100 cm, respectively. The duplicate measurements give also an indication of the spatial variability in measured mineral nitrogen data.

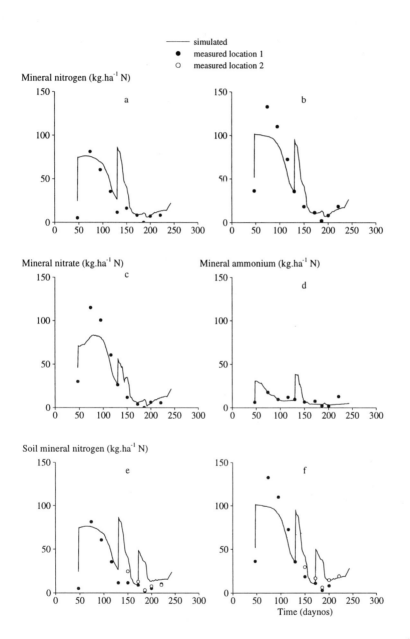

Fig. 8.10: Validation with winter wheat at Nagele: Soil mineral-N; fertilizer level 110 kg ha^{-1} N; soil layers: a) 0-40 cm, b) 0-100 cm. Nitrate (c) and soil mineral ammonium (d), soil layer 0-100 cm. Soil mineral-N, fertilizer level 150 kg ha^{-1} N; soil layers: e) 0-40 cm, f) 0-100 cm.

The simulated data appear to describe the temporal variation in mineral nitrogen in the soil following the fertilizer additions of 60 kg N at day-number 132 and of 40 kg N at day-number 172 correctly as compared with the measured data.

The effects of preferential drainage flow due to cracking in the soil profile has not been considered, which might partly explain the differences between measured and simulated mineral nitrogen in the soil. The formulation of partial anaerobiosis in the ANIMO model gives partially an explanation for the sharp reduction in mineral nitrogen in the soil profile after additions during the growing season. The ANIMO model did give a good simulation of the accumulated nitrogen uptake by the crop for all fertilizer treatments at the Nagele experimental fields.

8.2.5 Grazed grassland on clay soil

A research programme at the experimental farm Waiboerhoeve (Lelystad) aimed to analyze the influence of white clover on nitrate leaching and soil fertility in grazed grassland. An important feature of this research programme was to compare two complete dairy systems: a grassland and grassland/clover system. The experiments were conducted with different mixtures of grassland and white clover and during the period 1991-1994 the nitrate leaching was determined by incidental measurements of sub-surface drain water discharges and nitrate concentrations (Schils 1994).

To verify the preliminary results and to analyze the nitrate load on surface water by different grass/clover fields, an additional research programme was conducted to quantify the nitrate leaching during the season 1993/1994 (Kroes et al. 1996). Results of a comparison between simulated and measured nitrate sub-surface discharge from a grass/clover pasture are presented.

8.2.5.1 Method

The experimental fields at Lelystad can be characterized as:

- soil unit: clay;
- groundwater level: 0.5 - 2.0 m below soil surface;
- fertilization level: no artificial fertilizer and 220 kg.ha^{-1} N by animal manure.

Quantification of mass-discharges of nitrate by sub-surface drains was conducted by sampling of discharge proportional water volumes and analyses of nitrate (van den Toorn et al. 1994). The hydrological input required by the ANIMO model was generated by the SWAP model utilizing the physical soil properties derived from standard soil moisture retention curves (Wösten et al. 1994) for the four distinguished soil horizons.

A number of calibration runs were carried out with different values of the saturated hydraulic conductivity and the lower boundary condition until the measured and simulated data with respect to groundwater levels and drainwater discharges fitted well.

Nitrogen dynamics has been simulated using the ANIMO model version 3.4.2. Due to an insufficient number of replicates, no data were available for an independent field validation, so only the ability of the ANIMO model to reproduce the dynamic behaviour of nitrate leaching as influenced by climatical conditions could be verified. The empirical parameters p_1 and p_2 describing the vertical oxygen diffusion in the soil gas phase were adapted to obtain a good fit between simulated and measured values.

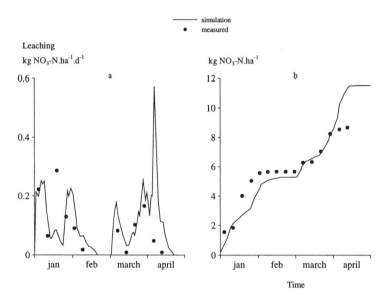

Fig. 8.11: Validation Lelystad: Measured and simulated nitrate drain discharge from a field covered by a mixture of grass and clover; daily (a) and cumulative (b); period January 1994-April 1994.

8.2.5.2 Results

Verification of nitrogen simulations took place by comparing the computed results with the measured data. A fair agreement between the simulated and the measured results was achieved for both the individual measurements (Fig. 8.11a) as well as for the cumulative results (Fig. 8.11b) The largest deviation occurred in the month of April 1994, where model simulations overestimated the water discharge. This deviation was most probably caused by an overestimation of the saturated hydraulic conductivity which incidentally caused an overestimation in simulated drain water discharge. Better results can be achieved by implementing preferential flow into both the hydrological and the nitrogen model.

8.2.6 Phosphorus leaching from grassland on a sandy soil

Intensive land use with high fertilizer levels resulted in phosphate saturated soils and as a consequence a high rate of leaching of phosphorous compounds to surface water and groundwater systems. During the years 1989-1994 the DLO-Winand Staring Centre carried out a research programme to evaluate several options to reduce phosphate leaching to surface waters. Within this framework Kruijne et al. (1996) validated the phosphate leaching as simulated by the ANIMO model.

8.2.6.1 Method

The phosphate adsorption formulations were based upon and parameterized by data originating from laboratory experiments (Schoumans 1995) and implemented in the ANIMO model version 3.5. The new formulation was calibrated using measured data of the penetration depth of the adsorbed phosphorus front and the phosphorus concentration in the liquid phase of the soil. Measured data were obtained from the experimental location in Putten (Schoumans and Kruijne 1995).

This field is characterized by:

- soil unit: 'Beekeerd' soil (humous sandy soil);
- groundwater level: 0.3 - 1.3 m below soil surface;
- P-fertilization: no artificial fertilizer and 45 kg.ha^{-1} P as animal manure.

The validation was carried out by comparing simulated phosphorus mass discharges towards a field ditch with measured data. Quantification of mass-discharges of P took place by analyses of P-concentrations in drainage water that was sampled proportionally to the drain discharge. During the first year the field equipment was installed and only water quality and adsorbed phosphate were measured. In the second and third year complete water balances and dissolved and adsorbed phosphate were measured.

A historical period of 45 years was simulated to initialize the ANIMO model and achieve a good approximation of the initial phosphorous sorption front. The final results of this initialization period were the starting-point for the calibration and validation phase. The hydrology was simulated with the SWAP model using the second year to calibrate and the third year to validate the simulated water balance on measured groundwater levels and water discharge to a open drainage ditch. Calibration was conducted by adaptation of the bottom boundary conditions, the drainage parameters and the soil physical parameters (Kruijne et al. 1996).

8.2.6.2 Results

The calibration phase resulted in an acceptable fit of model results to the measured data with respect to the sorbed phase and the liquid phase of the soil. Results of the sorbed phase are presented in Fig 8.12a as measured and simulated fractions of phosphate-sorption in relation to the depth in the upper 80 cm of the soil. Simulated and measured values are presented as averaged values for the year 1992. Calculation of standard deviation has been based on more than 20 measurements per sampling date. Results for the liquid phase, given as ortho-phosphate concentrations, have been presented in Fig. 8.12b as a function of depth.

The model validation was performed by comparison of calculated ortho-phosphate loads towards the ditch with measured values. Measured loads relate to loads as could be determined as P-discharges at the end of a field ditch. The results of this validation show in Fig. 8.13 a fair agreement between simulated and measured values. Other results have been presented by Kruijne et al. (1996).

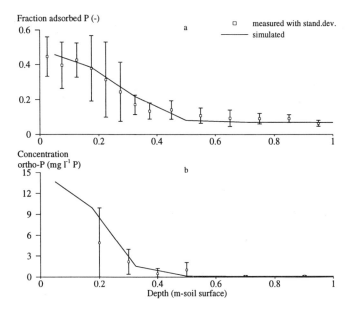

Fig. 8.12: Calibration Putten: measured and simulated fractions (-) of phosphate-occupation (a) and ortho-phosphate concentrations (mg l^{-1} P) in the liquid phase (b) in the upper 80 cm of the soil.

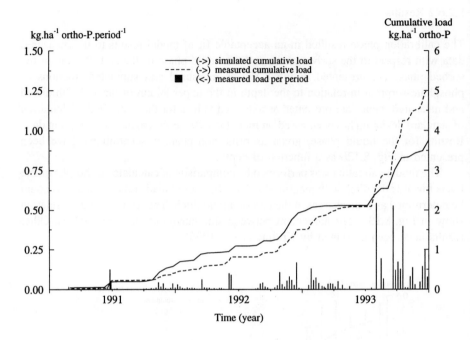

Fig. 8.13: Validation Putten: simulated (a) and measured loads (b) of ortho-phosphate (kg ha^{-1} P) discharged by a field ditch.

8.2.7 Flower bulbs on calcareous sandy soils

Flower bulb cultivation in the Netherlands can be characterized as an intensive form of agriculture with intensive use of fertilizers and pesticides. An important part of this cultivation is situated at vulnerable weakly humous calcareous sandy soils. Average phosphate surpluses during the eighties have been estimated at 220 kg.a^{-1} P$_2$O$_5$ (Steenvoorden *et al.* 1990). Environmental risks are considerable and research has been conducted to gain quantitative insight to assess these risks. Part of this research programme was a laboratory study of phosphate reactions in calcareous sandy soils (Schoumans and Lepelaar 1995) resulting in an alternative sorption and precipitation formulation to be implemented in the ANIMO model by an appropriate choice of the input parameters. The formulation was verified with data of laboratory experiments and validated under field conditions (Dijkstra *et al.* 1996).

8.2.7.1 Method

At first, laboratory research employing batch experiments yielded information on the sorption/precipitation mechanisms. The resulting model formulation has been verified in column leaching experiments with two soil samples. Field data on phosphate liquid concentrations in soil, groundwater and drainwater were used to validate the model under field circumstances.

Two different soil types were been identified as the most important groups to describe the phosphate behaviour in flower bulb cultivated sandy soils: mud-flat sands and coastal dune sands. Laboratory research yielded the parametrization of the Langmuir relation describing the equilibrium sorption. The data are presented in Table 8.3.

Table 8.3: Parametrization of the Langmuir equation describing the equilibrium sorption of phosphate in calcareous sandy soils.

Soil type	K_L (l.mg^{-1})	$X_{e,max,po4}$ (mmol.kg^{-1} P)	Explained variance (%)
Mud-flat sands	0.101	0.79	88
Coastal dune sands	0.144	0.43	85

Laboratory data supported the assumption that the sorption/diffusive precipitation reaction can be modelled according to a rate dependent Freundlich equation:

$$\frac{dX_P}{dt} = k_{ads}(K_F[Ca]^{N_2}[P]^{N_1} - X_P) \qquad (8.1)$$

where:
- $[Ca]$ = calcium concentration (mol·dm^{-3})
- $[P]$ = phosphate concentration (mol·dm^{-3})
- X_P = content sorbed/precipitated phosphate (mmol P·kg^{-1})
- k_{ads} = rate coefficient (d^{-1})
- K_F = sorption/precipitation coefficient (mmol P/kg · (mol P/l)$^{-N_1}$ · (mol Ca/l)$^{-N_2}$)
- N_1 = P-concentration exponent (-)
- N_2 = Ca-concentration exponent (-)

Seven soil samples were treated in a series of experiments. The quantity of rate dependent phosphate sorption/precipitation was determined from the total P-fixation minus the calculated quantity assigned to the equilibrium sorption phase. For the assessment of the parameters describing the rate dependent formulation, no distinction could be made between the two soil types. The resulting parametrization reads: $K_F = 600908$, $k_{ads} = 0.02529$, $N_1 = 1.024$ and $N_2 = 0.732$ and the percentage explained variance was 89.5 % (Schoumans and Lepelaar 1995).

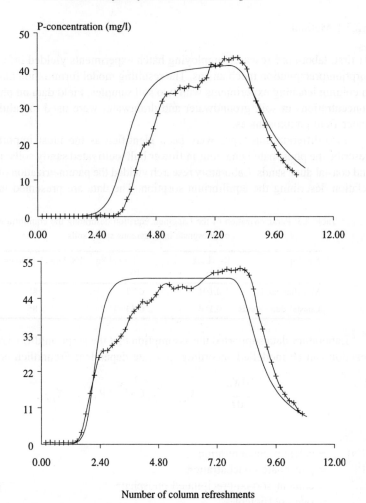

Fig. 8.14: Breakthrough curves of phosphate leaching experiments conducted with two soil samples originating from 0-25 cm depth (bottom) and 52-72 cm depth (top); + = measurement, full drawn line simulation.

Verification of these parameters was performed by comparing simulated and measured effluent concentrations of a leaching experiment. An example of the breakthrough curves of these phosphate leaching experiments is presented in Fig. 8.14. Two soil samples have been loaded by flushing with a phosphate influent concentration of 50 mg.dm^{-3} P (pH = 8). After saturation, the equilibrium sorption phase was removed by leaching with a zero-phosphate concentration. This experiment was simulated by the ANIMO model using the data concerning the experimental circumstances as input data and assuming that the influence of the organic phosphate cycle could be neglected.

The rate constant refers to the adsorption reaction, whilst the backward reaction was assumed to be zero.

The results show a fairly good performance of the model. The simulated concentrations agree reasonably well with the measured values. It can be concluded that the proposed process formulation and the derived set of parameters are able to describe the average phosphate concentrations in homogeneous calcareous sandy soils.

Field measurements included collection of hydrological data and data concerning nutrient concentrations in soil moisture and drain water. The hydrological conditions of the experimental field are characterized by a groundwater level at 0.6 - 1.0 m depth, a drain spacing of 5 m and sprinkling during spring and summer season. Once in four years, a six week inundation period of the soil profile is established to remedy biological soil contamination. No animal manure is used for fertilization, but organic compost is applied to prevent wind erosion of soil particles. During winter time, the fields are covered by straw to protect young flower bulbs from freezing.

In this study, the soil management, the material inputs and the nutrient extraction by the flower bulb crops were quantified and the initial condition with respect to the phosphate content of the soil was estimated from soil analyses.

8.2.7.2 Results

Simulation runs for validation purposes were carried out for a two-year period. Hydrological input data to the ANIMO model were generated by the SWAP model using measured moisture retention curves and measured hydraulic properties. The lower boundary of the model was described by the drainage characteristics as found in the field. The ANIMO model was calibrated on the basis of a qualitative comparison between simulated and measured values. Results of the phosphate concentrations of the soil water phase at different depths in one of the two soil profiles at the St. Maartensbrug site have been depicted in Fig. 8.15. Only minor adaptations of the initial distribution between the quantities present in the equilibrium sorption phase and the rate dependent sorption/precipitation phase were applied.

The dynamics of the simulated ortho-phosphate concentrations at 40 and 60 cm below soil surface agrees fairly well with the measured values, but dynamics of the ortho-P concentrations at 100 and 140 depth could not be reproduced. The organic-P concentrations are described satisfactorily by the model.

The field validation was carried out by an evaluation of the simulated drainwater concentrations (Fig.8.16). Both the measured and the simulated concentrations are within the same range, while the measured concentrations show a greater variability. The dynamic behaviour of the simulation run is mainly caused by the dynamics of dissolved organic phosphate. This agrees with the dynamics of the ortho-P concentration in the upper groundwater zone.

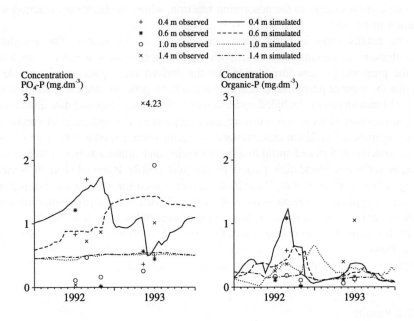

Fig. 8.15: Measured and simulated ortho-phosphate and dissolved organic phosphate concentrations in soil water at 40 cm, 60 cm, 100 cm and 140 cm depth.

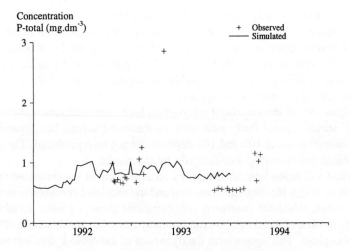

Fig. 8.16: Measured and simulated total-P concentrations in drainwater at a St.Maartensbrug flower bulb field plot.

The model was unable to simulate the measured short term dynamics of phosphate concentrations in drainage water. For the purpose of long term predictions, taking into account the soil heterogeneity, the model performance is judged to be useful. Validation on the concentrations of another drain situated at a distance of 200 m relative to the experimental drain considered in this section resulted to a worse result. This may be due to the soil heterogeneity within the catchment area of the drain tubes and due to the uncertainty of the initial occupation of the complex with phosphate and the initial distribution of sorbed phosphate.

References

Belmans, C., Wesseling, J.G. and Feddes, R.A. 1983. Simulation model of the water balance of a cropped soil: SWATRE. *Journal of Hydrology*, **63**, 271-286.
Boesten, J.J.T.I. 1986. *Behaviour of herbicides in soil: simulation and experimental assessment* Ph.D. Thesis, Agricultural University: Wageningen, The Netherlands.
Boesten, J.J.T.I. 1994. Simulation of bentazone leaching in sandy loam soil from Mellby (Sweden) with the PESTLA model. *J. Environ. Sci Health*, A **29** (6), 1231-1253.
Boesten, J.J.T.I. and Linden, A.M.A. van der, 1991. Modelling the influence of sorption and transformation on pesticide leaching and persistence. *J. Environmental Quality*, **20**, 425-435.
Bosch, H. van den, and Boesten, J.J.T.I. 1994. *Validation of the PESTLA model: Field tests for leaching of two pesticides in a humic sandy soil in Vredepeel (The Netherlands)*. Report **82**. SC–DLO: Wageningen, The Netherlands.
Brock, B.J. van den, Elbers, J.A., Huygen, J., Kabat, P., Wesseling, J.G., Dam, J.C. van, and Feddes, R.A. 1994. *SWAP93, Input instructions manual*. Interne Mededeling **288**. SC–DLO: Wageningen, The Netherlands.
Crum, S.J.H. and Deneer, J.W. (ed), 1995. *Development of the TOXSWA model for predicting the behaviour of pesticides in surface water; Proceedings of a Workshop*. Report **105**. SC–DLO: Wageningen, The Netherlands.
Dam, J. van, Huygen, J., Wesseling, J.G., Feddes, R.A., Kabat, P., Walsum, P.E.V. van, Groenendijk, P. and Diepen, C.A. van, 1997. SWAP version 2 Theory; Simulation of water flow, solute transport and plant growth in the soil-water-atmosphere-plant environment. *Technical Document* **45**. SC–DLO: Wageningen, The Netherlands.
Dijk W. van, Schröder, J., Holte, L. ten, and Groot, W.J.M. de, 1995. *Effecten van winter gewassen op verliezen en benutting van stikstof bij de teelt van snijmais, Verslag van onderzoek op ROC Aver-Heino tussen voorjaar 1991 en najaar 1994*. Verslag **201**. PAGV: Lelystad, The Netherlands.
Dijkstra, J.P. and Hack-ten Broeke, M.J.D. 1995. Simulation of different management options within integrated arable farming affecting nitrate leaching. In: *Scenario studies for the rural environment; proceedings of Symposium held in Wageningen, 12-15 Sept.1994*, (ed.: J.F.Th. Schoute, P.A. Finke, F.R. Veeneklaas and H.P. Wolfert), Environment & Policy, Volume **5**, pp. 329-334. Kluwer Academic Publishers: Dordrecht, The Netherlands.

Dijkstra, J.P., Groenendijk, P. Boesten, J.J.T.I. and Roelsma, J. 1996. *Emissies van bestrijdingsmiddelen en nutriënten in de bloembollenteelt. Modelonderzoek naar de uitspoeling van bestrijdingsmiddelen en nutriënten*. Rapport **376.5**. SC–DLO: Wageningen, Netherlands.

Feddes, R.A., Kowalik, P.J. and Zaradny, H. 1978. *Simulation of field water use and crop yield.* Simulation Monographs, PUDOC: Wageningen, The Netherlands

Fonck H. 1982a. *Stikstofconcentraties in bodemvocht en grondwater onder grasland op zandgrond in afhankelijkheid van runderdrijfmest- en kunstmeststikstofdosering.* Nota **1337**. Institute Land and Water Management Research: Wageningen, The Netherlands.

Fonck H. 1982b. *Stikstofconcentraties in bodemvocht en grondwater onder grasland op zandgrond in afhankelijkheid van runderdrijfmest- en kunstmestdosering (2e onderzoeksjaar 1981/1982).* Nota **1407**. Institute Land and Water Management Research: Wageningen, The Netherlands.

Fonck H. 1986a. *Stikstofconcentraties in bodemvocht en grondwater onder grasland op zandgrond in afhankelijkheid van runderdrijfmest- en kunstmestdosering (3e onderzoeksjaar 1982/1983).* Nota **1707**. Institute Land and Water Management Research: Wageningen, The Netherlands.

Fonck H. 1986b. *Stikstofconcentraties in bodemvocht en grondwater onder grasland op zandgrond in afhankelijkheid van runderdrijfmest- en kunstmestdosering (4e onderzoeksjaar 1983/1984).* Nota **1685**. Institute Land and Water Management Research: Wageningen, The Netherlands.

Fonck H. 1986c. *Stikstofconcentraties in bodemvocht en grondwater onder grasland op zandgrond in afhankelijkheid van runderdrijfmest- en kunstmestdosering (5e onderzoeksjaar 1984/1985).* Nota **1690**. Institute Land and Water Management Research: Wageningen, The Netherlands.

Groot J.J.R., Willigen P. de, and Verberne E.L.J., 1990. *Data-set. Workshop 5-6 June 1990. Nitrogen turnover in the soil crop ecosystem: modelling of biological transformations, transport of nitrogen and nitrogen use efficiency.* 97 pp. Institute for Soil Fertility: Haren, The Netherlands.

Hack-ten Broeke M.J.D, and Dijkstra, J.P. 1995. Effects of management options for grazing cattle within dairy farming. In: *Scenario studies for the rural environment; proceedings of Symposium held in Wageningen, 12-15 Sept.1994*, (ed.: J.F.Th. Schoute, P.A. Finke, F.R. Veeneklaas, H.P. Wolfert), Environment & Policy, Volume **5**, pp. 335-340. Kluwer Academic Publishers: Dordrecht, The Netherlands.

Jansen, E.J. 1991a. Results of simulations with ANIMO for several field situations. In *Soil and Groundwater Research Report II: Nitrate in soils*, pp. 269-280. Comm. Eur. Communities: Brussels, Belgium.

Jansen, E.J., 1991b. *Nitrate leaching from non-grazed grassland on a sandy soil: experimental data for testing of simulation models.* Report **26**, SC–DLO: Wageningen, The Netherlands.

Kroes, J.G. 1988. ANIMO *version 2.0, User's guide.* Nota **1848**. Institute for Land and Water Management Research: Wageningen, The Netherlands.

Kroes, J.G. and Boesten, J.J.T.I. 1993. *Vergelijking van de uitspoeling berekend met de modellen* TRANSOL *en* PESTLA. Rapport **238**. SC–DLO: Wageningen, The Netherlands.

Kroes, J.G. and Rijtema, P.E. 1996. TRANSOL, *a dynamic simulation model for transport and transformation of solutes in soils.* Report **103**. SC–DLO: Wageningen, The Netherlands.

Kroes, J.G., Groot, W.J.M. de, Pankow, J. and Toorn, A. van den, 1996. *Resultaten van onderzoek naar de kwantificering van de nitraatuitspoeling bij landbouwgronden.* Rapport **440**. SC–DLO: Wageningen, The Netherlands.

Kruijne, R., Wesseling, J.G. and Schoumans, O.F. 1996. *Onderzoek naar maatregelen ter vermindering van de fosfaatuitspoeling uit landbouwgronden. Ontwikkeling en toepassing van één- en twee-dimensionale modellen.* Rapport **374.4**. SC–DLO: Wageningen, The Netherlands. .

Reiniger, P, Hutson, J., Jansen, H., Kragt, J., Piehler, H., Swarts, M. and Vereecken, H. 1990. Evaluation and testing of models describing nitrogen transport in soil: a European project. In: *Transaction of 14th ICSS Kyoto, Japan.* Volume **I**, 56-61.

Rijtema, P.E. and Kroes, J.G. 1991. Some results of nitrogen simulations with the model ANIMO. *Fertilizer Research* **27**, 189-198.

Schils, R.L.M., 1994. Nitrate losses from grazed grassland and grass/clover pasture on clay soil. In: *Meststoffen* **1994**: 78-84.

Schoumans, O.F. 1995. *Beschrijving en validatie van de procesformulering van de abiotische fosfaatreacties in kalkloze zandgronden*, Rapport **381**, SC–DLO: Wageningen, The Netherlands.

Schoumans, O.F. and Kruijne, R. 1995. *Onderzoek naar maatregelen ter vermindering van de fosfaatuitspoeling uit landbouwgronden. Eindrapport.* Rapport **374**. SC–DLO: Wageningen, The Netherlands.

Schoumans, O.F. and Lepelaar, P. 1995. *Emissies van bestrijdingsmiddelen en nutriënten in de bloembollenteelt. Procesbeschrijving van het gedrag van anorganisch fosfaat in kalkrijke zandgronden.* Rapport 374. SC–DLO: Wageningen, The Netherlands.

Schröder, J., Holte, L. ten, Dijk, W. van, Groot, W.J.M. de, Boer, W.A. de, and Jansen, E.J. 1992. *Effecten van wintergewassen op de uitspoeling van stikstof bij de teelt van snijmaïs.* Verslag **148**, PAGV: Lelystad, The Netherlands.

Steenvoorden, J.H.A.M., Kolk, J.W.H. van der, Rondaij, R., Schoumans, O.F. and Waal, R.W. de, 1990. Bestrijdingsmiddelen en eutrofierende stoffen in bodem en water. Verkenning van de problematiek in het landinrichtingsgebied Bergen-Schoorl. Rapport 78. SC–DLO: Wageningen, The Netherlands.

Toorn, A. van den, Pankow, J. and Hooyer, O.M. 1994. *Nitraatuitspoeling van grasklaverpercelen op proefbedrijf de Waiboerhoeve in de Flevopolder. Veldonderzoek in de periode 1993 - 1994.* Interne Mededeling **314**. SC–DLO: Wageningen, The Netherlands.

Wadman, W.P. and Sluysmans, C.M.J. 1992. *Mestinjectie op grasland. De betekenis voor de bodemvruchtbaarheid en risico's voor de nitraatuitspoeling; Ruurlo 1980-1984.* Instituut voor Bodemvruchtbaarheid: Haren, The Netherlands.

Wösten, J.H.M., Veerman, G.J. and Stolte, J. 1994. *Waterretentie- en doorlatendheids karakteristieken van boven- en ondergronden in Nederland: de Staringreeks. Vernieuwde uitgave 1994.* SC–DLO: Wageningen, The Netherlands.

CHAPTER 9

REGIONAL MODEL APPLICATIONS

9.1 Regional application of ANIMO

The aim of the ANIMO model is to serve as a tool for evaluation of fertilization strategies and water management measures on nutrient leaching to groundwater and surface water systems at a regional scale. Soon after the first field validations became available in 1987 and 1988, the model was applied at a regional scale (Drent *et al.* 1988). Regional applications nearly always included comparisons between measured field data and simulated values, but at such a scale only a tentative verification was possible, rather than a thorough validation due to the enormous requirement of measured field data. The model was applied for different Dutch conditions with respect to climate, soils, cropping pattern and fertilization strategies on a regional scale which varied from the whole of the Netherlands to specific catchments. Fig. 9.1 refers to some of the regions where the model ANIMO was applied.

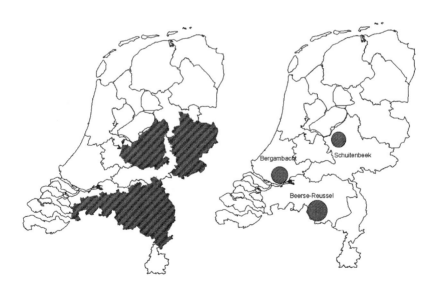

Fig. 9.1: Four regional applications of the model ANIMO in the Netherlands: nutrient leaching in the regions Bergambacht, Schuitenbeek and Beerze-Reusel-Rosep (right); phosphorus leaching from sandy soils (left).

In this chapter, three examples of regional model applications will be discussed:

- National study on leaching of nutrients to groundwater and surface water in the framework of a water management policy analysis in the Netherlands.
- Regional study on leaching of phosphate from P-saturated soils in the Eastern, Central and Southern sand districts of the Netherlands (Fig. 9.1, left).
- Regional study on the load on groundwater and surface water systems as influenced by fertilization levels differentiated to different types of land use from the viewpoint of physical planning of urban and rural areas and water management in the Beerze-Reusel catchment in the Southern part of the Netherlands (Fig. 9.1, right).

Other regional studies pertain to the Southern-Peel region (Drent et al. 1988), the Schuitenbeek catchment (Kruijne and Schoumans 1995) and the peat pasture area in Bergambacht polder (Hendriks et al. 1994), but they are not discussed in this chapter.

9.1.1 Leaching of nitrogen and phosphorus from rural areas to surface waters in the Netherlands

In the framework of the Policy Analysis of Water Management in the Netherlands the model ANIMO was applied twice: for the Third (NPD3) (Kroes et al. 1990, Uunk, 1991) and for the Fourth (NPD4, Boers et al. 1997, Kroes et al. 1997) National Policy Document. In both applications the model was used to analyze the effects of different scenarios of fertilizer management on nitrogen and phosphorus leaching from rural areas into Dutch surface waters. Both studies were initiated and funded by the Institute for Inland Water Management and Waste Water Treatment (RIZA) and aimed at a reduction of eutrophication of surface waters. The leaching studies were carried out by DLO-Winand Staring Centre (SC-DLO) in close cooperation with RIZA, and Delft Hydraulics. The results from these studies have been used by Delft Hydraulics and RIZA as input information in models describing the distribution of surface water and the quality of major surface water systems in the Netherlands.

9.1.1.1 Methodology

The model instrument, presented in Fig 9.2, consisting of a schematization procedure, a static model for fertilizer additions and dynamic models for water transport in soil (DEMGEN) and nutrient leaching to groundwater and surface water (ANIMO) was developed and applied. The fertilizer distribution model predicted the impact of policy alternatives on the short term and long term field additions. Dynamic models for water and nutrient behaviour in soils were inevitably required because of the combined impact of seasonal variations in meteorology, hydrology and timing of fertilizer applications, which is essential for the leaching of N and P to surface waters. The first step in the application of the model instrument was the determination of a spatial schematization. Basic data related to meteorology, geo-hydrology and drainage conditions originating from RIZA and data related to soil physics, soil chemistry and land use originating from SC-DLO were used to arrive at a set of calculation units (plots).

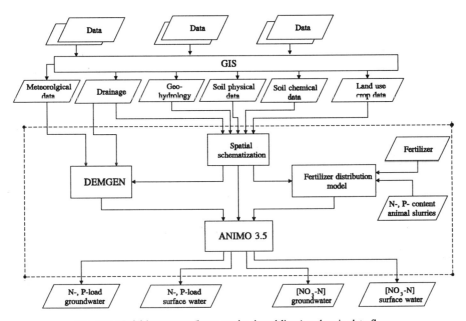

Fig. 9.2: Model instrument (between the dotted lines) and main data flow.

Once the spatial schematization was established, the hydrology of the soil system for each plot was simulated by the DEMGEN model. The fertilizer distribution was assessed by DELFT HYDRAULICS. Results of both types of computations were transmitted to SC-DLO, who applied the nutrient leaching model to simulate the behaviour of N and P in the soil. These simulations generated N and P mass balances of each plot. The N and P loads towards surface water systems were derived on the basis of these results.

The ANIMO model requires a good estimate of the initial distribution of N, P and C compounds in solid and liquid phase of the soil system, because poor initial conditions will cause error propagation. However, it is almost impossible to estimate the initial penetration of N and P fronts in the soil and the distribution of compounds over different pools of organic matter (including characteristic C/N and C/P ratios and decay rates). Therefore the decision was made to have the initial conditions determined by the model itself, through simulation of a historical period from 1940 to the year which served to evaluate the present situation. Evaluation years of the present situation were 1985 for NPD3 and 1993 for NPD4. Results of the evaluation year were used to verify the initial conditions by making a tentative comparison between measured and simulated data and were utilized as initial conditions for future scenarios (until

2045). Results from the hydrological model for the period 1971 to 1985 were used as input for nutrient simulations of history and scenarios. The fertilizer distribution model produced types and level of annual fertilizer applications. Data with respect to the time of application and the kind of fertilizer management were collected by SC-DLO and processed to generate yearly input files. Finally results were analyzed using general statistics, in the most recent NPD4-study within a GIS environment.

9.1.1.2 Results

This section describes results of the most recent study executed in the framework of the Fourth National Policy Document on Water Management in the Netherlands (NPD4). Validation of initial simulations to assess the environmental pollution and the store of minerals in soils in the evaluation year was conducted by comparing simulated and measured concentrations in groundwater and surface water systems. Special attention was paid to the validation of the discharge to surface water systems. Therefore three regions with different soil units were selected: a clay, a peat and a sandy region. The nutrient leaching model does not include the hydraulic, chemical and biological processes taking place in surface water systems. During the winter period, low temperatures will prevail and process rates will be low. Relatively high discharges in winter cause low residence times and will minimize the influence of processes in the surface water. Consequently, a comparison of winter discharges to the surface water system with measured concentrations in the surface water system can be made more safely than a comparison of summer discharges. An example of a validation result of the nitrogen concentration in the catchment of the river Schuitenbeek located in a sandy region in the central part of the Netherlands is given in Fig. 9.3.

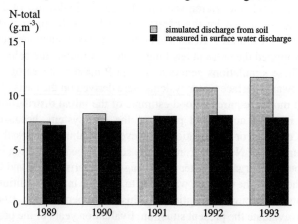

Fig. 9.3: N concentrations (g.m^{-3} N) in water leaching during the winter period from the soil to surface waters (simulated) and concentrations in surface waters (measured) in the region 'Schuitenbeek'.

Five fertilizer scenarios were simulated. Only three will be briefly discussed here: two extreme scenarios and one scenario representing, as good as possible, the present policy as formulated by the Dutch Government in 1995 (scenario: *policy*). One extreme scenario represents a continuation of the fertilizer use in 1993 (scenario: *present*), the other extreme scenario represents a prohibition of fertilizer use (scenario: *zero*). The scenario *policy* is aimed at reaching a balance between fertilizer levels, crop uptake and an acceptable loss of nutrients. In the fertilizer distribution model these acceptable losses were converted into fertilizer levels by adding estimates for crop uptake.

Scenario results pertaining to nitrogen and phosphorus leaching from soils to surface water systems in the Netherlands are presented in Fig. 9.4. To minimize the fluctuations caused by different meteorological conditions, the results are presented as a running average over periods of 15 years.

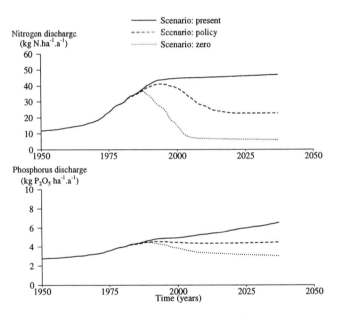

Fig. 9.4: Nitrogen (kg N.ha^{-1}.a^{-1}) and phosphorus (kg P$_2$O$_5$.ha^{-1}.a^{-1}) leaching from soils of rural areas to surface water systems in the Netherlands as calculated for the scenarios: *present*, *policy* and *zero*.

The difference between the nitrogen discharge of the scenarios *present* and *zero* is large. Results from scenario *present* for the year 2045 show a leaching of nitrogen which is six times higher than the leaching from *zero*. The same scenarios show that the leaching of phosphorus in scenario *present* is about twice as high as that of scenario *zero*. These results indicate the maximum reduction that can be achieved if fertilization is stopped. In the year 2045, the scenario *policy* resulted in a leaching of nitrogen of about 50% of the leaching that resulted from scenario *present*. The phosphorus leaching from scenario *policy* is about 70% of the leaching from scenario *present*. The phosphorus leaching from the scenario *policy* is almost constant from 1985 onward.

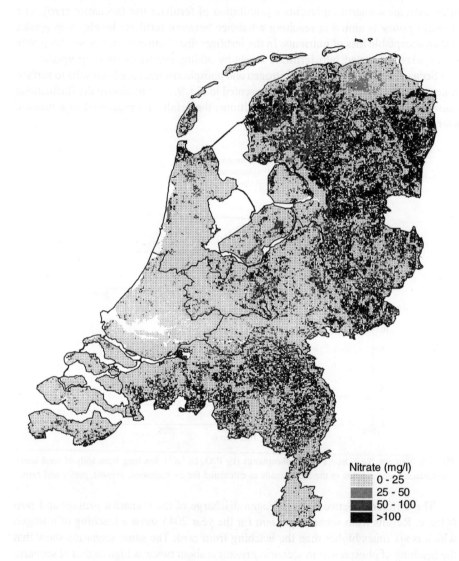

Fig. 9.5: Nitrate concentrations in groundwater in 1985 as a result of fertilization according to the scenario *present*.

Computed nitrate concentrations in groundwater as a result of the scenario *present* and the scenario *policy* are given in Fig. 9.5 and Fig. 9.6, respectively. The European nitrate directive for drinking water is exceeded in about 30% and 10% of the total area when the fertilization scenario *present* and scenario *policy*, respectively are implemented.

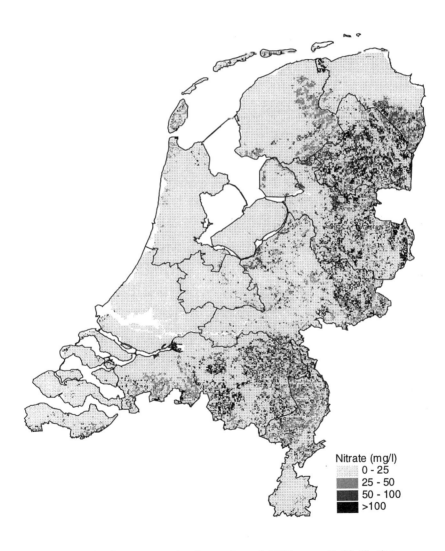

Fig. 9.6: Nitrate concentrations in groundwater in 2015 as a result of fertilization according to the scenario *policy*.

9.1.2 Simulation of phosphate leaching in catchments with phosphate saturated soils

Some rural regions in the Netherlands show an intensive agricultural activity with very high livestock densities. Especially on sandy soils which are characterized by a small buffer capacity for phosphate, high doses of animal slurries and manure lead to substantial phosphate surpluses. As a result, present phosphate leaching in these areas exceeds threshold values for surface water quality and adverse effects on the surface water quality occur. In the framework of a policy analysis, the effects of a reduction in phosphate fertilization was considered as one of the management measures. To investigate the reduction in phosphate leaching by limiting the allowed P surplus, a modelling study was carried out to evaluate the effects of different fertilization scenarios (Groenenberg et al. 1996).

9.1.2.1 Methodology

To estimate the phosphate leaching from soils in the area with sandy soils and a manure surplus in the Netherlands, simulations were carried out with dynamic models. The hydrology was simulated with the simple, two layer dynamic WATBAL model (Berghuis-van Dijk 1985). Nutrient related processes were simulated with the ANIMO model. Computations were made for about 2100 units unique in soil chemistry, hydrology, historic phosphate loads and land use. After initialization of the models with the historic phosphate loads, two different scenarios were run for a period of 60 years.

The methodology and models were verified on the Schuitenbeek catchment. In this catchment, measurements of phosphate leaching and phosphate concentrations were available for the period 1990-1993 which allow a comparison of simulated values with measured data. The comparison was hampered by the fact that the hydrologic years 1990-1993 were not in the meteorological data set used to compute the hydrology.

Results of the verification show that the simulated phosphate saturation as a function of soil depth was within the range of measured values. The results, however, also indicated that the simulated phosphate front is sharper than the measured one which causes leaching and concentrations of phosphate that are somewhat lower than the measured data, especially for groundwater regime class V/V*. A definition of the groundwater regime classes used in the Netherlands is given in Table 9.1.

9.1.2.2 Results

In the first scenario, a phosphate surplus of 10 kg.ha^{-1} P_2O_5 was used whereas in the second one a surplus of 40 kg.ha^{-1} P_2O_5 was used. The effects of the scenarios was evaluated on some selected plots. From these calculations it was concluded that leaching

Table 9.1: Groundwater regime classes, as defined by the mean highest groundwater table, the mean lowest groundwater table and the groundwater fluctuation in regions with pleistocene sands, marine clay soils and dune sands.

Groundwater regime class	Mean highest groundwater table (cm)	Mean deepest groundwatertable (cm)	Mean groundwater fluctuation (cm)
I	-5 ± 4	38 ± 7	43 ± 5
II wet	7 ± 3	66 ± 4	60 ± 3
II* dry	32 ± 7	67 ± 11	36 ± 10
III wet	17 ± 1	103 + 3	86 ± 10
III* dry	32 ± 3	102 ± 4	70 ± 3
IV	56 ± 3	104 ± 4	49 ± 3
V wet	17 ± 3	135 ± 5	118 ± 4
V* dry	32 ± 3	142 ± 4	110 ± 3
VI	61 ± 1	155 ± 2	94 ± 2
VII wet	101 ± 2	190 ± 3	90 ± 2
VII* dry	185 ± 3	281 ± 4	97 ± 3

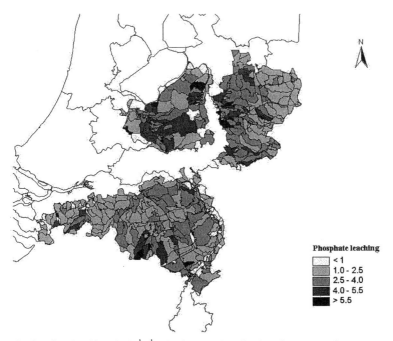

Phosphate leaching
- < 1
- 1.0 - 2.5
- 2.5 - 4.0
- 4.0 - 5.5
- > 5.5

Fig. 9.7: Phosphate leaching (kg.ha^{-1}.a^{-1}.P$_2$O$_5$) from sandy soils of rural areas to surface water systems as resulting from a scenario with 50 years of phosphate surplus of 10 kg.ha^{-1}.a^{-1}.P$_2$O$_5$.

at the level of the mean highest groundwater table tends to reach the value of the added surplus. Results of the scenario runs for the entire sandy area showed that almost all evaluated units, with groundwater regime classes I up to V^*, have a phosphate saturation higher than 25% which means that these units are, environmentally, considered as phosphate saturated. This is in accordance with results presented by Groenenberg et al. (1996).

Leaching fluxes towards surface waters increased in time in the *+40 kg*-scenario for both simulated concentration and quantity leached. In the *+10 kg*-scenario the leaching fluxes were stabilized and both phosphate leaching and phosphate concentration hardly increased between year 20 and year 50 of the scenario run. The phosphate leached, after 50 years simulation with this scenario, is shown in Fig. 9.7.

A first tentative assessment was made of the effects of a chemical treatment to reduce the leaching of phosphate, by assuming that this treatment was applied on all strongly P saturated soils in each catchment. In field experiments (Schoumans and Kruijne 1995) it was shown that this treatment reduced P leaching by about 70%. Results showed that such a measure can only reduce P leaching on catchment level if a scenario with a low P surplus for the other soils (with a P saturation < 75%) was used. In that case a reduction of maximal 20-30% in phosphate leaching was calculated. Higher reduction percentages are possibly only feasible in small (sub) catchments. In case of a larger surplus (40 kg), the positive effects of a chemical treatment for very strongly P saturated soils are diminished by an increased P leaching from other soils.

9.1.3 Beerze, Reusel and Rosep catchments

Prohibition of groundwater withdrawal for sprinkling of grassland is considered as a combative measure to the regional lowering of groundwater levels. However, a rise of the water table possibly implies undesirable environmental impacts. Also, the raise of surface water levels which lead to swamp conditions are considered as possible measures to stimulate the evolution of former agricultural soils to wet nature areas. However, this type of water management can yield an undesirable increase of nutrient leaching to surface water systems.

In the framework of a regional study in the Beerze, Reusel and Rosep catchment in the southern part of the Netherlands, the impact of raised water levels has been studied. The study was aimed at the quantification of the differences in nutrient discharges to groundwater and surface water as a consequence of the water management interference.

9.1.3.1 Methodology

Nitrogen and phosphorus balances and nutrient loads on surface water systems were calculated for different types of water management measures with integrated regional models for water management and groundwater flow. Fig. 9.8 gives a schematic presentation of the connection between different data files, data streams and required models that were used in this study for the evaluation of regional environmental impacts of diffuse sources on both groundwater and surface water quality.

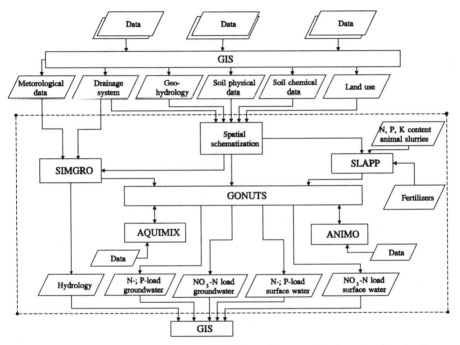

Fig. 9.8: Survey of data files, models and their interactions as used in the regional study of water pollution in the Beerze, Reusel and Rosep catchments.

The regional groundwater flow model SIMGRO (Querner, 1988; Querner and van Bakel 1989) was used to simulate the weekly water balances of 189 sub-regions for a meteorological series of 15 years. The SIMGRO model is an integrated hydrological model which comprises the unsaturated zone, the saturated zone and the surface water system. The saturated zone module consists of a quasi-three-dimensional finite element model using an implicit calculation scheme. The saturated zone is subdivided into aquitards with vertical flow and aquifers with horizontal flow. In the top layers of the saturated zone different drainage systems can be introduced. The unsaturated zone is modelled as two reservoirs: one for the root zone and one for the subsoil. The storage of water in the root zone is considered along with extractions and inflows. From the subsoil water balance, the phreatic surface elevation is calculated using a storage coefficient. The unsaturated zone is related to land use on a sub-regional level. Sub-regions are chosen to have relative uniform soil properties and hydrologic conditions.

The discharge from the saturated zone is influenced by the hydraulic properties of the surface water system by means of a storage function and a discharge function which relates the drainage level to the discharge of a sub-region.

The input data for the historical and future applications of animal slurries and inorganic fertilizers were obtained from calculations with the model SLAPP, optimizing the nutrient distribution for different forms of land use on basis of input obtained from a GIS-system yielding data on land use and intensity of animal husbandry for districts. This model is developed by Van Walsum (1988) for the generation of fertilization scenarios in regions with intensive animal husbandry. The model SLAPP (SLurry APPlication) translates the animal slurry production per district into actual fertilization data per type of land use per sub-region. Various restrictions in fertilization level per type of land use and in the required emission-poor application techniques can be introduced. The model calculates the over-production of animal slurry in terms of nitrogen, phosphate and potassium and indicates the required reduction in animal intensity for different scenarios.

The model ANIMO (Rijtema et al. 1997) was used in this regional study for the calculation of compound fluxes and compound balances within a sub-region per calculation unit. A calculation unit is defined as a unique combination of soil type, hydrological conditions and land use. ANIMO calculated the processes in the unsaturated zone and the top layer of the saturated zone for the determination of the local discharge of nutrients to surface water by the drainage systems present in the sub-region. The time scale for this type of calculation was relatively short and varied generally from 1 to 7 days depending on the problem to be analyzed. The results of the ANIMO calculations are summed up per type of land use and per sub-region for a period of half a year to one year and saved in files. The depth of the boundary between the local top-system and the deep regional groundwater system depends on the depth of the deepest streamline that discharges water to the local drainage systems. This depth depends on the drain distances, the hydraulic conductivity of the different layers in the soil profile and the pattern and magnitude of the regional groundwater flow. This depth was calculated for each sub-region with the output of SIMGRO.

Since the ANIMO model only describes the nutrient leaching as a local flow through a vertical column, the set of models was extended by a nutrient model, simulating the transport of nutrients in deeper regional flow systems. Interactions between sub-regions via the groundwater system are accounted for by linking the ANIMO model to the model AQUIMIX. The relation between ANIMO and the regional model AQUIMIX is determined by leakage and seepage fluxes between the sub-regional top system and the regional aquifer system, as is shown in Fig. 9.9.

This model AQUIMIX (van der Bolt et al., 1995) is developed for the transport of N-compounds and phosphate in the aquifer system. It considers the reduction of the nitrate concentration in the aquifer under anaerobic conditions through oxidation of organic matter and pyrite.

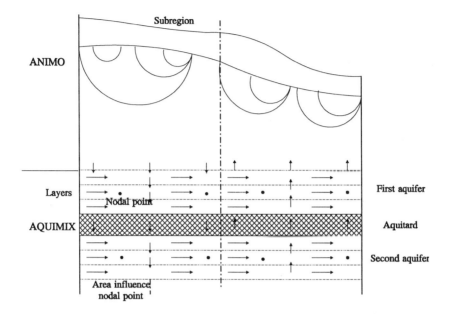

Fig. 9.9: Linking of the ANIMO model to the groundwater quality model AQUIMIX to account for regional interactions between infiltration areas and exfiltration areas.

The model AQUIMIX calculates per nodal point and per layer the regional solute transport and the geo-chemical processes in the saturated zone with time steps of ½ to 1 year. The calculated concentrations in the top layer of AQUIMIX at the end of the time step are the input concentrations for seepage in ANIMO.

The programme GONUTS (Geographical Oriented NUTrient Studies) (van der Bolt et al. 1995) organizes the data streams between the different models, executes pre-calculations and makes input data files for ANIMO. Interactions between sub-regions via the groundwater system are accounted for by linking both models. This procedure has been formalized in the programme GONUTS. GONUTS reads data, executes pre-calculations with data, writes the input data to files and calls the programmes ANIMO and AQUIMIX as subroutines. The following assumptions are made in the model:

- the model uses the geo-hydrological schematization and element network of the regional groundwater flow model SIMGRO;
- the smallest horizontal unit is the sphere of influence of a nodal point; the smallest vertical length is user defined;
- the length of a times step is user defined depending on layer thickness and geo-hydrological properties of the layers;
- all geo-chemical reactions are described as first order processes.

In the Beerze-Reusel-Rosep study, each sub-region was covered by a maximum of 10 different types of land use. The results of the SIMGRO model were verified with a limited number of groundwater elevation time series and estimated annual water discharges of some sub-catchments. A qualitative verification was conducted by a comparison of simulated and measured groundwater elevation maps for two points in time. The results were used as hydrological input to the ANIMO model.

This model configuration has been used to evaluate the environmental impacts with respect to reductions of permitted fertilization levels, the creation of new nature areas by the abolishment of fertilization (topic related to the physical planning of urban and rural areas), and water management strategies such as the prohibition of groundwater withdrawal for sprinkling purposes and the raise of the groundwater elevation to stimulate the generation of new nature areas. The research project was carried out in four phases:

- Data acquisition, schematization of the region and simulation of hydrology (van der Bolt *et al.* 1996a).
- Estimation of fertilizer distribution and simulation of historical mineral charging of soils and present environmental pollution by nitrogen and phosphorus compounds (van der Bolt *et al.* 1996b).
- Prediction of environmental impacts as caused by the reduction of fertilization levels. Effects of differentiated permitted fertilization levels with respect to the land use types as defined by the physical planning of urban and rural areas by national and provincial authorities were studied (van der Bolt *et al.* 1996c).
- Prediction of the nutrient load on groundwater and surface water systems as influenced by a set of water management measures (Groenendijk and van der Bolt 1996).

9.1.3.2 Results

Results presented below are related to the fourth phase of the research project which studied the impact of hydrological interferences. One of the hydrological scenarios contained the prohibition of groundwater withdrawal for sprinkling purposes. In the reference situation, only dry sensitive pastures were irrigated during drought periods. On sandy soils which are sensitive to drought, the sprinkling prohibition resulted in a rise of the groundwater. A complete stop of sprinkling, whilst equal fertilization doses are applied, will lead to an increase of the average nitrate concentration in groundwater in areas characterized by a relatively high sprinkling requirement (Table 9.2). The moisture deficit causes a decrease of crop dry matter production and as a consequence also a diminished mineral uptake by plant roots. The extra nitrogen surplus leaches to the groundwater zone and causes a slight increase of the nitrate concentration.

Table 9.2: Average nitrate concentration of groundwater in areas covered by grassland, classified by the average annual sprinkling requirement.

Sprinkling requirement (mm.a^{-1})	Nitrate concentration (mg.l^{-1} NO$_3$-N)	
	Sprinkled	Sprinkling prohibited
0 - 50	3.6	3.5
50 - 100	14.3	14.7
> 100	16.9	20.2

In the brook valleys with shallow groundwater levels, the nitrate concentrations were already low and no change is expected. It was concluded that fine tuning of farm management and fertilization strategies on water management measures are necessary to avoid adverse environmental impacts. Another water management scenario was defined to study the impacts of raising groundwater levels by water management. Within the future nature areas, the bottom of field ditches were raised and in the agricultural areas water conservation measures were applied. Water conservation was assumed by weir management during the spring and the summer seasons. These types of hydrological interferences are relevant as a measure to prevent negative effects on ecology.

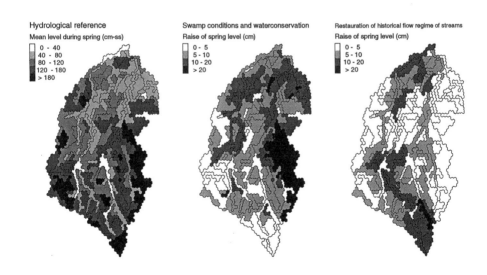

Fig. 9.10: Mean groundwater level during spring (left) raise of spring level resulted from the raise of third order drainage level and the weir management during summer (middle) and the additional raise of the spring level caused by the restoration of the historical flow regime of streams.

A special variant of this hydrological scenario comprised the restoration of the historical flow regime properties of semi-natural streams. The bottom of the streams were raised and the hydraulic resistances enlarged by an increased length of the stream bed. Results of the simulated groundwater levels during spring time are depicted in Fig. 9.10. Compared to the package of hydrological measures to raise groundwater elevations by less favourable drainage conditions, the restoration of the historical stream properties leads to an additional raise of groundwater levels. Conveyance of the precipitation surplus through the drainage system will be hampered and will lead to an additional raise of water levels in canals and field ditches.

The raise of groundwater levels induced wetter soil conditions which resulted in an increase of denitrification and in a decrease of mineralization. The additional increase of the groundwater elevation by the restoration of the historical flow regimes of natural streams leads to an additional decrease of nitrate concentrations (Fig. 9.11).

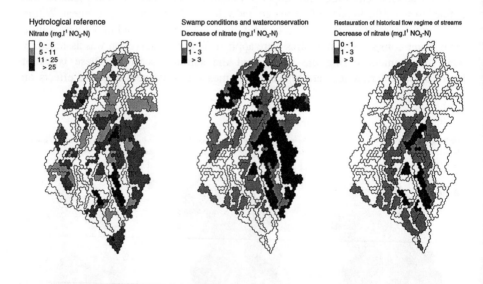

Fig. 9.11: Nitrate concentration in the reference situation predicted for areas covered by grassland (left) decrease of nitrate concentration as resulted from the raise of third order drainage level and the weir management during summer (middle) and the additional decrease of the nitrate concentration caused by the restoration of the historical flow regime of streams (right).

Based on the results of each sub-area, the results with respect to the P-leaching to surface water have been aggregated to sub-catchment level. All sub-catchments show an increase of P-discharge to surface water (Fig. 9.12). The phosphate load on surface water in these areas is 5 times as high as the P-discharge outside the stream valleys.

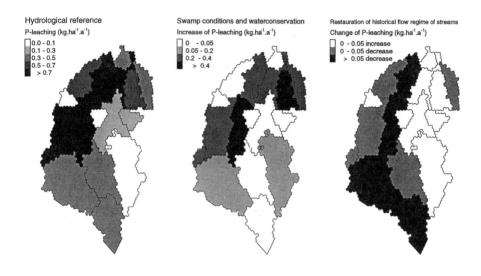

Fig. 9.12: P-leaching to surface water in the reference situation (left), P-discharge as resulted from the raise of third order drainage level and the weir management during summer (middle) and the raise of the spring level caused by the restoration of the historical flow regime of streams (right).

The raise of water tables due to the nature oriented water management will lead to a considerable increase of P-discharge in the future nature areas. Higher groundwater elevations lead to an increase of water discharge through shallow phosphate rich soil layers. The raise of groundwater generated adverse conditions from the point of view of surface water eutrophication by phosphates. The additional raise of the groundwater elevation which is caused by the retardation of surface water discharge leads in some sub-catchments to a slight decrease of phosphate leaching, compared to the phosphate discharge which results from the water management package without the adjustment of the flow regime of the streams, while in other sub-catchments a slight increase will occur. The average P-load on surface water systems showed an increase varying from 1.0 kg.ha^{-1}.a^{-1} to 1.4 kg.ha^{-1}.a^{-1} as a result of the water management measures, while the additional raise of groundwater elevations caused a counter balancing effect, and the average P-load amounted to 1.2 kg.ha^{-1}.a^{-1}.

Effects of increased mobilization of dissolved phosphate compounds induced by soil chemical reactions which will occur under these anaerobic circumstances have not been taken into account. Therefore, the calculated increase of phosphate leaching has to be considered as a prudent prediction.

9.2 Regional application of TRANSOL

9.2.1 Introduction

The aim of the TRANSOL model is to serve as a tool for the evaluation of strategies and water management measures to protect groundwater and surface water systems against pollution on a regional scale. The first version of the model TRANSOL originated in 1989. During most of the applications in the Netherlands, the model TRANSOL was used to simulate the discharge of pesticides to surface water systems. The model was also applied in some water quality management projects under arid conditions in the Arabic Republic of Egypt and in India to simulate the effects of different water management strategies on salinization.

The first regional application of the model was in a policy analysis of measures to reduce the load of pesticides to the surface waters in the catchment of the river Drentse Aa. The river water is used as a source for municipal water supply (van Bakel et al. 1992). Contamination of the river water was due to, respectively, spray drift, surface runoff and sub-surface discharge from fields with a shallow groundwater table. The introduction of pesticide free protection zones of about 10 % of the area of the catchment along the river reduced the pesticide load to the river water to acceptable levels, causing concentrations in the river water below the quality criteria for municipal water supply.

A second application of the model TRANSOL in combination with the model ANIMO was an analysis of the emission of pesticides and nutrients as a result of flowerbulb cultivation. Flowerbulb cultivation in the Netherlands is done at specialized holdings which are for the majority (about 70 %) situated at sandy soils in the coastal region. These sandy soils have a relatively high $CaCO_3$ content, an organic matter content that is maintained at 1.5 % and generally a very good water management. In comparison with other crops the quantity of pesticides and nutrients used is high in flowerbulb cultivation. The main types of pesticide used are soil disinfectants, fungicides and insecticides. Herbicides and disinfectants used in bulb treatment represent a minor part. Field measurements in monitoring programmes showed that the concentration of pesticides in surface waters exceeded the environmental water quality criteria in regions with a high concentration of flowerbulb farms. Dijkstra et al. (1995) used the model TRANSOL in an analysis of measures to reduce the effects of the use of pesticides in flowerbulb cultivation. Model testing was performed by comparison of computed concentrations with the results obtained from field studies executed both at a flowerbulb farm with a high quality farm management, as well as on an experimental farm with an integrated farm management system. The computed concentration of different pesticides was in agreement with the results of field measurements. The concentration of the pesticides in the shallow groundwater at 1 m below the surface is generally less than 0.01 % of the applied dose. For bentazone at the farm with standard management an emission fraction of 0.5-0.8 % to the shallow groundwater was found. The load to the surface water by drainage was strongly dependent on the ratio between drainage and leakage to the aquifer. It was concluded by Dijkstra *et al.* (1995) that the contribution of drainage water to the pesticide load of the surface water was low compared to the load by spray drift and activities like bulb-disinfection near the farm-buildings.

Roest *et al.* (1993) incorporated the model TRANSOL into the model FAIDS for the calculation of the redistribution of salt in the unsaturated zone of soils due to irrigation and evapotranspiration. The model FAIDS was developed in the framework of the model SIWARE of the Reuse of Drainage Water Project, which was aimed at a policy analysis of the water management and the reuse of drainage water in the Nile Delta of the Arabic Republic of Egypt.

Due to swelling and shrinking of Egyptian clay soils, a moisture deficit dependent crack volume was introduced, as a kind of bypass with horizontal infiltration in each layer during irrigation. For the heavy clay soils in the northern part of the Nile Delta these cracks are most probably the major vertical transport path for irrigation water. Part of the cracks moves horizontally towards drainage systems and leaches part of the salts, which have accumulated on crack surfaces. In order to account for this leaching, part of the water flowing through the cracks to drains has been considered as passing to the soil matrix. Soil salinity simulations, based on TRANSOL, were performed by two separate modules in the FAIDS model:

- REDIS for the calculation of the redistribution of salts after application of irrigation water on cracked soils;
- SAMIA for the calculation of the redistribution of salt under influence of evapotranspiration.

The module SAMIA was also used to calculate the discharge of salts in the saturated zone by drainage as a consequence of excess irrigation and seepage.

Calibration and validation of the SIWARE model has been performed on the basis of discharges of water and salt from individual catchments of drainage canals and pumping stations in the Nile Delta. The results did show that the SIWARE model, including the FAIDS model, can be reliably used for the analysis and evaluation of different water management alternatives in the Nile Delta (DRI/SC-DLO 1995).

Bronswijk *et al.* (1994) incorporated the model TRANSOL in combination with the model EPIDIM in the model SMASS in a study of the physical and chemical processes in acid sulphate soils. This model was used for the evaluation of different reclamation and water management techniques for the improvement of acid sulphate soils.

9.2.2 Beerze-Reusel-Rosep catchments

In the catchments of the rivers Beerze, Reusel and Rosep different forms of land use are present. Next to sub-regions with intensive agriculture, sub-regions with relatively high values for nature are present. About 35, 45 and 54 %, respectively, of the Beerze catchment, the Reusel catchment and the Rosep catchment have been intended for nature development. As nature development is strongly related to the abiotic environmental quality of the surface water in the catchments it is necessary to have a good impression of the effects of pesticide use on surface water quality.

The main aim of the study was to analyze the use of pesticides in the catchments and the consequences of this use on groundwater and surface water quality.

9.2.2.1 Methodology

Balances for different pesticides and pesticide loads on surface water systems were calculated for different types of land use with integrated regional models for water management and ground water flow. Fig. 9.13 gives a schematic presentation of the connection between different data files, data streams and required models used in this study for the evaluation of diffuse pesticide sources on both groundwater and surface water quality. This study started when the nutrient emission study discussed in section 9.1.3 for the same catchments was finished, using the same hydrological schematization and distribution of land use.

The input data for the use of pesticides were obtained from calculations with the pesticide information system ISBEST (Merkelbach and Lentjes 1993, Lentjes and Denneboom 1996). The information system ISBEST calculated per sub-region and per crop species the quantity of each pesticide used. ISBEST is an information system in which geographically independent data are stored, such as physical and chemical information on pesticides, application concentration and quantities and the number of applications for different crop species under different meteorological conditions. ISBEST gives this information on pesticide use for 82 different crops in the Netherlands and distinguishes 14 different agricultural regions, mainly based on soil classification. The ISBEST parameters dose and frequency have a direct relation with the application of the active compounds under the existing farming practice. The crop pest-stress indicates the intensity of the pest and the relative area that needs to be treated. Different pesticides can be used to attack the pest. The choice of a pesticide depends on the efficacy of the active compound in relation to the combination of pest and crop, positive side-effects against other pests and the price of the product. This is given by the ISBEST parameter market position. The combination of total crop area per sub-region, pest-stress and market position gives the use of a certain pesticide per sub-region per crop.

The model TRANSOL version 2.9 and the model PESTLA version 2.9 have been combined to the model PESTLA version 2.99. This model combines the possibility of parallel calculation of different compounds or of one compound with metabolites.

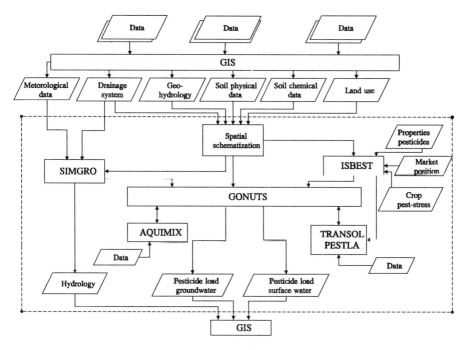

Fig. 9.13: Survey of data files, models and their interactions as used in the regional study of water pollution by pesticides in the Beerze, Reusel and Rosep catchments.

The transport schematization in PESTLA version 2.99 is similar to the schematization in TRANSOL version 2.9. The main processes considered in the model are: degradation, metabolite formation, crop uptake, precipitation of compounds and both equilibrium and kinetic sorption.

The programmes of the models GONUTS and AQUIMIX were both extended for the calculation of regional pesticide transport between sub-regions.

The following assumptions were made in this study (Aarnink *et al.* 1996):

- the model used the geo-hydrological schematization and element network of the regional groundwater flow model SIMGRO;
- the smallest horizontal unit was the sphere of influence of a nodal point; the smallest vertical length was user defined;
- the length of a time step was user defined depending on layer thickness and geo-hydrological properties of the layers;
- all geochemical reactions were described as first order processes;
- in the initial situation it was assumed that the soil is clean, which means that at the start of the calculations the concentration in soil moisture and the adsorbed quantity equal zero;

- degradation of pesticides was neglected below a depth of 1 m below surface;
- compound uptake by the crop was neglected;
- atmospheric deposition was neglected.

9.2.2.2 Results

The total area of the catchments equals 44 000 ha. The regional groundwater flow model SIMGRO (Querner, 1988; Querner and van Bakel 1989) was used to simulate the weekly water balances of 189 sub-regions for a meteorological series of 15 years. Each sub-region is divided into calculation units, which are defined as unique combinations of soil type, hydrological conditions and land use (crop type). The land use in the catchments is given in Table 9.3. Agricultural land use covers 59 % of the area, nature 31 % and urban land use 10 %. In the study 10 different types of land use were distinguished.

Table 9.3 gives also a brief review of the use of pesticides in the catchments, as calculated with ISBEST version 2.0. The non-agricultural use of pesticides is mainly restricted to herbicides, which are not only used in public gardens, kitchen gardens and lawns, but also on pavements, roads, parking places and tennis courts. The non-agricultural use in the catchments is, however, only 0.6 % of the total annual use of 131 821 kg active compounds. Therefore the study has been restricted to the agricultural use of pesticides. The use of soil disinfectants is mainly concentrated in vegetable growing and arboriculture. Though the use of soil disinfectants per ha in these land uses is high, the total area used for vegetable growing, flowerbulb growing and arboriculture is only 2.5 % of the total area. A selection of pesticides to be considered in the model study was based on the pesticides used in the catchments.

The following criteria for the selection of pesticides were used (Aarnink et al. 1996):

- the total use is more than 1000 kg;
- the physical-chemical properties of the pesticides are sufficiently known;
- the physical-chemical properties give a more or less representative picture of the 184 active compounds used in the catchments;
- the active compounds are used in the main crops.

Based on these criteria the following pesticides have been used as representative ones: MCPA, bentazone, atrazine, metolachlor, metamitron and fluazinam. To this selection desethyl-atrazine and hydroxy-atrazine, metabolites of atrazine, as well as lindane, have also been added.

Lindane is added to the selection because of the low degradation rate of this compound. In the study the emission of 9 different active compounds was analyzed, which represent, on the basis of their physical-chemical properties, the 184 active compounds used in the catchments.

Table 9.3: The distribution of land use and pesticide use in the catchments of the Beerze, Reusel and Rosep. Total area of the three catchments is 44 000 ha. The use of pesticides is given in kg active compound per ha (Aarnink et al. 1996).

Crop	Area (%)	Pesticide use in kg.ha^{-1} including soil disinfectants	excluding
Arable farming			
Maize	23.3	3.1	3.1
Potatoes	3.0	11.5	10.5
Sugar beets	3.1	6.4	6.1
Others	1.0	2.8	2.8
Stock farming			
Grass	26.3	0.6	0.6
Vegetable growing			
Carrots	0.67	49.9	6.6
Scorzonera	0.80	32.0	14.0
Strawberries	0.07	93.5	6.0
Leek	0.22	23.6	8.3
Asparagus	0.16	8.4	8.4
Others	0.21	7.5	7.5
Bulb growing			
Flowerbulbs	0.04	90.6	44.0
Arboriculture			
Ornamental shrubs	0.02	91.3	54.8
Fruit trees	0.04	67.0	29.9
Ornamental trees	0.16	50.2	13.6
Ornamental coniferae	0.16	48.7	10.8
Forest and hedge trees	0.05	45.6	7.3
Others	0.01	37.8	8.2
Nature			
Deciduous trees	7.0	-	-
Spruce trees	18.0	-	-
Others	6.0	-	-
Urban use			
Urban area	10.0	0.1	0.1

Soil disinfectants were used as representative active compounds, since 80 % of this type of pesticide was used in vegetable growing in the open and in arboriculture, covering only 2.5 % of the area spread over the catchments.

The physical chemical properties are important, since they determine the emission to groundwater and surface water to a great extent. The physical-chemical properties of the representative compounds are given in Table 9.4. The rate of degradation is given on the basis of half-life time, that is the number of days required to degrade or transform 50 % of the initial quantity of the active compound. The adsorption of pesticides to the soil matrix is given on the basis of the sorption to organic matter.

Table 9.4: The physical-chemical properties of the selected representative pesticides.

Pesticide	Half-life time (d)	Adsorption coefficient K_{om} (dm^3.kg^{-1})	Classification
Atrazine	50	70	moderately degradable slightly mobile
Desethyl-atrazine	45	18	moderately degradable moderately mobile
Hydroxy-atrazine	164	288	persistent immobile
Bentazone	48	0.4	moderately degradable very mobile
Fluazinam	107	5 330	persistent immobile
MCPA	15	29	degradable slightly mobile
Metamitron	30	100	moderately degradable slightly mobile
Metolachlor	100	103	persistent slightly mobile
Lindane	1 406	633	very persistent immobile

A fraction of the pesticides will be emitted to the environment by different transport pathways. The main pathways considered in this study are (Aarnink et al. 1996): spray-drift, leaching to groundwater and lateral drainage outflow.

The losses due to spray-drift have been estimated using data of measurements published by Porskamp et al. (1995). The mean downwind deposition is estimated as 5.6 % of the applied dose. The emission of spray-drift to surface water is calculated by introducing the surface water/land ratio in relation to the density of open field drains in the catchments. The load to the soil system is calculated as the applied dose minus the spray-drift.

The main transport of pesticides to the groundwater in the soil system is related to leaching, gas transport in the soil and tillage (deep ploughing). The model PESTLA version 2.99 calculated per calculation unit the compound fluxes, the concentration in the soil water and compound balances per layer and per soil profile. Table 9.5 gives the concentration of the pesticides in shallow groundwater at 1 m depth after 20 years of application. The concentrations and the cumulative fraction of the area are related to the area where the pesticide was applied. The results show that, after 20 years of application, the standard for municipal water supply of 0.1 $\mu g.dm^{-1}$ is not reached with desethyl-atrazine, hydroxy-atrazine, metamitron and lindane.

In the case of atrazine, MCPA and metolachlor, this water quality standard was exceeded in less than 5 % of the areal fraction, whereas it was exceeded in more than 95 % of the areal fraction in the case of bentazone. It must be concluded from the calculations that pesticides with a relatively high sorption to organic matter and/or a high degradation rate hardly reach the groundwater.

Table 9.5: The concentration of representative pesticides in shallow groundwater after 20 years of application in relation to the cumulative fraction of the area where these pesticides have been applied in the catchments of the rivers Beerze, Reusel and Rosep, derived from Aarnink *et al.*(1996).

Pesticide	Cumulative fraction of the area							
	0.00	0.05	0.10	0.20	0.40	0.60	0.80	0.95
	Concentration in shallow groundwater in $\mu g.dm^{-3}$							
Atrazine	> 4.5 10^{-1}	> 2.0 10^{-4}	> 4.2 10^{-4}	> 1.1 10^{-4}	> 6.8 10^{-5}	> 4.6 10^{-5}	> 1.2 10^{-5}	> 4.4 10^{-7}
Desethyl-atrazine	> 9.3 10^{-3}	> 4.2 10^{-4}	> 9.3 10^{-5}	> 7.7 10^{-5}	> 4.6 10^{-5}	> 3.9 10^{-5}	> 1.4 10^{-5}	> 1.4 10^{-6}
Hydroxy-atrazine	> 5.8 10^{-4}	> 9.3 10^{-5}	> 7.9 10^{-5}	> 6.0 10^{-5}	> 3.7 10^{-5}	> 2.5 10^{-5}	> 5.1 10^{-6}	> 3.0 10^{-7}
Bentazone	> 116.79	> 83.44	> 77.20	> 71.44	> 36.46	> 30.42	> 14.00	> 3.12
Fluazinam	> 1.0 10^{-9}	-	-	-	-	-	-	-
MCPA	> 5.8 10^{-1}	> 7.3 10^{-2}	> 9.7 10^{-4}	> 4.7 10^{-4}	> 9.3 10^{-5}	> 4.0 10^{-5}	> 1.4 10^{-5}	> 1.8 10^{-6}
Metamitron	> 8.8 10^{-5}	> 3.1 10^{-5}	> 2.4 10^{-5}	> 1.1 10^{-5}	> 4.3 10^{-6}	> 1.9 10^{-6}	> 4.2 10^{-7}	> 1.5 10^{-8}
Metolachlor	> 3.6 10^{0}	> 7.3 10^{-2}	> 8.1 10^{-3}	> 7.5 10^{-4}	> 1.5 10^{-4}	> 8.1 10^{-5}	> 5.0 10^{-5}	> 1.8 10^{-6}
Lindane	> 2.2 10^{-4}	> 5.2 10^{-5}	> 3.4 10^{-5}	> 2.4 10^{-5}	> 5.8 10^{-6}	> 1.0 10^{-6}	> 2.3 10^{-7}	> 7.9 10^{-9}

The load to the surface water is related to the lateral drainage outflow from the soil and the part of the spray-drift reaching the open water surface. It appears from Table 9.5 that the concentration of pesticides, with the exception of bentazone, in the shallow groundwater of the catchments is generally far below the water quality standard for municipal water supply.

This implies that the load to surface water of most of the representative pesticides by lateral drainage outflow, considering an annual discharge of between 250 and 350 m^3.ha^{-1}.a^{-1}, is less than 0.3 g.ha^{-1}.a^{-1}.

Table 9.6 gives the data of the annual load to surface water in the catchments, calculated by Aarnink *et al.* (1996). The load to the surface water by desethyl-atrazine and hydroxy-atrazine is mainly the result of lateral drainage outflow, since these compounds are formed as metabolites from atrazine in the soil. The very high surface water load by bentazone shows that there must be a considerable contribution to this load by lateral drainage outflow. However, it must be concluded from this analysis that spray-drift is generally the main source of surface water pollution by pesticides.

Table 9.6: The load of representative pesticides to surface water after 20 years of application in relation to the cumulative fraction of the area where these pesticides have been applied in the catchments of the rivers Beerze, Reusel and Rosep, derived from Aarnink *et al.* (1996).

Pesticide	Cumulative fraction of the area							
	0.00	0.05	0.10	0.20	0.40	0.60	0.80	0.95
	Load to surface waters in g.ha^{-1}.a^{-1}							
Atrazine	> 8.64	> 6.40	> 4.70	> 2.81	> 1.32	> 1.02	> 0.82	> 0.25
Desethyl-atrazine	> 2.8 10^{-3}	> 9.8 10^{-4}	> 5.9 10^{-5}	> 4.5 10^{-5}	> 2.5 10^{-5}	> 0.00	-	-
Hydroxy-atrazine	> 1.4 10^{-1}	> 1.3 10^{-2}	> 2.6 10^{-3}	> 0.00	-	-	-	-
Bentazone	> 65.17	> 44.34	> 37.60	> 26.22	> 9.96	> 1.35	> 0.22	> 0.00
Fluazinam	>3.58	> 2.85	> 2.36	> 1.50	> 0.67	> 0.51	> 0.42	> 0.12
MCPA	> 5.17	> 3.54	> 2.92	> 1.80	> 0.84	> 0.61	> 0.46	> 0.17
Metamitron	> 2.48	> 2.11	> 1.74	> 1.04	> 0.48	> 0.35	> 0.30	> 0.08
Metolachlor	> 9.74	> 5.24	> 3.28	> 1.87	> 0.94	> 0.68	> 0.56	> 0.16
Lindane	> 5.34	> 2.02	> 0.73	> 0.47	> 0.22	> 0.17	> 0.11	> 0.05

The distribution of the average bentazone load to the surface water per sub-region, expressed in kg.ha^{-1}.a^{-1} of the area where the pesticide was used in the sub-region after 20 years of application, is presented as an example in Fig 9.14 (left). The corresponding distribution of the bentazone load to shallow groundwater is presented in Fig 9.14 (right). This information was calculated for each pesticide used in the catchments of the rivers Beerze, Reusel and Rosep.

The load of sub-catchments in the rivers Beerze, Reusel and Rosep is obtained from adding up the loads of the sub-regions in those sub-catchments. The total load of pesticides to both surface water and groundwater per sub-region is calculated using the expression:

$$L_n^{sr} = \frac{\sum_{k=1}^{k_{\max}} (A_k^{cu(n)} \sum_{p=1}^{p_{\max}} L_p^{cu(n)})}{A_n^{sr}} \qquad (9.1)$$

where:
- $A_k^{cu(n)}$ = area of calculation unit k in sub-region n in ha
- A_n^{sr} = area of sub-region n in ha
- L_n^{sr} = total load of all pesticides in sub-region n in kg.ha^{-1}.a^{-1}
- $L_p^{cu(n)}$ = load per calculation unit in sub-region n of pesticide number p in kg.ha^{-1}.a^{-1}
- k = number of calculation unit (-)
- n = number of subregion (-)
- p = number of pesticide (-)

Load on surface water after 20 years
(kg.ha^{-1}.a^{-1})
- < 0.001
- 0.001 - 0.01
- 0.01 - 0.02
- 0.02 - 0.05
- no application

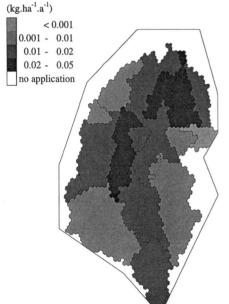

Load on groundwater after 20 years
(kg.ha^{-1}.a^{-1})
- < 0.001
- 0.001 - 0.01
- 0.01 - 0.1
- 0.1 - 1.0
- no application

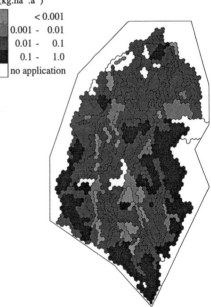

Fig. 9.14: The distribution of the load of bentazone on surface water after 20 years of application (left) and the distribution of the bentazone load to groundwater after 20 years of application (right) in the catchments of the rivers Beerze, Reusel and Rosep.
The loads are expressed in kg.ha.a^{-1} of the area where the pesticide was used.

References

Aarnink, W.H.B., Bolt, F.J.E. van der, Merkelbach, R.C.M. and Westein, E. 1996. *Belasting van grond- en oppervlaktewater met bestrijdingsmiddelen in de stroomgebieden van de Beerze, Reusel en Rosep.* Rapport 456. SC–DLO: Wageningen, The Netherlands.

Bakel, P.J.T. van, Elbers, J.A., Kroes, J.G., Leistra, M., Mourik, C.C.P. van, and Smelt, J.H. 1992. *Mogelijkheden tot reductie van bestrijdingsmiddelen in het stroomgebied van de Drentse Aa.* Rapport 200. SC–DLO: Wageningen, The Netherlands.

Berghuijs-van Dijk, J.T. 1985. *WATBAL: a simple water balance model for an unsaturated/saturated soil profile.* Nota 1670. Institute for Land and Water Management Research: Wageningen, The Netherlands.

Boers, P., Boogaard, H.L., Hoogeveen, J., Kroes, J.G., Menke, M.A., Noij, G.J., Roest, C.W.J., Ruijgh, E. and Vermulst, J.A.P.H. 1997. *De belasting van het oppervlaktewater met meststoffen nu en in de toekomst.* Rapport 97.013. Institute for Inland Water Management and Waste Water Treatment: Lelystad, The Netherlands; Rapport 532. SC–DLO: Wageningen, The Netherlands.

Bolt, F.J.E. van der, Walsum P.E.V. van and Groenendijk, P. 1996. *Nutriënten belasting van grond- en oppervlaktewater in de stroomgebieden van de Beerze, de Reusel en de Rosep. Simulatie van de regionale hydrologie.* Rapport 306.1. SC–DLO: Wageningen, The Netherlands.

Bolt, F.J.E. van der, Groenendijk, P. and Oosterom, H.P. 1995. *Nutriënten belasting van grond- en oppervlaktewater in de stroomgebieden van de Beerze, de Reusel en de Rosep. Simulatie van de nutriëntenhuishouding.* Rapport 306.2. SC–DLO: Wageningen, The Netherlands.

Bolt, F.J.E. van der, Groenendijk, P. and Oosterom, H.P. 1996. *Nutriënten belasting van grond- en oppervlaktewater in de stroomgebieden van de Beerze, de Reusel en de Rosep. Effecten van bemestingsmaatregelen.* Rapport 306.3. SC–DLO: Wageningen, The Netherlands.

Bronswijk, J.J.B. and Groenenberg, J.E. 1994. SMASS: A simulation model for acid sulphate soils: I. Basic principles. In: *Selected papers Saigon Symposium Acid Sulphate Soils*, (ed D. Dent and M.E.F. van Mensvoort), ILRI Publ. 52, Institute for Land Reclamation and Improvement: Wageningen, The Netherlands.

Bronswijk, J.J.B., Bosch, H. van den, Nugroho, K., Groenenberg, J.E. 1994. *SMASS - a simulation model of physical and chemical processes in acid sulphate soils. version 2.0.* Technical document 21. SC–DLO Winand Staring Centre: Wageningen, The Netherlands.

Dijkstra, J.P., Groenendijk, P., Boesten, J.J.T.I. and Roelsma, J. 1996. *Emissies van bestrijdingsmiddelen en nutriënten in de bloembollenteelt. Modelonderzoek naar de uitspoeling van bestrijdingsmiddelen en nutriënten.* Rapport 376.5. SC–DLO: Wageningen, The Netherlands.

Drent, J., Kroes, J.G. and Rijtema, P.E. 1988. *Nitraatbelasting van het grondwater in het zuidoosten van Noord-Brabant.* Rapport 26. Instituut voor Cultuurtechniek en Waterhuishouding: Wageningen, The Netherlands.

DRI/SC-DLO, 1995. *Reuse of drainage water in the Nile Delta; monitoring, modelling and analysis. Final Report Reuse of Drainage water Project.* Reuse Report 50. Drainage Research Institute: Kanater, Cairo, Egypt. SC–DLO: Wageningen, The Netherlands.

Groenenberg, J.E., Reinds, G.J. and Breeuwsma, A. 1996. *Simulation of phosphate leaching in catchments with phosphate saturated soils in the Netherlands.* Report 116 SC–DLO: Wageningen, The Netherlands.

Groenendijk, P. and Bolt, F.J.E. van der, 1996. *Nutriënten belasting van grond- en oppervlakte water in de stroomgebieden van de Beerze, de Reusel en de Rosep. Effecten van waterhuishoudkundige ingrepen.* Rapport **306.4**. SC–DLO: Wageningen, The Netherlands.

Hendriks, R.F.A., Kolk J.W.H. van der, and Oosterom, H.P. 1994. *Effecten van beheersmaatregelen op de nutriëntenconcentraties in het oppervlaktewater van peilgebied Bergambacht; Een modelstudie.* Rapport **272**. SC–DLO: Wageningen, The Netherlands.

Kroes, J.G., Roest, C.W.J., Rijtema, P.E. and Locht, L.J. 1990. *De invloed van enige bemestingsscenario's op de afvoer van stikstof en fosfor naar het oppervlaktewater in Nederland.* Rapport **55**. SC–DLO: Wageningen, The Netherlands.

Kroes, J.G., Boers, P.C.M., Boogaard, H.L. Ruijgh, E.F.W. and Vermulst, J.A.P.H. 1997. Impact of manure policy on leaching of nitrogen and phosphorus in the Netherlands. In *The Implementation of Nitrate Policies in Europe: Process of Change in Environmental Policy and Agriculture*,(ed. F Brouwer and W. Kleinhanss) Landwirtschaft und Umwelt; Schriften zur Umweltökonomik band14 pp. 253-269. Wissenschaftsverlag Vauk Kiel KG.

Kruijne, R., Wesseling J.G. and Schoumans, O.F. 1996. *Onderzoek naar maatregelen ter vermindering van de fosfaatuitspoeling uit landbouwgronden. Ontwikkeling en toepassing van één- en tweedimensionale modellen.* Rapport **374.4**. SC–DLO: Wageningen, The Netherlands.

Lentjes P.G. and Denneboom, J. 1996. *Data- en programmabeschrijving ISBEST versie 2.0.* Technical Document **31** SC–DLO: Wageningen, The Netherlands.

Merkelbach, R.C.M. and Lentjes P.G. 1993. *ISBEST: een informatiesysteem dat het bestrijdingsmiddelengebruik in Nederland beschrijft.* Internal report. SC–DLO: Wageningen, The Netherlands.

Porskamp, H.A.J., Michielsen, J.M.G.P., Huijsmans, J.F.M. and Zande, J.C. van de, 1995. *Emissiebeperkende spuittechnieken voor de akkerbouw. De invloed van luchtondersteuning, dopkeuze en teeltvrije zone op de emissie buiten het perceel.* Rapport **95-15**, IMAG-DLO: Wageningen, The Netherlands.

Querner, E.P. and Bakel, P.J.T. van, 1989. *Description of the regional groundwater flow model SIMGRO.* Report **7**. SC–DLO: Wageningen, The Netherlands.

Querner, E.P. 1988. Description of a regional groundwater flow model SIMGRO and some applications. *Agricultural Water Management*, **14**, 209-218.

Rijtema, P.E., Groenendijk, P, Kroes, J.G. and Roest, C.W.J. 1997. *Modelling the nitrogen and phosphorus leaching to groundwater and surface water; Theoretical backgrounds of the ANIMO model.* Report **30**. SC–DLO: Wageningen, The Netherlands.

Roest, C.W.J., Rijtema, P.E., Abdel Khalek, M.A., Boels, D., Abdel Gawad, S.T. and El Quosy, D.E. 1993. *Formulation of the on-farm water management model FAIDS.* Report **24**, Reuse of Drainage Water Project. DRI: Cairo, Egypt. SC–DLO: Wageningen, The Netherlands.

Schoumans, O.F. and Kruijne, R. 1995. *Onderzoek naar maatregelen ter vermindering van de fosfaatuitspoeling uit landbouwgronden. Eindrapport.* Rapport **374**. SC–DLO: Wageningen, The Netherlands.

Uunk, E.B.J. 1991 *Eutrophication of surface waters and the contribution of agriculture.* Proceedings **303**, The Fertilizer Society, Greenhill House, Thorpe Wood Peterborough

Walsum, P.E.V. van, 1988. *SLAPP: een rekenprogramma voor het genereren van bemestingsscenario's (betreffende dierlijke mest and stikstofkunstmest) ten behoeve van milieu-effectonderzoek, versie 1.0.* Nota **1920**. Instituut voor Cultuurtechniek en Waterhuishouding: Wageningen, Netherlands.

LIST OF SYMBOLS

Symbol	Description	Units
α	slope of the walls of open field drains	m m^{-1}
α	first order rate coefficient	d^{-1}
α	soil type dependent constant	cm^{-1}
α_{aq}	surface area of single aggregate	m^2
γ	chemical activity coefficient	-
Γ	residence time	d
Γ_r^*	residence time per m in radial zone	d m^{-1}
Γ_d^o	residence time above drain level	d
Γ^*	residence time per meter drain distance	d m^{-1}
Γ_r	residence time in radial flux field	d
Γ_{max}^r	maximum residence time in radial flux field	d
$\overline{\Gamma}_r$	average residence time in radial flux field	d
$\overline{\Gamma}^*$	average residence time per m drain distance in radial flux field	d m^{-1}
Υ	drainage resistance	d
Υ_{cpl}	compound uptake resistance in plants	d m^{-1}
Υ_{pl}	crop resistance for liquid flow	bar d m^{-1}
Υ_{so}	compound uptake resistance in soil	d m^{-1}
δ_r	slope of phreatic water level near drains	-
ε_g	volume fraction of gas filled macro pore space	m^3 m^{-3}
ε_s	volume fraction of macro porosity of the soil	m^3 m^{-3}
ε_t	volume fraction of total pore space	m^3 m^{-3}
ε_w	volume fraction of effective water filled pore space	m^3 m^{-3}
η	labyrinth factor	-
θ	volume fraction of water in soil	m^3 m^{-3}
θ_ψ	moisture fraction at suction ψ	m^3 m^{-3}
θ_e	volume fraction of water in macro pores	m^3 m^{-3}
θ_{fc}	volume fraction of water at field capacity	m^3 m^{-3}
θ_i	volume fraction of water of internal aggregate	m^3 m^{-3}
θ_r	moisture fraction in the root zone	m^3 m^{-3}
θ_{sat}	volume fraction of water at saturation	m^3 m^{-3}
$\overline{\theta}$	average volume fraction of water	m^3 m^{-3}
κ	light absorption coefficient	-
κ_{max}	maximum light absorption coefficient	-
Λ	dispersion length	m
λ_T	thermal conductivity	J m^{-1} d^{-1} °K^{-1}
λ_w	capillary conductivity	m d^{-1}
μ	storage coefficient	m^3 m^{-3}

Symbol	Description	Units
ξ	decomposition rate correction factor	-
ξ_θ	moisture dependent decomposition activity coefficient	-
ξ_{ae}	aeration dependent decomposition activity coefficient	-
ξ_N	reduction factor for decomposition depending on N	-
ξ_N^{fix}	N-dependent fixation activity coefficient	-
ξ_P	reduction factor for decomposition depending on P	-
ξ_{pH}	pH dependent decomposition activity coefficient	-
ξ_Q	nutrient dependent reduction factor for dry matter production	-
ξ_{so}	transpiration stream concentration geometry factor	-
ξ_T	temperature dependent decomposition activity coefficient	-
ρ_d	bulk density of dry soil	kg m^{-3}
ρ_{dm}	specific weight of dry matter	kg m^{-3}
ρ_s	specific weight of soil	kg m^{-3}
ρ_{sat}	bulk density of shrinking soil at saturation	kg m^{-3}
ρ_w	specific weight of water	kg m^{-3}
ρ_θ	bulk density of shrinking soil at moisture fraction θ	kg m^{-3}
σ_{cpl}	transpiration stream concentration uptake rate	m d^{-1}
σ_{cpl}^{max}	maximum transpiration stream concentration uptake rate	m d^{-1}
σ_{dm}	efficiency coefficient for dry matter production	-
σ_{pl}	selectivity coefficient for material uptake by plants	-
σ_{pl}^{max}	maximum selectivity coefficient for material uptake by plants	-
σ_r^{upt}	root uptake coefficient	-
σ_r^{uptmax}	maximum root uptake coefficient	-
σ_{so}	soil transpiration stream concentration factor	-
ς_n	locally used empirical parameter	kg m^{-3}
τ	fraction of a time step	d
τ_a	apparent age of organic material	a
υ_{ag}	volume of a single aggregate	m^3 m^{-3}
υ_c	volume fraction of clay minerals	-
υ_i	volume of aggregate class i	m^3 m^{-3}
υ_i^{an}	anaerobic volume fraction of aggregate class i	m^3 m^{-3}
υ_{om}	volume fraction of organic material	-
υ_q	volume fraction of quartz	-
υ_r	volume fraction of roots	m^3 m^{-3}
ϕ	phase shift	rad
Φ	latitude	°
χ	relative duration of bright sunshine	-
χ_{m-1}^m	ratio of mole mass metabolite/parent compound	-
ψ	soil moisture suction	bar
ψ_a	soil moisture suction at air entry point	bar
ψ_l	leaf water suction	bar
ψ_l^c	leaf water suction at which stomata start to close	bar
ψ_m	mean soil moisture suction in rootzone	bar

List of Symbols

Symbol	Description	Units
ψ_{max}	maximum soil moisture suction	bar
ψ_{min}	minimum soil moisture suction	bar
ψ_{os}	mean osmotic suction in root zone at field capacity	bar
ω	frequency of temperature wave	rad d^{-1}
Ω	radial resistance factor	d m^{-1}
Π_r	root density	m^{-2}
a	empirical constant	-
a	boundary constant	-
a	molar activation energy	J mol^{-1}
a	water transport coefficient in unsaturated soil	m bar$^{1.4}$ d^{-1}
A_γ	temperature dependent coefficient	mole$^{0.5}$ (dm^3)$^{0.5}$
A_{ae}	aerated area	m^2
A_{ag}	surface area of an aggregate	m^2
A_{cr}	horizontal surface area of soil cracks	m^2
A_{cr}^{rel}	relative horizontal surface area of cracks	(-)
$A_k^{cu(n)}$	area of calculation unit k in sub-region n	ha
A_{max}	maximum assimilation rate	kg CH$_2$O ha^{-1} h^{-1}
A_n^{sr}	area of sub-region n	ha
A_r	cross sectional area for radial flux	m^2 m^{-1}
b	boundary constant	-
b	empirical constant	-
B_γ	temperature dependent coefficient	nm^{-1} mole$^{-0.5}$ (dm^3)$^{-0.5}$
c	solute concentration	kg m^{-3}
c_a	concentration in free atmosphere	m^3 m^{-3}
c_{aq}	concentration in water phase	kg m^{-3}
c_e	concentration in macro pores	kg m^{-3}
c_{eq}	equilibrium concentration	kg m^{-3}
c_g	gas volume concentration in air-filled pores	m^3 m^{-3}
c_g	concentration in soil air	kg m^{-3}
c_g^{ox}	oxygen concentration in soil air	m^3 m^{-3}
c_i	internal mass concentration in soil aggregates	kg m^{-3}
c_{om}^i	concentration of inflow of dissolved organic material	kg m^{-3}
c_{org}^o	concentration of outflow of dissolved organic material	kg m^{-3}
c_{max}	concentration at maximum sorption	kg m^{-3}
c_{om}	concentration of organic material in solution	kg m^{-3}
c_{opt}	optimum compound concentration in soil	kg m^{-3}
c_{pl}	mineral compound concentration in plants	kg m^{-3}
c_{pr}	concentration in rain	kg m^{-3}
c_r	concentration at root surface	kg m^{-3}
c_r^{min}	minimum concentration at root surface	kg m^{-3}
c_{rmax}	concentration at $r = r_{max}$	kg m^{-3}
c_s	equilibrium or saturation concentration	kg m^{-3}

Symbol	Description	Units
c_{sat}	saturation concentration	kg m^{-3}
c_w	gas concentration in water-filled pores	kg m^{-3}
c_{we}	equilibrium gas concentration in water	kg m^{-3}
c_w^{ox}	oxygen concentration in water	kg m^{-3}
c^x	mass concentration in soil	kg m^{-3}
$\bar{c}$	time averaged concentration in bulk solution	kg m^{-3}
$\bar{c}_{dr}$	average drain water concentration	kg m^{-3}
$\bar{c}_e$	average concentration in macro pores during time step	kg m^{-3}
$\bar{c}_g$	average gas volume concentration	m^3 m^{-3}
$\bar{c}_i$	average internal mass concentration in the aggregates	kg m^{-3}
$\bar{c}_{id}$	average concentration infiltrating water from drain	kg m^{-3}
$\bar{c}_{org}$	average concentration of organic material in solution	kg m^{-3}
C_a	volumetric heat capacity of air	J m^{-3} °K^{-1}
C_{cl}	volumetric heat capacity of clay minerals	J m^{-3} °K^{-1}
CEC	cation exchange capacity	mole kg^{-1}
C_h	volumetric heat capacity	J m^{-3} °K^{-1}
C_o	volumetric heat capacity of organic materials	J m^{-3} °K^{-1}
C_q	volumetric heat capacity of quartz	J m^{-3} °K^{-1}
C_w	volumetric heat capacity of water	J m^{-3} °K^{-1}
C_w^ψ	differential soil water capacity	m^{-1}
C_γ	ion dependent parameter	mole^{-1} dm^3
d	diameter of hydrated ion	nm
D_a	diffusion coefficient in air	m^2 d^{-1}
D_a^{ox}	diffusion coefficient for oxygen in free atmosphere	m^2 d^{-1}
D_{dd}	coefficient for diffusion plus dispersion	m^2 d^{-1}
D_m	damping depth	m
D_n	numerical dispersion coefficient	m^2 d^{-1}
D_s	solute diffusion coefficient in soil	m^2 d^{-1}
D_s^{ox}	diffusion coefficient for oxygen in soil	m^2 d^{-1}
D_{sw}^{ox}	diffusion coefficient of oxygen in water	m^2 d^{-1}
D_T	thermal diffusivity	m^2 d^{-1}
D_w	diffusion coefficient in water	m^2 d^{-1}
E_{pr}	volumic mass of exudate and dead root material	kg m^{-3}
F	flux from soil air to soil water	kg d^{-1}
f_{ae}	aerated fraction	-
f_{ae}'	reduced aerobe fraction	-
$f_{dm}^{\psi=0}$	minimum dry matter fraction at $\psi = 0$	-
f_n	fraction number organic material	-
f_{pH}	pH dependent reduction factor	-
f_{pH}^{den}	pH dependent reduction factor for denitrification	-
f_{pH}^{min}	pH dependent reduction factor for mineralization	-
f_{red}	reduction factor	-

List of Symbols

Symbol	Description	Units
fr_{red}^{den}	reduction factor for mineralization due to anaerobiosis	-
f_{θ_*}	reduction factor for conductivity in dry soils	-
fr^*	steady state fraction	-
fr_{as}	biomass formation coefficient	-
fr_c	fraction of carbon in organic material	kg kg^{-1}
fr_{cl}	weight fraction of clay in mineral soil parts	kg kg^{-1}
fr_{dm}	ratio of dry matter weight over fresh weight	kg kg^{-1}
fr_{dm}^{min}	minimum ratio of dry matter weight over fresh weight	kg kg^{-1}
fr_j	volume fraction of aggregate class j	-
fr_l	fraction of grazing losses	-
fr_{linex}	fraction of linear extensibility of soils	-
$fr_{linex,wp}$	fraction of linear extensibility of soils to wilting point	(-)
fr_m	formation fraction of metabolite	-
fr_N	weight fraction of nitrogen	-
fr_{om}	weight fraction of organic material	-
fr_{om}^s	weight fraction of soluble organic material	-
fr_P	weight fraction of phosphor	-
h	pressure head above base of aquifer	m
h_{aq}	pressure head of the aquifer	m
$h_{d,i}$	pressure head in drain system i	m
h_m	pressure head in the middle between two drains	m
h_r	pressure head required for radial flux	m
h_w	water depth in drain	m
h_θ	matrix pressure head in unsaturated soil	m
$\overline{h}$	average pressure head at phreatic water level	m
H	thickness of the aquifer below drain level	m
H	Henry's distribution coefficient	-
H^*	relative aquifer thickness in relation to drain distance	-
H_c	Henry's law constant	mole Pa^{-1} m^{-3}
H_d	equivalent thickness of drainage layer	m
H_d^c	corrected layer thickness per meter drain distance	-
H_{wl}	water level in drain	m
i	number	-
I	infiltration: precipitation including irrigation	m d^{-1}
I	ionic strength	mole dm^{-3}
j	number	-
J_{cr}	sink term for compound uptake by crops	kg m^{-2} d^{-1}
J_e	exudate and dead root material production	kg m^{-2} d^{-1}
J_m^*	reduced net decomposition of organic material	kg m^{-2} d^{-1}
J_{NO3}	nitrate demand for denitrification	kg m^{-2} d^{-1}
J_{om}	net decomposition of organic material	kg m^{-2} d^{-1}
J_{om}^{an}	reduced net decomposition under anaerobic conditions	kg m^{-2} d^{-1}

Symbol	Description	Units
J_s	total solute flux	kg m^{-2} d^{-1}
k	number	-
k	rate constant	d^{-1}
k_{bio}	apparent death rate of biomass	d^{-1}
k_d	real death rate of biomass	d^{-1}
k_{dbio}	first order decomposition rate of dead biomass	d^{-1}
k_{ex}	first order decomposition rate for exudates	d^{-1}
k_h	hydraulic conductivity	m d^{-1}
k_i	initial decomposition rate	a$^{-0.6}$
k_r	death rate factor for roots	d^{-1}
k_r^c	rate constant for root mass consumption	d^{-1}
k_r^d	rate coefficient for dying roots	d^{-1}
k_w	rate of change in water content	d^{-1}
k_x	hydraulic conductivity in x direction	m d^{-1}
k_y	hydraulic conductivity in y direction	m d^{-1}
k_θ	change in moisture fraction	m^3 m^{-3} d^{-1}
k_0	zero order decomposition rate coefficient	kg m^2 d^{-1}
k_0	capillary conductivity at saturation	m d^{-1}
$k_{0,om}^{dis}$	zero order production rate of dissolved material	kg m^2 d^{-1}
k_0^S	zero order decomposition rate of sorbed material	kg m^2 d^{-1}
k_1	first order decomposition rate coefficient	d^{-1}
k_1^S	first order decomposition rate of sorbed material	d^{-1}
k_2	reaction factor	kg^n m^{-3n}
$\underline{k}$	average first order decomposition rate coefficient	d^{-1}
K	rate coefficient	m^3 kg^{-1} d^{-1}
K_a	dissociation constant of organic acid	mole dm^{-3}
K_{ad}	time dependent adsorption rate coefficient	d^{-1}
K_B^{sol}	Bunsen's coefficient of solubility	m^3 m^{-3}
K_c^M	complexation constant cation with organic acid	mole^{-1} dm^3
K_d	linear adsorption coefficient	m^3 kg^{-1}
K_{di}	material dissolution rate	m^3 kg^{-1} d^{-1}
K_F	Freundlich sorption coefficient	(m^3 kg^{-1})^n
K_G	Gapon selectivity coefficient	-
K_{GT}	Gaines-Thomas selectivity coefficient	-
K_{HE}	heterovalent selectivity coefficient	-
K_K	Kerr selectivity coefficient	-
K_L	coefficient in Langmuir sorption isotherm equation	m^3 kg^{-1}
K_n	nth order decomposition rate coefficient	kg$^{(1-n)}$ m$^{-3(1-n)}$ d^{-1}
K_N	chemical selectivity coefficient	-
K_{oc}	adsorption coefficient to organic carbon	m^3 kg^{-1}
K_{ow}	octanol/water partition coefficient	-
K_p^{DOC}	partition coefficient cations between DOC and soil solution	mol$_c$ kg^{-1}

List of Symbols

Symbol	Description	Units
K_{pr}	material precipitation rate	m^3 kg^{-1} d^{-1}
$\overline{K}_d$	average sorption coefficient in case of non-linear sorption	m^3 kg^{-1}
l	number	-
L	drain distance	m
L_{ai}	leaf area index	-
L_B	width of buffer zone	m
L_{dr}	length of drains	m
L_n^{sr}	total load in sub-region n	kg ha^{-1} a^{-1}
$L_p^{cu(n)}$	load per calculation unit of pesticide p	kg ha^{-1} a^{-1}
m	number	-
m	concentration of acidic functional groups in DOC	mol$_c$ kg^{-1}
m_r	geometry factor for the root system	bar
M	aeric mass	kg m^{-2}
M_{biom}	mass fraction of biomass	-
M_{cl}	mass fraction of clay	-
M_N	net mineralization of nitrogen per time step	kg m^{-2}
$M_{N\text{-}NO3}$	nitrate mass fraction	kg kg^{-1}
$M_{N\text{-}tot}$	weight fraction of total nitrogen	kg kg^{-1}
M_{oc}	mass fraction of organic carbon	-
M_P	net mineralization of phosphate per time step	kg m^{-2}
M_r	dry mass fraction of roots	-
M_s	dry mass fraction of shoots	-
n	number	-
n_c	number of grazing animals	cow ha^{-1}
n_{max}	maximum number of aggregate classes	-
N	Freundlich exponent	-
N	maximum number of organic fractions	-
N_{ag}	total number of aggregates per aggregate class	-
N_d	maximum number of drainage layers	-
N_s	nitrogen requirement in the root zone per kg dry matter	kg kg^{-1} ha^{-1}
p	number of pesticide	-
p	material constant	-
p	gas pressure	atm
p_1	soil dependent empirical constant	-
p_2	soil dependent empirical constant	-
pc_{ro}	dry matter partition coefficient to roots	-
pc_{sh}	dry matter partition coefficient to shoots	-
P	dry matter production rate	kg m^{-2} d^{-1}
P_a	actual production rate	kg CH$_2$O ha^{-1} d^{-1}
P_s^a	actual production rate for a full grown crop	kg CH$_2$O ha^{-1} d^{-1}
P_o^m	maximum production on overcast days	kg CH$_2$O ha^{-1} d^{-1}
P_c^m	maximum production on perfectly clear days	kg CH$_2$O ha^{-1} d^{-1}

P_{cst}	gross photosynthesis on perfectly clear days	kg CH_2O ha^{-1} d^{-1}
P_{ost}	gross photosynthesis on completely overcast days	kg CH_2O ha^{-1} d^{-1}
P_r	areic dry root mass present in living roots	kg m^{-2}
P_r	volumic root mass	kg m^{-3}
P_r^{gr}	gross production rate of roots	kg m^{-2} d^{-1}
$\bar{P}_r^{gr}$	mean gross production rate of roots	kg m^{-2} d^{-1}
P_{st}	gross photosynthesis of a standard crop	kg CH_2O ha^{-1} d^{-1}
$\bar{P}_{st}$	average standard dry matter production	kg ha^{-1} d^{-1}
q	water flux	m d^{-1}
q_d	total drainage flux	m d^{-1}
$q_{d,i}$	drainage flux to drain system i	m d^{-1}
$q_{d,r}$	regional drainage flux	m d^{-1}
q_{et}	evapotranspiration flux	m d^{-1}
q^i	incoming flux	m d^{-1}
q^i_d	incoming flux from the drainage system	m d^{-1}
$q_{i+½}$	discharge flux to next layer	m d^{-1}
$q_{i-½}$	discharge flux from previous layer	m d^{-1}
$q^i_{i+½}$	incoming flux from compartment $i+1$	m d^{-1}
$q^i_{i-½}$	incoming flux from compartment $i-1$	m d^{-1}
q^i_n	net upward flow through model bottom	d^{-1}
q^o	outgoing flux	m d^{-1}
q^o_d	outgoing flux to the drainage system	m d^{-1}
$q^o_{i+½}$	outgoing flux to layer $i+1$	m d^{-1}
$q^o_{i-½}$	outgoing flux to layer $i-1$	m d^{-1}
q_{max}^r	maximum water extraction rate by roots	m d^{-1}
q_{pr}	rainfall	m d^{-1}
q_r	flux for radial flow	m d^{-1}
q_r^{net}	net recharge	m d^{-1}
q_s	soil evaporation	m d^{-1}
q_{sr}	surface runoff	m d^{-1}
q_t	transpiration rate	m d^{-1}
q_{tp}	potential transpiration rate at optimum water supply	m d^{-1}
q_x	horizontal flux of water	m d^{-1}
Q	quantity of solid organic material	kg m^{-2}
Q_{add}	addition of solid material	kg m^{-2}
Q_{biom}	quantity of biomass	kg m^{-2}
Q_{biom}^{nl}	quantity of biomass present by unlimited growth	kg m^{-2}
Q_{dbiom}	quantity of dead biomass present in the soil	kg m^{-2}
Q_{dbiom}^{pr}	production rate of dead biomass	kg m^{-2} d^{-1}
Q_{ex}	quantity of exudates present	kg m^{-2}
Q_p	volumic quantity present as solid compound	kg m^{-3}
Q_{pl}^{prod}	total dry matter production of a crop	kg ha^{-1}

List of Symbols

Q_r	dry matter present in roots	kg ha^{-1}
Q_{sh}	dry matter present in shoots	kg ha^{-1}
Q_{sh}^N	nitrogen present in grass shoots	kg ha^{-1}
Q_{shmax}	maximum dry matter present in standing crop	kg ha^{-1} a^{-1}
r	radial distance	m
r_a	pore radius at air entry point	m
r_b	radial distance of the boundary from the open field drain	m
r_b^*	relative distance of the boundary per meter drain distance	-
$r_{f\theta}$	prescribed function of pressure head	-
r_{max}	radius of maximum sphere of influence	m
r_o	hydraulic radius of drain	m
r_p	radius of air filled soil pore	m
r_r	radius of a plant root	m
r_{sbl}	radius of a soil block	m
r_v	average radius of air-filled pores	m
r_{ve}	aggregate radius if $\varepsilon_g = \varepsilon_e$	m
r_ψ	pore radius at soil moisture suction ψ	m
R	radial distance from the centre	m
R_a	retardation factor	-
R_d	sink for decomposition	kg m^{-3}d^{-1}
R_g	gas constant	l atm °C^{-1} mole^{-1}
R_g	gas constant	J mol^{-1} °K^{-1}
R_p	source for production	kg m^{-3}d^{-1}
R_u	sink for plant uptake	kg m^{-3}d^{-1}
R_x	sink for lateral drainage	kg m^{-3}d^{-1}
S	source or sink function	m d^{-1}
S	water solubility	mole dm^{-3}
S	source or sink term	kg m^{-3} d^{-1}
S	extraction rate of water volume per unit volume of soil	m^3 m^{-3} d^{-1}
S_e	exudate production rate	kg m^{-3} d^{-1}
S_{ox}	oxygen production rate	kg m^{-3} d^{-1}
S_{ox*}	reduced oxygen production	kg m^{-3} d^{-1}
S_{ox}^C	oxygen demand for carbon oxidation	kg m^{-3} d^{-1}
S_{ox}^N	oxygen demand for nitrification	kg m^{-3} d^{-1}
S_w	volume of extracted water	m^3 m^{-3} d^{-1}
S_θ^r	water extraction by roots	m d^{-1}
t	time	d
t_c	day number of transition between growing periods	d
t_h	day number of harvest	d
t_0	time at beginning of time step	d
t_p	day number of planting time	d
T	temperature	°C

T_{am}	amplitude of temperature wave at soil surface	°C
T_{ref}	reference temperature	°C
T	average temperature at soil surface	°C
u	dry matter herbage intake during grazing	kg cow^{-1} d^{-1}
U	crop uptake rate	kg m^{-2} d^{-1}
U_i^{gr}	total gross dry matter intake by grazing cattle	kg ha^{-1} d^{-1}
U_r	sink term for compound uptake per m root length	kg m^{-1} d^{-1}
U_{cr}	compound uptake rate by crop	kg m^{-2} d^{-1}
U_{cr}^{max}	maximum compound uptake rate by crop	kg m^{-2} d^{-1}
U_{cr}^{opt}	optimum compound uptake rate by crop	kg m^{-2} d^{-1}
U_{cr}^{req}	required compound uptake rate by crop	kg m^{-2} d^{-1}
v_D	Darcian flow velocity	m d^{-1}
V	water storage	m
V_{ag}	total soil volume of an aggregate type and size	m^3
V_b	areic moisture volume below the root zone	m^3 m^{-2}
V_{cr}	volume of soil cracks	m^3
V_{dry}	volume of soil aggregates in dry state	m^3
V_g	gas volume	l
V_r	areic moisture volume in root zone	m^3 m^{-2}
V_s	volume of solids in soil	m^3 m^{-3}
V_{sat}	volume of saturated soil aggregates	m^3
V_θ	areic moisture volume	m^3 m^{-2}
x	horizontal distance from infiltration point	m
x_i	horizontal distance of infiltration point from water divide	m
x^*	relative distance in relation to drain distance	-
X_{ag}	total quantity of a compound present in aggregates	kg kg^{-1}
X_{eq}	quantity present in aggregates in equilibrium with c_e	kg kg^{-1}
X_e^s	mass fraction of equilibrium sorption to solid phase	kg kg^{-1}
$X_{m,l}^P$	mass fraction of precipitated parent compound	kg kg^{-1}
X_n^s	mass fraction of non-equilibrium sorption to solid phase	kg kg^{-1}
X_{on}	organic nutrient content in dry matter	kg kg^{-1}
X_P	mass fraction of precipitated solid compound	kg kg^{-1}
X_{pl}	mineral compound content in dry matter	kg kg^{-1}
X_{pl}^{N-NO3}	NO$_3$-N fraction in crop	kg kg^{-1}
X_{pl}^{N-tot}	total nitrogen-N fraction in crop	kg kg^{-1}
X_{pl}^P	P fraction in crop	kg kg^{-1}
X^s	adsorbed quantity	kg m^{-3}
X_{max}^s	maximum adsorbed quantity	kg m^{-3}
X_{ex}^s	sorbed quantity in equilibrium with external concentration	kg kg^{-1}
Y_{ox}	oxygen consumption rate	m^3 m^{-2} d^{-1}
Y_{ox}'	reduced oxygen demand	m^3 m^{-3} d^{-1}
z	depth	m

List of Symbols

z_r	depth in rootzone	m
Δz	layer thickness	m
Δz_r	layer thickness in the root zone	m
$\bar{z}_{sh}$	mean thickness of shoots	m
Z	depth of layer boundary	m
Z_{an}	depth at which anaerobiosis starts	m
Z_{cr}	critical depth for capillary rise	m
Z_d	drain depth	m
Z_{deep}	depth of deepest observed groundwater table	m
Z_{gt}	depth of groundwater table	m
Z_{max}	maximum depth	m
Z_{min}	minimum depth	m
Z_n	depth of layer boundary n	m
Z_r	depth of rooting	m
Z_r^{max}	maximum depth roots due to soil constraints	m
Z_{sat}	layer thickness of saturated soil	m

INDEX

acid sulphate soils, 113
acidification potential, 113
additions root materials, 146
    fresh materials, 146
adsorption, 93–94
    Freundlich equation, 99–101, 117–118, 120
    first order rate, 100
    instantaneous, 115
    intra-aggregate, 102–103
    isotherms, 98
    Langmuir equation, 99, 101, 124–126
    linear, 98
    multi-site, 9, 115, 118,
    non-equilibrium, 9, 100
    phosphorus, 260–261, 263–264
    time dependent, 115
aeration, 60, 155, 170–172, 175–176, 183, 185
agriculture, 2
alfalfa phosphorus uptake, 237
aluminium, 92, 120, 124
ammonification, 154–155, 169–170, 200
ammonium, 153–156, 160, 257
    production rate, 154–155
    uptake, 232
anaerobiosis, 156–157, 171–172, 175–176, 183, 186–188, 199
animal density, 231
animal grass uptake, 229
ANIMO, 7–9, 86–87, 127, 140, 145, 159, 168, 171, 273, 281–283
    model calibration, 249–250, 252–253, 261–262
    model validation, 246, 249, 251, 254–259, 261–262, 265

    regional application, 271
application technique slurry, 252, 254
AQUIMIX, 7–8, 86–87, 281–283, 291
arable crops nitrogen requirements, 236
arboriculture, 293–294
asparagus, 293
assimilation rate, 220
    factor, 138
atmospheric deposition, 2, 153, 161
atrazine, 292, 294–296
BALANCE model, 24–25, 29
barium, 92
barley residues nitrogen content, 237
Beerze, 271, 280–286, 290–297
beets phosphorus uptake, 239
bentazone, 289, 292, 294–297
Bergambacht, 271–272
biodegradation, 133–134
biological process, 188
biomass, 131
    assimilation factor, 138
    cell synthesis, 138
    death rate, 139, 150–151
    decomposition, 140, 149
    growth, 139, 148–151
    production, 148
    respiration, 138
    turnover, 139
biosynthesis, 138, 152
buffer zones, 85
bulb-disinfection, 289
Bunsen's coefficient, 181
cadmium, 119–121, 123
calcium, 92, 120
calculation unit, 284, 290, 297
capillary conductivity, 17
carbon cycle, 10, 141
carrots, 293

CASCADE model, 72
catch crop, 249
catchment Beerze, 271, 280, 283–286
  290–297
catchment Bergambacht, 271–272
catchment Drentse Aa, 288
catchment Reusel, 271, 280, 283–286
  290–297
catchment Rosep, 271, 280, 283–286,
  290–297
catchment Schuitenbeek, 271, 274, 278
cation exchange, 93
cattle slurry, 145, 152
cattle manure, 152
cell synthesis, 138
cereals phosphorus uptake, 239
chemisorption, 93
chromium, 120, 121, 123
classes groundwater regime, 279
clay, 18–19, 32, 197, 248, 258–259
  basin, 18–19, 32
  content, 11, 20, 60
  loam, 18–19, 32
  loam silty, 18–19, 27, 32,
  silty, 18–19, 27, 32
  thermal properties, 196–197
cobalt, 92, 119
complexation, 119
  organic acids, 114
compost, 145, 152
compound transport, 60
concentration phosphorus, 260–264,
  266
convection, 62–63, 72, 74
copper, 92, 119–121, 123
crack area, 23, 79
crack formation, 19
crack volume, 22
crop nutrient uptake, 231
crop phosphorus uptake, 237
crop residues, 227
  barley, 227
  beets, 227

cabbage, 227
  nitrogen content, 235
  maize, 227
  oats, 227
  peas, 227
  potato, 227
  rye, 227
  sprouts, 227
  wheat, 227
crop resistance, 212–213
crop soil cover, 218
crop uptake rate, 208–209, 212– 215
cyanazine, 118
Davies equation, 91–92
Debije-Hückel equation, 91
deciduous trees, 293
  dry matter production foliage, 238
  dry matter production roots, 238
  dry matter production stemwood,
  238
  litter, 138, 143–144, 150, 152
  nitrogen concentration foliage 238
  nitrogen concentration roots 238
  nitrogen concentration stemwood
  238
decision support system, 3
decomposition, 65, 133–134
  rate, 145–146
DEMGEN model, 16, 273
denitrification, 9, 60, 153, 156, 190
  200–201
desethyl-atrazine, 292, 294–296
diffuse sources, 3
diffusion, 206– 207
discharge relation, 53
dispersion, 62–63, 70–72, 74–75
dissolution, 104
  instantaneous, 111, 116–117
  non-equilibrium, 110
distribution coefficient, 76, 115
drain distance, 33
drain discharge, 259
  nitrate, 259

phosphorus, 260, 262, 266
drain layer equivalent thickness, 33
   finite thickness, 40
   infinite thickness, 36–38
drain line, 34–37, 40, 82–83
drain perfect, 33, 40, 80
drainage, 24
   water, 80
Drentse Aa, 288
drinking water, 1
dry matter, 219
   shoots, 219
   distribution, 220, 222
   fraction, 219–220
   production, 233
   production deciduous trees, 238
   production grass, 228
   production spruce trees, 238
emission, 80
EPIDIM, 7–8, 94, 97, 107, 289
ethoprophos 243–244
evaporation, 24
exchange gas, 107–108
exudate, 151, 223
   composition, 151
   decomposition, 147–148
   production, 147–148, 223–224, 226
FAIDS model, 24, 262
farmyard manure, 138 143–144, 150
field capacity, 17
finite elements, 16
fixation, 153, 158
FLOCR model, 23
flower bulbs, 248, 262–266, 288
fluazinam, 292, 294–296
flux distribution saturated domain, 46
foliage, 138, 143–144, 150
   deciduous trees, 238
      nitrogen concentration, 238
   spruce trees, 238
      nitrogen concentration, 238
   plants, 152

forage maize, 248–251
   nitrogen, 250–251
forest, 238, 293
   nitrogen requirement, 238
   plantation production, 238
   trees, 293
fruit trees, 293
fulvic compounds, 93, 119
fulvic acids, 114
functional model, 4
Gaines-Thomas equation, 95, 97
Gapon equation, 94–95
GONUTS model, 7–8, 281–283, 291
grass, 228–229
   animal uptake, 229
   continuous grazing, 228
   dry matter production, 228
   hay-winning, 228
   nitrogen content, 253–254
   nitrogen requirement, 235
   nitrogen uptake, 236
   phosphorus uptake, 237
   respiration, 229
   root production, 229–231
   root respiration, 229
   rotational grazing, 228
   shoot nitrate concentration, 236
   shoot production, 228 230
grassland, 248, 252–254, 258–262
   sprinkling, 280, 284–285
   utilization, 229
groundwater, 1
   flow, 15
   level, 284–285
   pollution, 85, 278, 285–287, 295–297
   regime classes, 279
growth reduction, 233
half life time, 76
heavy metals, 9, 92, 119–123
hedge trees, 293
Heino, 247–251
horizontal flow, 41–42

humic acids, 114
humic compounds, 93, 119
humous sand, 248, 260–262
hydrogen, 92, 120
hydrology, 284–286
    reference situation, 285–286
    schematization, 53
hydroxy-atrazine, 292, 294–296
immiscible fluids, 132
immobilization, 10, 153, 155, 161, 162
infiltration, 24
    point, 34
information system ISBEST, 7, 290–291
ion, 89–92
    activity, 90–91
    complexation, 90
    dependent parameters, 92
    speciation, 89
ionic strength, 91
iron, 92, 124
irrigation, 24
ISBEST information system, 7, 290–291
isochrones, 34
isotropy, 50
Kerr equation, 94
kinetics zero order rate, 65
kinetics first order rate, 65, 74
lateral outflow, 81
lateral drainage flux, 55
leaching, 153, 161
lead, 119–121, 123
leaf water potential, 221–222
leaf area index, 217–218
leaf thickness, 219
leek, 293
legiminosae, 9, 153, 158
Lelystad, 247–248, 258–259
lindane, 292–296
linear extensibility, 20–22
litter deciduous trees, 138, 143–144, 150, 152
litter spruce trees, 138, 143–144, 150, 152

loam, 18–19, 27, 32
    fine sandy, 18–19, 32
    loess, 18–19, 27, 32
    sandy, 18–19, 32
    silt, 18–19, 32
loamy sand, 18–19, 27, 32, 248, 252–254
    fine, 18–19, 32
    medium coarse, 18–19, 32
    medium fine, 18–19, 27, 32
magnesium, 92, 120
maize, 293
    phosphorus uptake, 237
    residues nitrogen content, 235
manganese, 92
market position, 290–291
mass conservation equation, 61–62, 65–67
MCPA, 292–296
mechanistic model, 4
metabolytes, 9, 133–134, 292, 294–296
    formation, 9
    leaching, 292, 294–296
metamitron, 292, 294–296
metolachlor, 292, 294–296
metribuzin, 118
mineral nitrogen, 250–251, 253–257
mineralization, 9, 153–155, 161–162, 190
MIXING CELL, 72
model, 35
    schematization, 35, 38, 45
    validation ANIMO, 246, 249, 251, 254–259, 261–262, 265
    validation TRANSOL, 241–244
    verification, 74
multi-component model, 5, 106
Nagele, 247–248, 255
nature restoration, 285–287
nickel, 92, 120–123
nitrate, 250, 253, 255, 257, 259
    concentration, 285–286
    grass shoots, 236

Index

drain discharge, 259
uptake, 234
nitrification, 154, 156, 169–170, 190, 200
nitrogen, 153, 155
  ammonification, 153, 155
  atmospheric deposition, 153
  balance, 153
  concentration, 152, 235, 238, 253–254
    barley residues, 235
    cattle slurry, 152
    cattle manure, 152
    compost, 152
    crop residues, 235
    deciduous trees, 238
    exudates, 152
    foliage plants, 152
    grass, 253–254
    litter deciduous trees, 152
    litter spruce trees, 152
    maize residues, 235
    peat, 152
    pig manure, 152
    potato residues, 235
    poultry manure, 152
    roots, 152
    sewage sludge, 152
    spruce trees, 238
    stubble, 152
    wheat residues, 235
  cycle, 153
  denitrification, 153, 156
  fixation, 9, 153, 158
  immobilization, 153, 155
  leaching, 153, 272, 274, 275, 280, 284
  mineralization, 153–155
  optimum concentration, 232
  organic, 153, 155
  requirement arable crops, 234
  requirement forest plantations, 238
  requirement grass, 235
  uptake, 229–230
  uptake forage maize, 250–251
  uptake grass, 236
  uptake pH, 232
  uptake wheat, 256
  volatilization, 153, 160
non-equilibrium model, 5
nutrient, 11, 60, 152, 233, 288
  availability, 11
  balance, 60
  immobilization, 152
  shortage, 233
octanol/water–partition–coefficient, 118–119
organic acids, 122
organic matter, 10, 135
  additions, 146
  compost, 145
  cattle slurry, 145
  decomposition, 135–138, 142–143, 197
    rate, 138, 143
    multi-component, 135–136, 142–143
    time-dependent, 135, 137–138, 142–143
  dissolved, 148
  farmyard manure, 138, 143–144
  foliage, 138, 143–144
  fraction, 20
  hypothetical material, 145
  litter, 138, 143–144
    deciduous trees, 138, 143–144
    spruce trees, 138, 143–144
  peat, 138, 143
  pig manure, 145
  poultry manure, 145
  sawdust, 138, 143–144
  sewage sludge, 145
  slurries, 145
  straw, 138, 143–144
  thermal properties, 196–197
organic micro-pollutants, 117, 170

organic nitrogen, 153, 155
organic–phosphorus, 161–162, 266–267
ornamental coniferae, 293
ornamental shrubs, 293
ornamental trees, 293
osmotic pressure, 221
oxygen, 11, 172–177, 187–188
    concentration, 172–176, 181
    consumption, 178
    demand, 173 178
oxygen diffusion, 172
    aggregated soils, 184–187
    coefficient, 174–175, 181
    temperature, 181
    water phase, 179–180
    water phase sand, 180–183
oxygen transport, 173–178
partition coefficient dry matter, 222
peat, 18–19, 27, 32, 138, 143, 152
    oxidation, 169
pest-stress, 288, 290–291
pesticides, 9, 170, 288, 290–296
pesticide leaching, 292–296
    atrazine, 292–296
    bentazone, 292, 294–296
    desethyl-atrazine, 292, 294–296
    fluazinam, 292, 294–296
    hydroxy-atrazine, 292, 294–296
    lindane, 292, 294–296
    MCPA, 292, 294–296
    metabolite, 292, 294–296
    metamitron, 292, 294–296
    metolachlor, 292, 294–296
pesticides physical-chemical properties, 291, 294
pesticide use, 293
pesticide urban use, 293
PESTLA, 243–246, 291
pH, 11, 60, 127, 199–201
    ammonification, 200
    denitrification, 200–201
    nitrification, 200

nutrient uptake, 232
phosphate, 123–128, 152
    fraction compost, 152
    fraction cattle manure, 152
    fraction litter deciduous trees, 152
    fraction litter spruce trees, 152
    fraction foliage plants, 152
    fraction pig manure, 152
    fraction peat, 152
    fraction poultry manure, 152
    fraction roots, 152
    fraction sewage sludge, 152
    fraction slurries, 152
    fraction stubble, 152
    precipitation, 123–128
    saturated soils, 271, 278–280
    sorption, 123–128
phosphorus, 260–264
    adsorption, 260–264
    atmospheric deposition, 161
    concentration, 260–266
    cycle, 161
    drain discharge, 260 262, 266
    immobilization, 161–162
    leaching, 161, 271–272, 275, 278–280, 286–287
    mineralization, 161, 162
    optimum concentration, 232
    organic, 266–267
phosphorus uptake, 232
    alfalfa, 237
    maize, 237
    cereals, 237
    crops, 237
    grass, 237
    sugar beets, 237
photosynthesis, 11, 215–217
    clear sky, 216–217
    gross, 215, 217–218
    latitude, 216–217
    net, 215
    overcast sky, 216–217
    standard crop, 216

physical planning, 284–287
physical-chemical properties pesticides, 291, 294
pig manure, 145, 152
plant resistance, 212–213
plant uptake rate, 212, 214– 215
planting time, 224
ploughing, 162
point sources, 3
policy analysis, 272, 275–277, 284
pollution, 1, 85, 278, 285–287, 295–297
potassium, 92
potatoes, 235, 293
poultry manure, 145, 152
precipitation, 9, 104, 106, 110–111, 116–117
    instantaneous, 111, 116–117
    non-equilibrium, 9, 110
    time dependent, 106
production forest plantations, 238
production reduction, 233
Putten, 247–248, 260–262
radial flow, 42–43
rainfall, 24, 187–188
REDIS model, 289
redistribution, 60
redox reaction, 111–112
regional, 51
    application ANIMO, 271
    application TRANSOL, 288
    drainage, 51
    flow system, 52
    water balance, 51
    water transport, 49
residence time, 34–41, 44–47
respiration, 138, 223
restoration nature, 285–287
retardation factor, 74
Reusel, 271, 280, 283–286, 290–297
Richard equation, 62
risk analysis, 2
roots, 152, 205

concentration, 207–211, 213
density, 205–211, 213, 226
depth, 221, 224
    beet, 224–225
    effective, 226
    lettuce, 224–225
    cabbage, 224–225
    maize, 224–225
    peas, 224–225
    potato, 224–225
    spring cereals, 224–225
    winter cereals, 224–225
    development, 205
dry matter, 225
    production, 222
    deciduous trees, 238
    spruce trees, 238
growth, 222
mass consumption, 223–224
materials, 146
exudate production, 147
distribution, 147
production, 147, 221, 223
hairs, 147
nutrient transport, 205–207
nutrient uptake, 212
penetration resistance, 226
radius, 205–209
specific weight, 226
sphere of influence, 207–208, 226
water transport, 206–207
root-zone, 25
Rosep, 271, 280, 283–286, 290–297
Ruurlo, 247–248, 252–254
salt redistribution, 289
SAMIA model, 289
sand, 248–251, 262
    coarse, 18–19, 32
    medium coarse, 18–19, 26–27, 32
    medium fine, 18–19, 32
    fine, 18–19, 32
sanitary landfill, 122
saturated domain, 6, 33

saturated zone, 6
saturation, 105
sawdust, 138, 143–144, 159
scenario analysis, 272, 275–277, 284
Schuitenbeek, 271, 274, 278
scorzonera, 293
sewage sludge, 145, 152
shoots, 219
    dry matter production, 222
    thickness, 219
shrinkage, 19–21
    layer thickness, 23
    soil, 289
silty loam, 248, 255
SIMGRO model, 6–7, 16, 281, 283, 291–292
SIWARE model, 289
SLAPP model, 6–7 281–282
slurry application, 160, 252, 254
SMASS model, 289
sodium, 92
soil, 16
    aeration, 165
    cracks, 79, 174–175, 184–186
    disinfectants, 293–294
    dry bulk density, 22
    heat capacity, 190, 194–195
    heterogeneity, 50
    moisture, 11, 167–168
        characteristic, 17
        content, 221
        stress, 11, 219–220
        potential, 60
        suction, 168–170
    physical properties, 17
        basin clay, 18–19, 32
        clay, 18–19, 32
        clay light, 18–19, 32
        clay silty, 18–19, 27, 32
        clay loam, 18–19, 32
        clay loam sandy, 18–19, 32
        clay loam silty, 18–19, 27, 32
        loam, 18–19, 32
        loam loess, 18–19, 27, 32
        loam fine sandy, 18–19, 32
        loam sandy, 18–19, 32
        loam silt, 18–19, 32
        loamy sand fine, 18–19, 32
            medium coarse, 18–19, 32
            medium fine, 18–19, 27, 32
        peat, 18–19, 27, 32
        sand coarse, 18–19, 32
            medium, 18–19, 26–27, 32
            medium fine, 18–19, 32
            fine, 18–19, 32
    salinity, 222
    shrinking, 19, 174–175, 184–187
    structure, 22
    swelling, 19, 172, 184–187
    standard, 17
    temperature, 11, 190–194
    thermal properties, 194–196
solubility constant, 105
sorption, 98
    first order rate, 100
    Freundlich–equation, 99–101, 117–118, 120
    instantaneous, 115
    intra-aggregate, 102–103
    isotherms, 98
    Langmuir–equation, 99, 101, 124–126
    linear, 98
    multi-site, 115, 118
    non-equilibrium, 100
    time dependent, 115
Southern Peel, 271–272
spray-drift, 289, 295
sprinkling grassland, 280, 284–285
spruce trees, 293
    dry matter foliage, 238
    dry matter roots, 238
    dry matter stemwood, 238
    litter, 138, 143–144, 150, 152
    nitrogen concentration, 240

foliage, 240
roots, 240
stemwood, 240
St. Maartensbrug, 247–248, 262–265
stability control, 73
strategy water management, 284–287
straw, 138, 143–144, 150
strawberries, 293
streamline, 34–37
structural model, 4
stubble, 152
sub-region, 284, 290, 297
sugar beets, 293
sulphate-acid soils, 2
sunshine duration, 217
surface water pollution, 1, 80, 85, 278, 285–287, 295–297
surface storage, 76–78
surface runoff, 76–78
swamp conditions, 285–287
SWAP model, 17, 248, 252, 255, 260
SWATRE model, 16
swelling, 19, 21
swelling soil, 289
synthetic organic compounds, 9
temperature, 11, 60, 91, 188, 190
    amplitude, 191
    damping depth, 191 197
    mineralization, 190
    nitrification, 190
    phase shift, 191
    wave, 191
thermal properties, 194
transformation process, 60
TRANSOL model, 7–9, 86–87, 132, 134, 168, 170, 291
    regional application 288–289
    validation, 243–246
transpiration, 24
    actual, 219
    potential, 219
    stream concentration factor, 211
    stream concentration rate, 214

transport multi-component, 69
transport single-component, 64
TREND model, 6–7
unsaturated zone, 5–6
unsaturated domain, 16
uptake, 210, 232
    deficit, 233
    nitrogen, 232
        optimum, 232–233
    nutrients optimum, 232–233
    phosphorus, 232
    requirement, 233
    specific rate, 210
validation, 243 246
    ANIMO, 246, 249–251, 254–265
    TRANSOL, 243–246
Vanselow equation, 96–97
volatilization, 153, 160
WATBAL model, 24, 278
water, 2
    waste, 2
    balance, 24, 79
    balance definition sketch, 28
    conservation, 285–287
    management strategy, 284–287
wheat, 248, 255–257
    nitrogen, 256
    residues nitrogen content, 234
zinc, 92, 119–121, 123